Solar Thermal Desalination Technologies for Potable Water

H. Sharon · K. Srinivas Reddy

# Solar Thermal Desalination Technologies for Potable Water

## Exploring Viable Options for Reliable and Sustainable Water Production

Ane Books Pvt. Ltd.

Springer

H. Sharon
Indian Institute of Petroleum and Energy
Visakhapatnam, India

K. Srinivas Reddy
Indian Institute of Technology Madras
Chennai, India

ISBN 978-3-031-49154-2 ISBN 978-3-031-49155-9 (eBook)
https://doi.org/10.1007/978-3-031-49155-9

Jointly published with Ane Books Pvt. Ltd.
The print edition is not for sale in South Asia (India, Pakistan, Sri Lanka, Bangladesh, Nepal and Bhutan) and Africa. Customers from South Asia and Africa can please order the print book from: ANE Books Pvt. Ltd.
ISBN of the Co-Publisher's edition: 978-81-19160-54-9



This Springer imprint is published by the registered company Springer Nature Switzerland AG
The registered company address is: Gewerbestrasse 11, 6330 Cham, Switzerland

Paper in this product is recyclable.

*To the Creator of the Sun and the Earth*

# Preface

*The world its course maintains through life that rain unfailing gives;*
*Thus rain is known the true ambrosial food of all that lives*
*—Thirukkural*

Fresh water demands in various sectors are rising rapidly due to increasing population, improved life style, and continuous industrial and agricultural expansion. "World Water Day" an annual United Nations (UN) observance day will be held on 22 March to highlight the importance of fresh water. Due to uneven fresh water distribution, rising water pollution, and lack of proper wastewater treatment infrastructures at least two billion people have no access to clean and safe drinking water. Desalination technology seems to be a promising option in addressing water scarcity problems and is thus expanding rapidly all over the globe. However, desalination is an energy-intensive process, thereby low-income nations with no/little fossil fuel resources will not be able to procure and operate large-scale desalination systems in an economical way to meet their rising water demands. In addition, the entire globe is driving towards energy transition from conventional to clean fuels to limit global warming and its adverse impacts. Seventeen sustainable development goals have been set by the United Nations (UN) for its member countries to achieve a sustainable society by 2030. Enhancing access to affordable clean energy and clean water to all are some among the important goals.

Learning from the nature and utilizing nature seems to be the best way to solve water scarcity in a sustainable way. Hydrological cycle occurring in nature can be replicated in thermal desalination systems with the aid of solar energy to produce potable water. Numerous researches have been conducted in this area around the globe to develop efficient, economical, and highly productive desalination systems. This book consisting of 13 chapters makes an attempt to transfer knowledge to the scientific community and general public about the various aspects of solar thermal desalination systems. Chapters 1 and 2 deal with the basics of water-related issues and solar energy fundamentals, respectively. Chapters 3 to 10 deal with thermodynamic modeling and performance improvement approaches of various solar thermal desalination systems. Chapter 11 deals with large-scale hybrid solar chimney-assisted desalination systems. Moreover, the suitability of solar thermal distillation process for wastewater treatment and improved sanitation has been presented in Chap. 12. Apart from these, a separate chapter (Chap. 13) on economics, environmental footprints, and sustainability assessment of solar thermal desalination systems has been included. We believe that this book will be of great companion to the undergraduate, graduate researchers, entrepreneurs, and policy-makers interested in the area of solar energy, desalination, water treatment, energy transition, clean technologies, rural upliftment, sanitation, and sustainability.

At most care has been taken to present the readers with accurate information from the available literatures. However, it is impossible to have a human work without mistakes and omissions of some technologies due to certain restrictions. Hence, suggestions, critics, and directions for further improvements of this book are highly welcomed from the peers, student community, researchers, and other interested readers.

Visakhapatnam, India H. Sharon
Chennai, India K. Srinivas Reddy

# Acknowledgements

*The help given without weighing the return, when weighed, outweighs the sea*

*—Thirukkural*

The authors would like to thank the contributors of various peer-reviewed research articles in the area of solar thermal desalination systems without which the contents of this book are highly impossible. Thanks are also extended to Prof. Tanongkiat Kiatsiriroat, Chiang Mai University, Thailand; Prof. Sarada Kuravi, New Mexico State University, USA; and Prof. Hiroshi Tanaka, National Institute of Technology, Kurume College, Japan for permitting us to use figures from their research works. The authors are thankful to the publishers, namely, Elsevier, John Wiley & Sons, and Taylor & Francis for permitting them to reuse the necessary figures and data from their database. The authors are highly thankful to Mr. Rathinam, General Manager, ANE Books Pvt. Ltd., India for coordinating this book work and publisher Springer-Nature for agreeing to publish this book work.

H. Sharon would like to acknowledge the continuous encouragement and moral support of his parents J. Hilary Doss and N. Thangam and brother H. John. R. Sunil during the course of this book work. He also remembers and extends his thanks to Prof. K. S. Reddy and Prof. Ligy Philip of Indian Institute of Technology Madras. Thanks are also extended to the authorities of Indian Institute of Technology, Madras; National Institute of Technology, Andhra Pradesh; and Indian Institute of Petroleum and Energy, Visakhapatnam for providing access to useful online resources for this book preparation. Last but not the least, the author would like to remember and thank the Lord Almighty for providing sufficient strength, good health, knowledge, and wisdom to complete this book.

K. S. Reddy is grateful to the Curriculum Development Cell of Centre for Continuing Education (CCE), Indian Institute of Technology, Madras for providing him financial support to carry out this book work. He also acknowledges the encouraging atmosphere in the Heat Transfer and Thermal Power Laboratory, Department of Mechanical Engineering, Indian Institute of Technology, Madras that favors this process.

H. Sharon
K. Srinivas Reddy

# Contents

# About the Authors

**H. Sharon** is an assistant professor in the mechanical engineering department of the Indian Institute of Petroleum and Energy, Visakhapatnam. Prior to this position, he worked as an ad-hoc faculty and assistant professor at the National Institute of Technology Andhra Pradesh and Karunya Institute of Technology and Sciences Tamil Nadu, respectively. He has a Ph.D., in mechanical engineering with a specialization in solar energy from the Indian Institute of Technology Madras. He received his Bachelor's in mechanical engineering and Master's in energy engineering from the Government College of Engineering Tirunelveli and Anna University of Technology Tirunelveli, respectively. His research interests include solar photovoltaic & thermal energy systems, desalination, waste valorization, and sustainability assessment of energy technologies. He has published about 26 peer-reviewed high-impact journal articles, 3 book chapters, and 1 Indian patent.

**K. Srinivas Reddy** obtained his Ph.D. from the Indian Institute of Technology Delhi in Energy Studies in 1999. Presently, Dr. Reddy is a Professor of Mechanical Engineering at the Indian Institute of Technology Madras and he is also an honorary professor at the University of Exeter, UK. Professor Reddy is a Fellow of the National Academy of Engineering (FNAE) and a Fellow of the World Society of Sustainable Energy Technologies (FWSSET). He is a specialist in renewable energy with special research interests in concentrating solar power, energy efficiency, and the environment. He has 9 patent applications and more than 300 publications with more than 8000 citations and an h-index of 50. He has secured research funding of about USD15 Million as Principal Investigator (PI) and Co-PI in more than 30 research projects. Dr. Reddy has received the WSSET Innovation Award, the Mid-Career Level Institute Research and Development Award (IRDA), the Shri J. C. Bose Patent Award, and twice, the Bhagyalakshmi and Krishna Ayengar Awards. As a part of human resource developmental activities, he has supervised over 200 UG, PG, and Ph.D. students.

# Chapter 1
# Introduction

## 1.1 Water Distribution

Water is one of the unique resources which are available above and below the earth's surface. However, this essential resource is not uniformly distributed throughout the earth. For example, India holds nearly 16.0% of the world's total population but has only 4.0% of the global renewable water. **Renewable water** refers to the water reserves that are continuously replenished by natural water cycle. Moreover, water quality also varies from location to location. Nearly, 97.5% of the water available on the earth is saline in nature and the remaining 2.5% is fresh water. All the available fresh water reserves cannot be accessed easily because around 68.7% of the total fresh water reserves are trapped in glaciers and ice caps. Nearly, 30.0% of the total fresh water reserves are ground water and the remaining 1.2% is available as surface water in lakes and rivers (Manju & Sagar 2017). Fresh water is an essential ingredient in agricultural, industrial, and energy sectors. Moreover, fresh water is also essential for quenching the thirst of living beings in Mother Earth (Reif & Alhalabi 2015). Current global water demand in agricultural sector, industrial sector (including power generation sector), and domestic sector is distributed as 69.0%, 19.0%, and 12.0%, respectively. By 2050, the global water demand is expected to increase by about 20.0 to 30.0% from the current demand (United Nations 2019).

## 1.2 Water Scarcity and Water Stress

Significant population in most of the developing and under developed regions of the globe has no access to safe drinking water. Around 1.1 billion people over the globe consume unsafe water for their daily survival. When a single person has no access to safe and clean water to meet all his daily water requirements ranging from drinking to washing, the person is said to be **water insecure**. Regions where groups of people or community are water insecure for a significant or prolonged period



H. Sharon and K. Srinivas Reddy, *Solar Thermal Desalination Technologies for Potable Water*, https://doi.org/10.1007/978-3-031-49155-9_1

are called **water-scarce regions**. Water scarcity can be quantified by estimating the water requirement by an individual to meet all his/her needs and the amount of water actually available. Minimum water requirement of an individual in a year has been estimated to be about 1700.0 $m^3$ and it also includes the per capita demands in all sectors (domestic, industrial, energy, and agricultural sectors). Nations which are not capable of providing this minimum water to the citizens come under **water stress zones**. If water supply in nations falls below 1000.0 $m^3$/capita/year and 500.0 $m^3$/capita/year then those nations are said to be under **water scarcity** and **absolute water scarcity**, respectively (Rijsberman 2006).

Water scarcity is of two types, namely, physical and economical water scarcity. **Physical water scarcity** refers to the non-availability of water resources or disability of the countries to meet the required water demands even after taking necessary actions to tackle water stress through economical and technical means. North African countries, Middle East, China, Pakistan, and some portions of the United States of America and India suffer from physical water scarcity. **Economical water scarcity** refers to the lack of necessary water treatment or water supply-related infrastructures to make the water resources available for peoples' use. Most of the countries in the African continent and some portion of the Indian sub-continent are under economical water scarcity. Global physical and economical water scarcity zones are shown in Fig. 1.1. Most of the water-scarce zones in the globe have poor rainfall and/or densely populated such that the per capita water supply per year falls below 1000 $m^3$.

Estimated least water requirement for satisfying the domestic needs per capita is around 50.0 L/d, which is just, 20.0 $m^3$/capita/year and is lower than the limit set for water scarcity (1000.0 $m^3$/capita/year). Hence, it should be understood that the

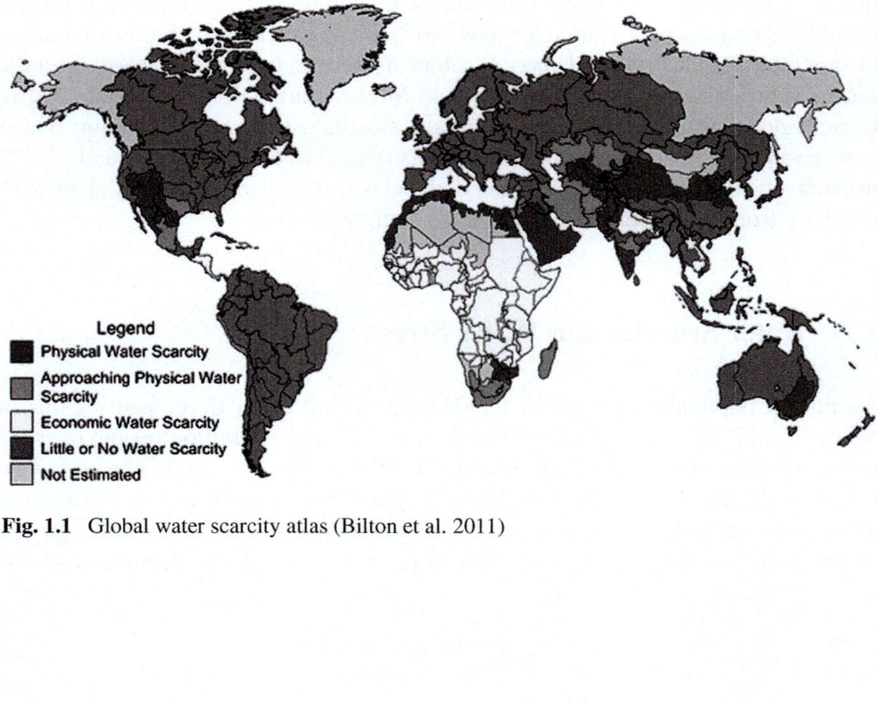


**Fig. 1.1** Global water scarcity atlas (Bilton et al. 2011)

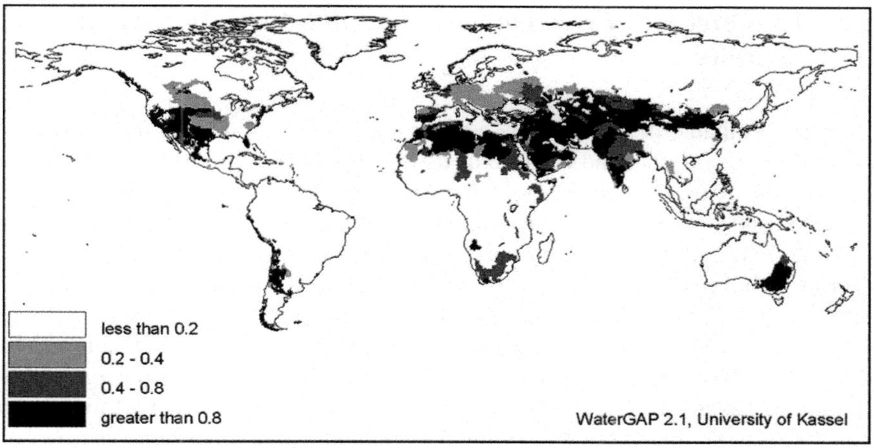


**Fig. 1.2** Global water stress zones based on water withdrawal to availability ratio (Alcamo et al. 2003)

water scarcity not only refers to the lack of water for meeting the domestic water requirements of people (Rijsberman 2006). Only 54.0% of the urban population and 16.0% of the rural population in India receive tap water supply to their houses. Water supply to urban India is much lower than any other developing countries in the globe. In addition, piped water supply in India has fallen from 10–15 h/d in 1980 to 1.5 h/d in 2006 (Tortajada 2016). According to the United Nations Organization, most of the people on the globe are living in severe water scarcity zones. Moreover, the lack of clean and safe drinking water is a major reason for the spread of most of the communicable diseases (Nandi et al. 2017).

Water stress zones are generally identified by calculating the criticality ratio. **Criticality ratio** is defined as the ratio of annual water withdrawal to the annual available water resources and is generally expressed in terms of percentage. Water withdrawal refers to the water withdrawn from the rivers, lakes, and groundwater reserves to meet the water demands of mankind. If criticality ratio of a region lies within 20% to 40%, the region is said to be under **water stress**. If criticality ratio lies above 40%, the region is said to be under **high water stress**. Global water stress zones are shown in Fig. 1.2. Regions that are slightly above and around tropic of cancer are under water stress. Nearly, 54.0% of the Indian states fall under high to extremely high water stress zones (Alcamo et al. 2003).

## 1.3 Desalination: Effective Solution for Addressing Water Scarcity

Naturally available fresh water reserves are not capable of meeting the rising fresh water demands of mankind because of its less availability and manmade pollution (Gude et al. 2010). Moreover, large quantities of water received by rainfall are not collected and stored properly because most of the Asian countries lack finance for construction of dams. Hence, the occurrence of floods and drought in same location is always noticed (Rijsberman 2006). In addition, due to climate change, rainfall pattern is also not uniform; as a result, most of the over-exploited ground water reserves are not replenished. Hence, the available ground water reserves are not able to meet the growing water demands in agricultural, industrial, domestic, and energy sectors (Manju & Sagar 2017). This situation can be tackled only if mankind finds some other ways to produce fresh water from the available saline water reserves in oceans and polluted wastewater. Luckily, desalination technology which was developed long back resembling natural hydrological cycle has the capability to tackle this problem.

**Desalination** is the process of separating salts and minerals from saline water (Reif & Alhalabi 2015). The salinity of potable water must be around 500 ppm as per World Health Organization's (WHO) drinking water standards (Reif & Alhalabi 2015). Ocean water has salinity above 35,000 ppm making it non-potable but it can be desalinated to recover potable water (Reif & Alhalabi 2015). Commercial biological wastewater treatment plants are found to be ineffective in treating wastewater up to the required standards set by the water agencies (Kaur et al. 2012). Mostly, industrial and domestic wastewaters are saline in nature, hence desalination technology can be utilized for treating wastewater too (Rebhun 2004).

Desalination capacity in different regions of the globe along with the type of feed water used is displayed in Table 1.1. Desalination technology is widely adopted in Middle East, Arab countries, North America, Asia, Europe, Africa, Central America, South America, and Australia to meet their drinking water and process water demands. Seawater is used as a major feed for desalination plants in Gulf and North African countries. American continent and European countries use brackish water as major feed water for their desalination process. Industrial wastewater has been majorly used as feed for desalination systems in Asian countries. However, from the reported data, it could be seen that the wastewater desalination process is less practiced and is at the emerging stage in most of the countries. The available data revealed that desalination is widely practiced in Asian continent and least practiced in Australia.

**Table 1.1** Global desalination capacity based on feed water type (Escobar & Schafer, 2009)

| Continent | Region/country | Desalination feed water | | | Cumulative desalination capacity ($m^3$/d) | Distribution of global desalination capacity | |
|---|---|---|---|---|---|---|---|
| | | Seawater (%) | Brackish water (%) | Waste water (%) | | Region wise (%) | Continent wise (%) |
| Australia | Australia-Pacific | 0.80 | 1.10 | 5.40 | 4,61,266 | 1.05 | 1.05 |
| Asia | Japan, Korea and Taiwan | 1.20 | 3.20 | 12.10 | 27,39,212 | 6.21 | 60.45 |
| | East Asia | 3.40 | 2.80 | 21.00 | 19,10,142 | 4.33 | |
| | West Asia | 2.20 | 4.40 | 2.30 | 10,57,340 | 2.40 | |
| | Gulf countries | 61.40 | 21.10 | 19.00 | 1,93,60,853 | 43.89 | |
| | Middle East (Rest) | 3.10 | 4.50 | 2.70 | 15,95,763 | 3.62 | |
| Europe | Northern Europe | 1.70 | 4.40 | 4.30 | 15,99,545 | 3.63 | 13.62 |
| | Southern Europe | 10.60 | 12.80 | 9.70 | 4,40,5024 | 9.99 | |
| Africa | North Africa | 7.30 | 3.20 | – | 23,59,769 | 5.35 | 6.31 |
| | Africa (Rest) | 1.20 | 2.40 | 0.40 | 4,24,160 | 0.96 | |
| South America | South America | 2.20 | 3.10 | 0.90 | 9,05,687 | 2.05 | 2.05 |
| North America | Caribbean | 2.40 | 0.60 | – | 7,23,487 | 1.64 | 16.53 |
| | North America | 2.50 | 36.40 | 22.20 | 65,68,659 | 14.89 | |

## 1.4 Conventional Desalination Technologies

The most commonly adopted large-scale desalination technologies are Multi-stage Flash (MSF), Multi-effect Distillation (MED), Vapor Compression Distillation (VC), Reverse Osmosis (RO), and Electro-dialysis (ED) (Ali et al. 2011). RO, MSF, MED, ED, and other desalting systems occupy about 65.0, 21.0, 7.0, 3.0, and 4.0% of the total global desalination capacity (Mito et al. 2019).

### *1.4.1 Multi-stage Flash Desalination (MSF)*

Heated feed water is allowed to flash in successive vessels maintained at low pressure, thereby vapors are generated and are then condensed with the help of condensers to obtain condensate. Feed water after flashing in one vessel enters the next vessel which is maintained at a pressure lower than the previous one and the same process continues in a number of vessels. The vapor formed in one vessel is collected as condensate in the same vessel.

### *1.4.2 Multi-effect Distillation (MED)*

Feed water is sprayed inside the first vessel containing heat exchanger tubes carrying hot heat transfer fluid. The sprayed water evaporates within the vessel to form hot vapor which is utilized as heat source for generation of vapors in the next successive vessel. The successive vessels are maintained at pressures lower than the previous vessel. Feed water is sprayed inside each vessel separately. The vapor formed in one vessel is collected as condensate in the next successive vessel.

### *1.4.3 Vapor Compression Desalination (VC)*

In vapor compression desalination, the vapors generated in one vessel are compressed with the help of mechanical or thermal compressor to raise its temperature and this high-temperature vapor is used as heat source for the feed water to be evaporated in the next vessel. The vapor formed in one vessel is collected as condensate in the next successive vessel.

### 1.4.4 Reverse Osmosis (RO)

In reverse osmosis process, feed saline water pressurized above the osmotic pressure is allowed to pass through semi-permeable membrane thereby the fresh water permeates across the membrane and the highly concentrated brine is rejected.

### 1.4.5 Electro-Dialysis (ED)

The electro-dialysis system consists of anode, cathode, and a number of anion and cation exchange membranes which are placed alternatively to form a number of chambers. The chambers are filled with saline water to be treated. When direct current is applied across the electrodes, the negative ions in feed water pass through the anion exchange membrane and the positive ions pass through the cation exchange membrane and the feed water in the chamber becomes free from these ions and is taken out as treated water. The ions get accumulated in the feed water of the chambers adjacent to its both sides. This concentrated brine is rejected as brine.

The energy requirements and greenhouse gas emissions of conventional fossil fueled commercial desalination technologies along with the quality of distillate produced by them are tabulated in Table 1.2. The minimum thermodynamic energy required for desalting saline water of salinity 33000 ppm at 25 °C is about 0.77 kWh/m$^3$. However, the actual energy consumed for desalination process will be about 5 to 26 times higher than this theoretical limit due to the requirement of extra energy for sustaining the process at a finite rate. The energy requirement and greenhouse gas emissions of thermal energy-based systems are very much higher than membrane-based systems (Ali Al-Karaghouli & Kazmerski 2013). **Recovery rate** is the ratio of distillate produced to the feed water supplied and is about 35 to 50% for RO system and around 10 to 25% for thermal systems (Moosa et al. 2022). Low recovery ratio is preferred in thermal systems to mitigate the corrosion effect and high energy consumption. However, distillate quality obtained by thermal desalination is very much superior to the distillate quality obtained by membrane-based separation process. In addition, feed water pre-treatment requirement is also nearly nil for thermal desalination systems because it can tolerate feed water of high salinity which is not possible with membrane-based separation systems. Moreover, for treating feed water of salinity above 50000 ppm, high pressure of at least 80 bar is required in RO membrane systems. The commercial RO membranes cannot sustain this high pressure as of now (Ahmed et al. 2021).

Low energy consumption of RO systems in comparison to thermal-based desalination systems is due to the better developments in membranes, pumps, and better scale formation control techniques. Feed water generally used in desalination systems are, namely, seawater (20000 to 50000 ppm), brackish water (3000 to 20000 ppm), river water (500 to 3000 ppm), brine (greater than 50000 ppm), and wastewater. Fresh water (less than 500 ppm) is also desalinated for obtaining double distilled

**Table 1.2** Energy requirements, harmful gas emissions, and distillate quality of conventional large-scale desalination technologies desalting seawater (Ali Al-Karaghouli & Kazmerski 2013; Gude et al. 2010)

| Plant characteristics | Desalination technology | MSF | MED | TVC | RO | ED |
|---|---|---|---|---|---|---|
| Plant capacity | ($m^3$/d) | 50000–70000 | 5000–15000 | 10000–30000 | Up to 128000 | 2–145000 |
| Thermal energy | (MJ/$m^3$) | 190–282 | 145–230 | 227 | – | – |
| Electrical energy | (kWh/$m^3$) | 2.5–5.0 | 2.0–2.5 | 1.6–1.8 | 4.0–6.0 | 2.7–5.5 |
| Electrical equivalent of consumed thermal energy | (kWh/$m^3$) | 15.83–23.5 | 12.2–19.1 | 14.5 | – | – |
| Total electricity consumption | (kWh/$m^3$) | 19.58–27.25 | 14.45–21.35 | 16.26 | 4.0–6.0 | 2.64–5.5 |
| Harmful gas emissions | (kg of $CO_2$/$m^3$) | 13.5–22.5 | 7.2–18.1 | 9.9–10.8 | 4.5–8.1 | 2.3–5.0 |
| Distillate quality | (ppm) | <10 | <10 | <10 | 400–500 | 150–500 |
| Recovery rate | (%) | 22.0 | 25.0 | | 42.0 | |
| Feed water pre-treatment requirement | | No | No | No | Yes | Yes |

water for industrial and chemical applications (Moosa et al. 2022). RO and ED are suitable for brackish water desalination. ED is more suitable and cost-effective for desalting feed water of salinity less than 5000 ppm (Ali Al-Karaghouli & Kazmerski 2013).

## 1.5 Emerging Desalination Technologies

Emerging desalination technologies are adsorption desalination, membrane distillation, forward osmosis, and desalination battery (Ahmed et al. 2021).

### *1.5.1 Adsorption Desalination (AD)*

In adsorption desalination process, vapors generated by heating feed water at low pressure are adsorbed by the adsorbent material which is held at ambient/low temperature. The adsorbed water is then recovered from the adsorbent material by heating it by circulating hot fluid. It is a low-temperature process, hence waste heat/solar heat can be effectively utilized for this process (Moosa et al. 2022). The specific electric energy consumption of this system is in the range of 38.0 to 64.0 kWh/m$^3$ of distillate produced (Sharon & Reddy 2015).

### *1.5.2 Membrane Distillation (MD)*

In membrane distillation process, a thermal gradient is maintained across the hydrophobic membrane separating the heated feed water and permeate. The temperature difference creates partial pressure difference, thereby vapors travel through the membrane and get condensed in the permeate side. Membrane distillation is more compact, requires only less feed water heating and its energy consumption is nearly independent of feed water salinity. Some drawbacks of this system are low distillate productivity and high energy consumption (Ahmed et al. 2021).

### *1.5.3 Forward Osmosis (FO)*

In forward osmosis system, natural osmosis phenomenon is utilized. Water molecules from the saline feed water move towards the draw solution (hypertonic glucose, dimethyl ether, etc.) which is at a higher concentration than the feed water through the semi-permeable membrane. The draw solution is then heated using waste/solar heat energy to recover the water molecules trapped within it. Forward osmosis is more energy efficient and has specific energy consumption of about 0.25 to 0.84 kWh/m$^3$ (Moosa et al. 2022; Sharon & Reddy 2015).

### *1.5.4 Desalination Battery (DB)*

Desalination battery is an extended version of capacitive deionization (CDI). In CDI, electric field is applied to saline water such that the positive and negative ions migrate towards the oppositely charged porous electrodes and the feed water becomes free from salt. The electrodes can be regenerated by desorption of the ions by reversing the potential difference. The salt removal efficiency mainly depends on the geometry and material of the electrode. Desalination battery consists of sodium ion, chlorine ion storage electrodes, and seawater. During charging, $Na^+$ and $Cl^-$ ions of saline feed water get stored in the electrode material by chemical bonding, thereby the feed water becomes desalinated water. The electrodes can be regenerated by discharging the battery. The energy consumption of the desalination battery for desalting seawater to produce desalinated water of 192 ppm is about 4.91 kWh/m$^3$ which is comparable to the energy consumption of RO system (Ahmed et al. 2021).

## 1.6 Utilization of Solar Energy for Desalination

Desalination is an energy-intensive process. It has been estimated that at least 10000 tons of oil are required every year to produce 1000 $m^3/d$ of desalinated water (Kalogirou 2005). Fossil fuel-operated conventional desalination plants have contributed to significant greenhouse gas (GHG) emissions to the environment which has been shown earlier in Table 1.2. Available global commercial desalination plant capacity can address only 7.5% of the global drinking water demand. Therefore, there is an urgent need to install additional fossil fuel-powered desalination systems to address the rising fresh water demands. These additional plants may contribute to increased pollution and rapid environmental degradation (Gude et al. 2010). Hence, replacement of fossil fuel with renewable energy in desalination sector is highly essential to mitigate these harmful gas emissions and to improve the sustainability of desalination sector (Subramani et al. 2011).

Per capita availability of fresh water in arid, semi-arid, rural, and remote regions are declining at a faster rate and the supply of piped water to these regions from the centralized desalination systems is projected to be very expensive. Renewable energy-driven desalination systems are highly suitable for these regions where no other mode of power supply is possible (Chaibi 2000). Renewable energy commonly employed for desalination is solar, wind, and geothermal energy. Among these, solar energy occupies nearly 57.0% of the global renewable energy-based desalination market. Wind energy and other renewable energy occupy 24.0% and 19.0% of the global renewable energy-based desalination market (Eltawil & Zhengming 2009). Arid, semi-arid, rural, and remote regions are blessed with abundant solar radiation, hence solar energy-based community-scale desalination systems can be a good option to tackle water scarcity in these zones in a sustainable way (Shatat et al. 2013).

The thermal and electrical energy requirement of desalination systems can be obtained from the incident solar energy in a single step which is not possible with other renewable energy sources. Moreover, solar thermal energy can be used effectively for both desalination and detoxification. Fossil fuel-rich Middle East and Arab countries have opened their doors towards exploring the possibilities of utilizing solar energy in their desalination sectors to improve sustainability by minimizing carbon emissions (Abutayeh & Yogi Goswami 2009). Global horizontal solar radiation intensity over the globe is shown in Fig. 1.3. Comparing the global water stress atlas (Fig. 1.2) and global horizontal solar radiation intensity atlas (Fig. 1.3), it can be seen clearly that the regions under water stress are blessed with abundant solar radiation potential of at least 4.6 $kWh/m^2d$. Thus, solar energy can be justified as an effective option for powering desalination systems in water-scarce zones. In addition, solar thermal collectors employed for feed water heating are very well developed (Rodriguez et al. 2002).

Adoption of solar energy for desalination process will help nations to achieve 4 out of 17 sustainable development goals (SDGs) set by the United Nations (UN) to achieve sustainability by 2035. The possible achievable SDGs by adopting solar energy in desalination sector are, namely, SDG 6—clean water and sanitation, SDG

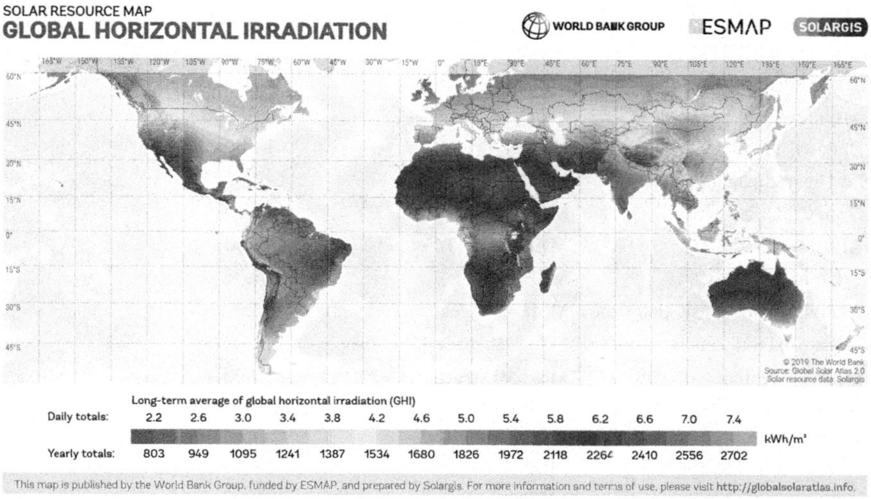


**Fig. 1.3** Atlas showing global horizontal solar radiation potential (Solargis 2022)

11—sustainable cities and communities, SDG13—climate action, and SDG 16—peace, justice, and strong institutions. UN has already reported and warned about water-related war/conflicts in near future due to rising water stress combined with problems associated with the share of nearly 286 international rivers and 592 trans-boundary aquifers by 153 countries (Xiao et al. 2013). Solar desalination process can produce significant amount of drinking water for population, thereby it can help to maintain peace among communities hence achieving SDG 16 is highly possible.

## 1.7 Commercial Solar Thermal Desalination Systems

The selected list of successfully operating commercial solar thermal desalination systems at various parts of the globe is given in Table 1.3. Mostly, MSF and MED systems are preferred for large-scale solar thermal energy-operated desalination systems. However, selection of proper/suitable desalination technology is very important for flexible and maintenance-free operation in arid, remote, and rural areas (Shatat et al. 2013).

The commercialization status of various solar desalination systems is listed in Table 1.4. It could be seen that high-capacity solar PV-powered RO system, solar-powered MED system, and low-capacity solar still are at the application stage. Moreover, it can be observed that the solar still is relatively economical for small-scale applications where water demand is within 100.0 L/d. The required quantity of water for activities like drinking, cooking, bathing, washing, and ablution is about 3.0, 5.0, 15.0, 7.0, and 10.0 L/capita/d, respectively. The minimum daily water demand of a

**Table 1.3** List of well-known commercial solar thermal desalination systems in operation (Bundschuh et al. 2021)

| Location, country | Process | Capacity ($m^3$/ d) |
|---|---|---|
| Arabian gulf | MED | 6000.00 |
| El Paso, Texas, USA | MSF | 19.00 |
| La Desired island, France | MED–14 effects | 40.00 |
| Gran Canaria, Spain | MSF | 10.00 |
| Almeria, Spain | MED-TVC–14 effects | 72.00 |
| Berken, Germany | MSF | 10.00 |
| Al Azhar University, Palestine | MSF–4 stages | 0.20 |
| Takami Island, Japan | MED–16 effects | 16.00 |
| La Paz, Mexico | MSF-10 stages | 10.00 |
| Islands of Cape Verde, Cape Verde | Auto-flash | 300.00 |
| Margarita de Savoya, Italy | MSF | 50.00–60.00 |
| University of Ancona, Italy | MED | 30.00 |
| Lampedusa Island, Italy | MSF | 0.30 |
| Tunisia | MSF | 0.20 |
| Abu Dhabi, UAE | MED–16 effects | 16.00 |
| Dead Sea, UAE | MED–18 effects | 120.00 |
| Safat, Kuwait | MSF | 10.00 |
| Kuwait | MSF | 100.00 |
| Dead Sea, Jordan | MED | 30.00 |
| Hzag, Tunisia | Solar still | 0.10–0.35 |

**Table 1.4** Commercialization status of various solar-powered desalination technologies (Bundschuh et al. 2021)

| Type of solar powered desalination unit | Capacity ($m^3$/d) | Potable water production cost (USD/$m^3$) | Commercialization status |
|---|---|---|---|
| PV-Seawater RO unit | <100.00 | 11.70 to 15.60 | Under application |
| PV-Electro-dialysis | <100.00 | 10.00 | Advanced R&D |
| Membrane distillation | 0.15 to 10.00 | 10.40 to 19.50 | Advanced R&D |
| Multi-effect humidification and de-humidification unit | 1.00 to 100.00 | 2.60 to 6.50 | Advanced R&D |
| Concentrated solar power-based MED | >5000.00 | 2.00 to 2.50 | Under application |
| Solar still | <0.10 | 1.40 to 6.50 | Under application |

**Fig. 1.4** Solar still at Las Salinas, Chile (4700 $m^2$ aperture area, Distillate-23000 L/d) (Hirschmann 1975)

**Fig. 1.5** Solar still at Awania Village, India (1867 $m^2$ aperture area, Distillate–5000 L/d) (Natu et al. 1979)

person living in rural and remote region is estimated to be about 40 L/d. Hence, solar still can be considered as a better option for application in rural, remote, and coastal regions. Large-scale single-slope solar still of capacity 23,000 L/d has operated successfully for continuous 40 years in mines of Chile whose photograph is shown in Fig. 1.4. Double slope solar still shown in Fig. 1.5 of capacity 5000 L/d has been used to treat fluorine-contaminated well water of Awania Village in India. Photograph of solar still named "**water pyramid**" shown in Fig. 1.6, which has the capability to

**Fig. 1.6** Water pyramid developed by aqua aero water systems, BV Netherlands (650 $m^2$ aperture area, Distillate—1250 L/d, rain water harvest—600 $m^3$/year) (Tukker et al. 2010)

produce distillate of about 1250 L/d and collect 600 $m^3$ of rain water/year in tropical climate conditions has been designed and successfully implemented in various parts of the globe by Aqua Aero Water Systems, BV Netherlands.

## 1.8 Classifications of Small-Scale Solar Thermal Desalination Systems

In the past few decades, researchers throughout the globe have significantly contributed to enhance solar thermal desalination systems' distillate productivity. Moreover, compact, high productive, and hybrid-type solar thermal desalination systems have also been developed. Solar thermal desalination systems (non-membrane type) can be broadly classified into direct and indirect type. In direct type, solar radiation enters the aperture of the system directly and initiates evaporation of saline water kept within it to produce potable water. In indirect type, solar radiation is utilized by solar collectors to produce hot heat transfer fluid/hot feed water which is then transferred to the desalination system for initiating water evaporation. Generally, in indirect type, the aperture of the system is not exposed to direct solar radiation. Solar thermal desalination systems can also be further classified based on their shape, orientation, and mode of operation. General classification of compact/small-scale solar thermal desalination systems is shown in Fig. 1.7. The type of solar thermal desalination systems most suitable for various locations based on solar radiation potential and ambient temperature is listed in Table 1.5.

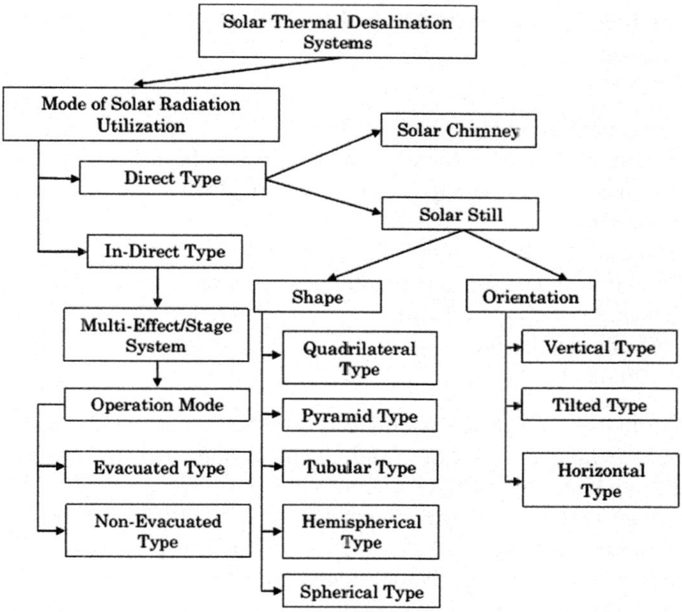


**Fig. 1.7** General classification of compact solar thermal desalination systems

**Table 1.5** Suitable solar thermal desalination systems based on climatic conditions (Xiao et al. 2013)

| Climatic conditions | | Suitable solar thermal desalination system |
|---|---|---|
| Yearly cumulative solar radiation potential ($MJ/m^2d$) | Yearly average ambient temperature (°C) | |
| 4200–5000 | ~4.5 to 22.2 | Reflector or solar collector assisted system |
| 5000–5400 | ~5.8 to 22.2 | Heat energy storage assisted system |
| 5400–6700 | ~5.2 to 12.9 | System with increased evaporating surface area |
| >6700 | ~6.5 to 14.4 | System integrated with enhanced condensing surfaces, Multi-effect/stage systems |

## 1.9 Features of Small-Scale Solar Thermal Desalination Systems

(a) Low-cost and high-quality distillate production.
(b) Can be constructed using locally available materials.
(c) Life time nearly equivalent to commercially available large-scale thermal desalination systems.
(d) Can be easily installed at roof tops.
(e) Suitable for arid, coastal, remote, and rural regions due to requirement of low maintenance and operation cost.
(f) No brine disposal problem.
(g) No environmental threat.

Solar thermal desalination systems can also be utilized in hospitals, telephone exchanges, automobile workshops, and chemical industries for producing double distilled water which can be used for various industrial processes, sterilizing medical appliances, and refilling batteries. Even though solar still has significant benefits, their main limitations which prevent its wider public acceptability are low distillate production rate and huge ground area occupancy. For sizing area of the simple solar still to meet the daily water requirement, daily water demand and minimum distillate production of the still per $m^2$ must be known. The maximum area required for installing solar still to meet the daily fresh water demand can be estimated by (Tiwari and Somwanshi 2018),

$$\text{Maximum area required} = \frac{\text{Daily demand}}{\text{Minimum yield of the still per m}^2} \tag{1.1}$$

## References

Abutayeh, M., & Yogi Goswami, D. (2009). Solar flash desalination under hydrostatically sustained vacuum. *Journal of Solar Energy Engineering, 131*, 031016-1–7.

Ahmed, F. E., Khalil, A., & Hilal, N. (2021). Emerging desalination technologies: Current status, challenges and future trends. *Desalination, 517*, 115183.

Alcamo, J., Doll, P., Henrichs, T., Kaspar, F., Lehner, B., Rosch, T., & Siebert, S. (2003). Global estimates of water withdrawals and availability under current and future "business-as-usual" conditions. *Hydrological Sciences, 48*(3), 339–348.

Ali Al-Karaghouli, A., & Kazmerski, L. L. (2013). Energy consumption and water production cost of conventional and renewable-energy-powered desalination processes. *Renewable and Sustainable Energy Reviews, 24*, 343–356.

Ali, M. T., Fath, H. E. S., & Armstrong, P. R. (2011). A comprehensive technoeconomic review of indirect solar desalination. *Renewable and Sustainable Energy Reviews, 15*, 4187–4199.

Bilton, A. M., Wiesman, R., Arif, A. F. M., Zubair, S. M., & Dubowsky, S. (2011). On the feasibility of community scale photovoltaic powered reverse osmosis desalination systems for remote locations. *Renewable Energy, 36*, 3246–3256.
Bundschuh, J., Kaczmarczky, M., Ghaffour, N., & Tomaszewska, B. (2021). State of the art of renewable energy sources used in water desalination: Present and future prospects. *Desalination, 508*, 115035.
Chaibi, M. T. (2000). An overview of solar desalination for domestic and agriculture water needs in remote arid areas. *Desalination, 127*, 119–133.
Eltawil, M. A., & Zhengming, Z. (2009). Wind turbine-inclined still collector integration with solar still for brackish water desalination. *Desalination, 249*(2), 490–497.
Escobar, I., & Schafer, A. (2009). *Sustainable water for future: Water recycling versus desalination*. Elsevier.
Garcia-Rodriguez, L., Palmero-Marrero, A. I., & Gomez-Camacho, C. (2002). Comparison of solar thermal technologies for applications in seawater desalination. *Desalination, 142*, 135–142.
Gude, V. G., Nirmalakhandan, N., & Deng, S. (2010). Renewable and sustainable approaches for desalination. *Renewable and Sustainable Energy Reviews, 14*, 2641–2654.
Hirschmann, J. R. (1975). Solar distillation in Chile. *Desalination, 17*, 17–30.
Kalogirou, S. (2005). Seawater desalination using renewable energy sources. *Progress in Energy and Combustion Science, 31*, 242–281.
Kaur, R., Wani, S. P., Singh, A. K., Lal, K. (2012). Wastewater production, treatment and use in India. *Presented at the Second Regional Workshop of the Project 'Safe Use of Wastewater in Agriculture'* (pp. 16–18). New Delhi.
Manju, S., & Sagar, N. (2017). Renewable energy integrated desalination: A sustainable solution to overcome future fresh-water scarcity in India. *Renewable and Sustainable Energy Reviews, 73*, 594–609.
Mito, M. T., Ma, X., Albuflasa, H., & Davies, P. A. (2019). Reverse osmosis (RO) membrane desalination driven by wind and solar photovoltaic (PV) energy: State of the art and challenges for large-scale implementation. *Renewable and Sustainable Energy Reviews, 112*, 669–685.
Moosa, B., Trivedi, P., Saleem, H., & Zaidi, S. J. (2022). Desalination in the GCC countries-a review. *Journal of Cleaner Production, 357*, 131717.
Nandi, A., Megiddo, I., Ashok, A., Verma, A., & Laxinarayan, R. (2017). Reduced burden of childhood diarrheal diseases through increased access to water and sanitation in India: A modeling analysis. *Social Science & Medicine, 180*, 181–192.
Natu, G. L., Goghari, H. D., & Gomkale, S. D. (1979). Solar distillation plant at Awanta, Gujarat, India. *Desalination, 31*, 435–441.
Rebhun, M. (2004). Desalination of reclaimed wastewater to prevent salinization of soils and groundwater. *Desalination, 160*, 143–149.
Reif, J. H., & Alhalabi, W. (2015). Solar-thermal powered desalination: Its significant challenges and potential. *Renewable and Sustainable Energy Reviews, 48*, 152–165.
Rijsberman, F. R. (2006). Water scarcity: Fact or fiction? *Agricultural Water Management, 80*, 5–22.
Sharon, H., & Reddy, K. S. (2015). A review of solar driven desalination technologies. *Renewable and Sustainable Energy Reviews, 41*, 1080–1118.
Shatat, M., Worall, M., & Riffat, S. (2013). Economic study for an affordable small scale solar water desalination system in remote and semi-arid region. *Renewable and Sustainable Energy Reviews, 25*, 543–551.
Solargis (2022). https://solargis.com/maps-and-gis-data/download/world.
Subramani, A., Badruzzaman, M., Oppenheimer, J., & Jacangelo, J. G. (2011). Energy minimization strategies and renewable energy utilization for desalination: A review. *Water Research, 45*, 1907–1920.
Tiwari, A. K., & Sonwanshi, A. (2018). Techno-economic analysis of mini solar distillation plants integrated with reservoir of garden fountain for hot and dry climate of Jodhpur (India). *Solar Energy, 160*, 216–224.

Tortajada, C. (2016). Policy dimensions of development and financing of water infrastructure: The cases of China and India. *Environmental Science & Policy, 64*, 177–187.
Tukker, J. J., Anema, K., Nitzsche, M. (2010). Water pyramids: innovative design provides sustainable water delivery. *Journal AWWA*, 26–28.
United Nations (2019). Leaving no one behind. The United Nations world water development report.
Xiao, G., Wang, X., Ni, M., Wang, F., Zhu, W., Luo, Z., & Cen, K. (2013). A review on solar stills for brine desalination. *Applied Energy, 103*, 642–652.

# Chapter 2
# Solar Radiation Potential: Modeling, Measurement, and Utilization

## 2.1 Solar Energy

Solar energy is clean, non-toxic, and it cannot be exhausted and thereby it falls under renewable energy category. Moreover, solar energy is the primary source for most of the renewable energy sources (wind energy, biomass, hydro-energy, and tidal energy) available in Mother Earth. Even fossil fuel is also considered a stored form of solar energy in early ages. However, due to high greenhouse gas emissions, air pollution, and limited reserves of fossil fuel, there is always an urgent call for incorporating renewable energy in various sectors of nations around the globe. Among all sectors, energy and water sectors are more essential for sustaining the improved life style of mankind and they also play an important role in enhancing socio-economic betterment of nations. Solar energy can be utilized effectively to address the demands in energy and water sectors. Global annual energy demand is about $4.1 \times 10^{20}$ Joules and the amount of solar energy reaching earth's surface per hour is about $4.3 \times 10^{20}$ Joules. Moreover, cumulative annual solar energy reaching earth's surface is estimated to about ten times higher than the total energy content of available fossil and nuclear fuel reserves in Mother Earth.

## 2.2 Structure of the Sun

The sun is a middle-aged star of diameter, mass, and luminosity of $1.36 \times 10^6$ km, $2.0 \times 10^{30}$ kg and $4.0 \times 10^{26}$ W, respectively. It is a nuclear fusion reactor containing high-pressure plasma (ionized gases) retained by its own gravitational forces. The sun is mainly made of hydrogen atoms and it currently has about 70 to 73 wt% hydrogen atoms, 25 wt% helium atoms, and remaining 2 to 5 wt% other elements. The sun consists of a central core which is then followed by radiation zone, convection zone, and solar atmosphere (Garg and Prakash 2009).



H. Sharon and K. Srinivas Reddy, *Solar Thermal Desalination Technologies for Potable Water*, https://doi.org/10.1007/978-3-031-49155-9_2

### 2.2.1 Core

The core of the sun extends to a radius of about $0.2R$ and it has a density, pressure, and temperature of about 15,000 kg/m$^3$, $1.5 \times 10^7$ K, and $250 \times 10^9$ bar. The high pressure is due to high gravitational force. Approximately, the core contains about 40 to 50% of the total mass of the sun. Nuclear fusion reaction occurs inside the core where four hydrogen atoms combine to form one helium atom. Mass of the formed helium atom is about 0.72% less than combined hydrogen atoms. This loss in mass is converted into tremendous energy as per Einstein's equation ($E = mc^2$). Rate at which energy is released from the sun is about $4.0 \times 10^{26}$ J/s.

### 2.2.2 Radiation Zone

It lies between $0.2R$ and $0.7R$ and in this zone energy released from the core is transmitted by radiation. The temperature and density of this zone are about $5 \times 10^6$ K and at least 100 kg/m$^3$, respectively. It takes at least 1000 years for the energy released from the core to leave this zone.

### 2.2.3 Convection Zone

This zone extends between $0.7R$ and $R$ and its temperature and density are about $2 \times 10^6$ K and $10^{-5}$ kg/m$^3$, respectively. Turbulent motion of hot gases occurs in this zone where hot gases rise, irradiate their energy, gets cooled off, and sink down in a cyclic manner.

### 2.2.4 Solar Atmosphere

The atmosphere of the sun consists of photosphere, chromosphere, and the corona.

- *Photosphere*: It is the lowest and denser ($10^{-4}$ kg/m$^3$) part of the solar atmosphere and is about 300 km thick. It is the only visible surface of the sun (to the naked eye) and its temperature is about 4600 K (top) to 6400 K (bottom). Photosphere is usually considered as the surface of the sun below which it is not possible to view through.
- *Reversing layer*: It is formed of a several hundred km thick cooler gases and it lies between the photosphere and chromosphere.
- *Chromosphere*: It is about 2000 km thick and it lies between the corona and the reversing layer. The chromosphere can be seen as a reddish line during total solar

eclipse day. The reddish line is due to the burning of superheated hydrogen present in this hot layer.

- *Corona*: It is a very thin and outermost layer of the sun's atmosphere whose temperature is about $10^6$ K. It has very low density (of about $10^{-3}$ kg/m$^3$) and it appears as white plumes of ionized gases flowing into the space which can be seen only on total solar eclipse day. The sudden high temperature in this region is still a mystery and the heating process causing this temperature rise is believed to be due to the magnetic activity occurring in the sun's atmosphere.

Each latitude of the sun rotates at slightly different speed. It takes about 25 earth days and 36 earth days for completion of one rotation by the sun's surfaces near its equator and the poles, respectively, due to the fact that the sun is not a solid body like the earth. Moreover, the sun also revolves around the center of the Milky Way Galaxy at a speed of 8,00,000 km/h. It takes about 250 million years for the sun to complete one revolution.

## 2.3 Sun-Earth Geometric Relations

The earth has a mass and diameter of about $5.97 \times 10^{24}$ kg and $12.7 \times 10^3$ km, respectively. The earth completes one rotation about its own axis in 24 h and completes one revolution about the sun in about 365 days. Earth revolves around the sun in an elliptical path with an eccentricity of only about 0.0167 (closer to circular path). Thereby, the distance between the center of the sun and the center of the earth varies throughout the year and its maximum and minimum value is about $15.2 \times 10^9$ km (occurs during early July) and $14.7 \times 10^9$ km (occurs during early January), respectively. The position in which earth is closer to the sun is called perihelion and the position in which earth is far from the sun is called aphelion. The earth is closer to the sun by about 4.0% in perihelion than the aphelion. The mean earth-sun distance is about $14.9 \times 10^7$ km and is called one Astronomical Unit (1 AU). The angle subtended by the sun's disk over any point in the earth is about 0.54° or 32′. The earth-sun distance ($E_o$ in AU) for each day in a year can be estimated by the following equation (Foster et al. 2010):

$$E_0 = 1 + \left(0.033 \cos\left[\frac{360n}{365}\right]\right) \tag{2.1}$$

where "$n$" is the day number of the year. For January 30, $n = 30$; February 29, $n = 60$.

The earth's axis of rotation is tilted by about 23.45° from the perpendicular to the ecliptic plane. The rotational axis of the earth in North Pole always faces the North Star Polaris and earth's tilt angle does not vary during its rotation and revolution. The tilt angle of the earth is a major reason for the occurrence of different seasons like winter, spring, summer, and autumn. The position of the earth around the sun during

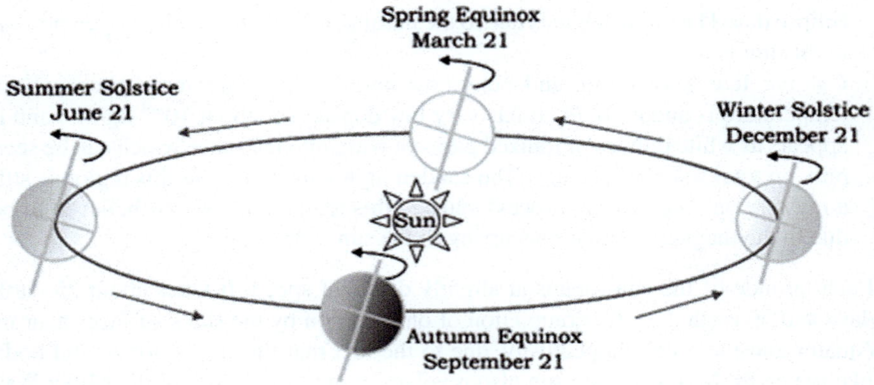


**Fig. 2.1** Position of the earth around the sun in various seasons

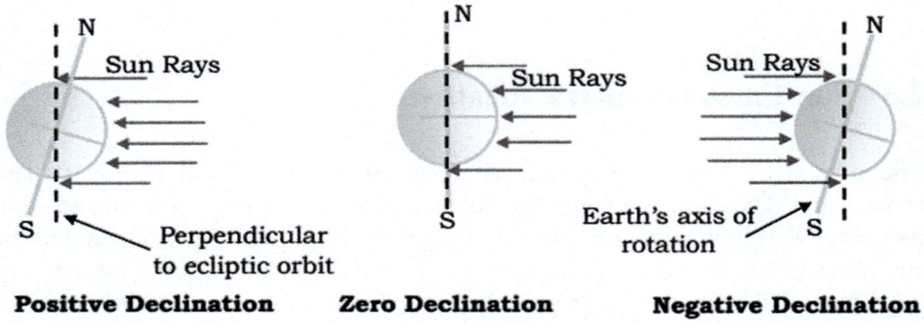


**Fig. 2.2** Impact of declination angle on sun ray's incidence angle over the earth's surface

various seasons is shown in Fig. 2.1. The angular displacement of the sun from the earth's equator is called **declination angle** ($\delta$). Declination angle can also be defined as the angle between the line joining the centers of the earth and the sun to the earth's equatorial plane. Declination angle varies with seasons and it ranges from +23.45 to −23.45°. The impact of declination angle on sun ray's angle of incidence over the earth's surface is shown in Fig. 2.2. The variation of declination angle for different days is estimated by (Foster et al. 2010)

$$\delta = 23.45 \sin\left[\frac{360}{365}(284 + n)\right] \tag{2.2}$$

### 2.3.1 *Summer Solstice*

Summer solstice occurs in June 21 for locations in northern hemisphere of the earth. During summer solstice, the North Pole of the earth is tilted towards the sun and the regions between 66.5°N and 90°N latitude receive sun light for 24 h. The sunshine hours in other northern latitudes will be higher than 12 h. Moreover, sun rays will be perpendicular to the locations at 23.5°N latitude. Declination angle is +23.45° during summer solstice in northern hemisphere.

### 2.3.2 *Winter Solstice*

Winter solstice occurs in December 21 for locations in northern hemisphere of the earth. During winter solstice, the North Pole of the earth is tilted away from the sun and the regions between 66.5°N and 90°N latitude will be in total darkness for 24 h. The sunshine hours in other northern latitudes will be lesser than 12 h. The sun rays will be perpendicular to the locations at 23.5°S latitude. Declination angle is −23.45° during winter solstice in northern hemisphere.

### 2.3.3 *Equinox*

During equinox the earth's axis is neither tilted towards nor tilted away from the sun. Spring equinox and autumn occur in March 21 and September 21, respectively. All the locations of the earth receive sun rays for 12 h. Sun rays will be perpendicular to the locations in the equator on both the equinox days. Declination angle is 0° during both the equinoxes.

Summer solstice, autumn equinox, winter solstice, and spring equinox fall on December 21, March 21, June 21, and September 21, respectively. High glancing angle of the sun rays in southern hemisphere than in northern hemisphere makes solar radiation less concentrated in southern hemisphere. In addition, the earth also will be about 4.0% further away from the sun during winter in southern hemisphere. Hence, southern winter will be colder than northern winter.

## 2.4 Earth's Co-ordinate System

Any location on the earth's surface is indicated by its latitude ($L$) and longitude ($\phi$). The schematic of the earth indicating latitude, longitude, and declination angle is shown in Fig. 2.3.

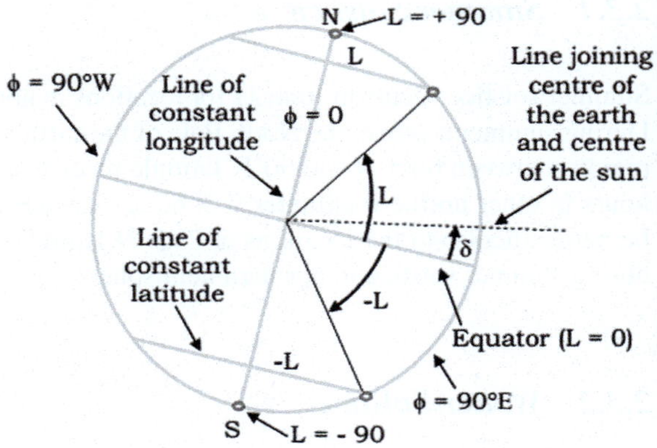


**Fig. 2.3** Schematic of the earth indicating its latitude ($L$), longitude ($\phi$), and declination angle ($\delta$)

## 2.4.1 *Latitude* (L)

It is defined as the angle between the hypothetical line formed by joining the center of the earth and any point on the earth's surface to the hypothetical line's projection over the earth's equatorial plane. Latitude varies between 0° and 90° in both northern and southern hemispheres. At the equator, latitude corresponds to 0°. Northern latitudes are indicated as $+L°$ or $L$°N while southern latitudes are indicated as $-L°$ or $L$°S.

## 2.4.2 *Longitude* (ϕ)

Longitude is also called meridian. Longitude is defined as an imaginary line which runs over the earth's surface and connects both the poles. Longitude passing through the old royal astronomical observatory in Greenwich, England has been designated as 0° longitude and is called the prime meridian. Longitudes to the east of prime meridian are indicated as $\phi$°E and longitudes to the west of prime meridian are indicated as $\phi$°W. Longitudes vary between 0 to 180°E and 0 to 180°W.

Geographical location of Chennai, Tamil Nadu is 13°N, 80°E which indicates that Chennai is 13° north to the equator and 80° east to the prime meridian.

Geographical location of Austin, Texas is 30.2°N, 97.7°W which indicates that Austin is 30.2° north to the equator and 97.7° west to the prime meridian.

Geographical location of Sydney, New South Wales is −33.8°, 151.2°E which indicates that Sydney is 33.8° south to the equator and 151.2° east to the prime meridian.

## 2.5 Apparent Position of the Sun in the Sky

For an observer in the earth, the sun appears to rise in the east and set in the west and this is mainly due to the eastward rotation of the earth. To locate the position of the sun in the sky at any instant solar altitude angle, solar zenith angle, solar azimuth angle, and hour angle are required. The schematic representing the apparent path of the sun in the sky along with the associated angles for a location in northern hemisphere is shown in Fig. 2.4.

### 2.5.1 *Solar Altitude Angle* ($\alpha_s$)

It is the angle between the sun's ray and its projection on the horizon and is estimated by

$$\sin(\alpha_s) = \sin(L)\sin(\delta) + \cos(L) + \cos(\delta)\cos(\omega) \tag{2.3}$$

Arcsine of a negative number does not exist thereby solar altitude angle can be estimated by (Foster et al. 2010)

$$\cos(\alpha_s) = \sqrt{1 - \sin^2 \alpha_s} \tag{2.4}$$

$$\alpha_s = \tan^{-1}\left(\frac{\sin(\alpha_s)}{\cos(\alpha_s)}\right) \tag{2.5}$$

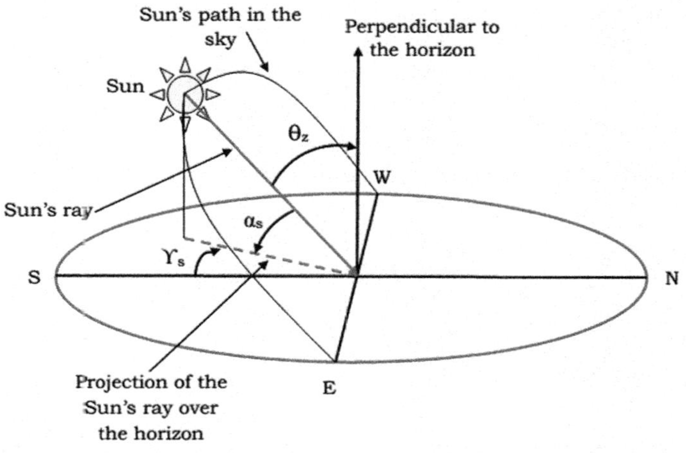


**Fig. 2.4** Apparent position of the sun in the sky

### 2.5.2 Solar Zenith Angle ($\theta_z$)

It is the angle between the sun's ray and the line perpendicular to the horizon and is estimated by

$$\cos(\theta_z) = \sin(L)\sin(\delta) + \cos(L) + \cos(\delta)\cos(\omega) \tag{2.6}$$

It can be seen from Fig. 2.4

$$\alpha_s + \theta_z = 90^\circ$$

Thereby, $\sin(90 - \alpha_s) = \cos(\theta_z)$.

### 2.5.3 Solar Azimuth Angle ($\Upsilon_s$)

It is the angle between due south (for northern hemisphere locations) and the projection of the sun's ray on the horizon.

The solar azimuth angle is estimated by (Foster et al. 2010)

$$\gamma_s = C_1 C_2 \gamma_s + C_3 \left(\frac{1 - C_1 C_2}{2}\right) 180 \tag{2.7}$$

The constant "$C_1$" indicates whether the sun lies within the first or fourth quadrant and above the horizon.

$C_1 = 1$; if $\left[\cos(\omega) > \frac{\tan(\delta)}{\tan(L)}\right]$ else use $C_1 = -1$ for the particular hours;

$C_2 = 1$; if ($L = 0^\circ$ (or) $L = \delta$ (or) $|L| > |\delta|$) else $C_2 = -1$

The constant "$C_3$" indicates whether the sun has crossed the local meridian or not.

$C_3 = 1$; if ($\omega \leq 0$) else $C_3 = -1$

The pseudo-solar azimuth angle "$\gamma_s$" is estimated by

$$\sin(\gamma_s) = \frac{\sin(\omega)\cos(\delta)}{\sin(\theta_z)} \tag{2.8}$$

### 2.5.4 Hour Angle ($\omega$)

It is defined as the angle through which the earth must be rotated to bring the local meridian (longitude) directly in line with the sun's ray. The time when the sun is at the highest position in the sky in a day is called solar noon. Earth needs 24 h to complete one full rotation (i.e., 360°). Therefore, in 1 hour, the earth must have rotated by about 15°. At solar noon, solar time is 12:00 and hour angle is 0°. Other than solar noon, hour angle during forenoon and afternoon hours is taken as negative and positive, respectively.

At solar time of 11:00 A.M., $\omega = -15°$.
At solar time of 12:00 P.M., $\omega = 0°$.
At solar time of 13:00 P.M., $\omega = +15°$.
At solar time of 14:30 P.M., $\omega = +37.5°$.
(60 min = 15°; 30 min = 7.5°).

From Fig. 2.4, it can be seen that during the sun set or sun rise, solar zenith angle ($\theta_z$) is 90°. Substituting, the same in Eq. (2.6), helps to estimate sun set hour angle which is given as

$$\omega_{\text{sunset}} = \cos^{-1}(-\tan(L)\tan(\delta)) \tag{2.9}$$

Hours between solar noon to sun set are estimated by dividing "$\omega_{\text{sunset}}$" by 15° (as earth rotates by 15° in 1 h).

The total sunshine hours in a day ($N_s$) can be then estimated as

$$N_s = \frac{2}{15}\omega_{\text{sunset}} \tag{2.10}$$

## 2.6 Extraterrestrial Solar Radiation Intensity Estimation

The solar radiation intensity measured exterior to the earth's atmosphere is called **extraterrestrial solar radiation intensity** ($I_o$). The amount of solar radiation energy (over all wavelengths) received by a surface of unit area kept perpendicular to the sun's ray at the earth's mean distance from the sun outside the earth's atmosphere is called **solar constant** ($I_{sc}$). Solar constant values are measured continuously using space crafts, high-altitude balloons, satellites, and aircrafts. Average of the readings obtained for 30 plus years has been suggested as solar constant value. Solar constant value recommended by World Radiation Center (WRC) is 1367 W/m$^2$. As the earth-sun distance varies day by day, the amount of solar radiation received by the surface kept exterior to the earth's atmosphere also varies as per inverse-square law.

The extraterrestrial solar radiation intensity received by the surface of unit area kept perpendicular to the sun's ray for different days can be estimated by (Foster et al. 2010)

$$I_o = I_{sc}\left[1 + 0.033\cos\left(\frac{360n}{365}\right)\right] \tag{2.11}$$

In the absence of earth's atmosphere, the solar radiation intensity reaching the horizontal surface kept on earth's surface is estimated as

$$I_{ho} = I_o \cos(\theta_z) \tag{2.12}$$

## 2.7 Terrestrial Solar Radiation Intensity Estimation Over Horizontal and Tilted Surfaces

Solar radiation intensity measured on the earth's surface in the presence of earth's atmosphere is called **terrestrial solar radiation intensity**. Earth's atmosphere consists of gas molecules, suspended dust particles, tiny solids, liquid particulates, and various clouds. Because of the presence of these components in the earth's atmosphere, the intensity of solar radiation gets depleted while passing through it. The depletion occurs due to scattering, absorption, and reflection of the solar radiation by these components in the earth's atmosphere (Garg and Prakash 2009).

If the sky is clear the depletion occurs due to

- *Selective absorption*: Solar radiation of certain shorter wavelength below 0.29 $\mu$m gets absorbed by $O_2$, $O_3$ and $CO_2$.
- *Rayleigh scattering*: Scattering of solar radiation by particles size smaller than the wavelength of the radiation.
- *Mie scattering*: Scattering of solar radiation by particles larger than the size of the wavelength of the radiation.

Around one-half of the scattered radiation has escaped to the space and the remaining reaches the earth's surface from various directions as diffuse solar radiation. If the sky is cloudy, part of solar radiation gets absorbed, some gets reflected, and the remaining is transmitted to the earth's surface as diffuse solar radiation.

### *2.7.1 Albedo*

The fraction of the total incident solar radiant energy reflected back to space by the cloud reflection, atmospheric scattering, and earth's surface reflection is called albedo. In general, albedo for the earth is about 0.30.

The radiation received by the surface on the earth will be lesser than the solar radiation intensity in the space due to the above-mentioned physical processes. Any surface on the earth will receive direct, diffuse, and reflected solar radiation as shown in Fig. 2.5.

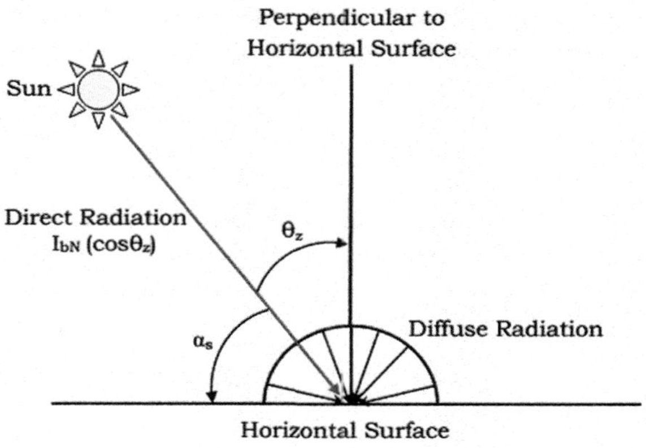


**Fig. 2.5** Schematic representing direct and diffuse radiation reaching horizontal surface kept on the earth's surface

### *2.7.2 Direct Radiation*

The solar radiation whose direction does not change while passing through the earth's atmosphere is called direct or beam solar radiation. Direct solar radiation intensity reaching the earth's surface decreases by about 10% on clear days and 100% on dark and cloudy days.

### *2.7.3 Diffuse Radiation*

The scattered solar radiation reaching the earth's surface is called diffuse solar radiation.

### *2.7.4 Reflected Radiation*

The solar radiation reflected to the surface by the surroundings is called reflected solar radiation.

### *2.7.5 Air-Mass Ratio*

It is the dimensionless path length of the sunlight through the earth's atmosphere. In mathematical form, it can be represented as the ratio of the transverse path length to the vertical path length of the sunlight through the earth's atmosphere. Vertical path length will be shorter than the transverse path length. Air-mass ratio ($m$) can be related to the solar zenith angle and solar altitude by

$$m = \frac{1}{\cos\theta_z} = \frac{1}{\sin\alpha_s} \tag{2.13}$$

The above equation works well for solar zenith angle between 0° to 70°. For a wide range of solar zenith angles, the recommended formula which works even when the sun is in horizon ($\theta_z = 90°$) is given as (Wald 2021)

$$m = \frac{1}{\cos\theta_z + 0.50572(6.07995° + 90° - \theta_z)^{(-1.634)}} \tag{2.14}$$

If the sun is overhead then solar altitude angle and solar zenith angle become 90° and 0°, respectively, thereby the air-mass ratio becomes one. Air-mass ratio equals to zero outside the earth's atmosphere and it is about two for solar zenith angle of 60° or solar altitude angle of 30°.

In the above-mentioned formulas, air-mass ratio is estimated at sea level where the atmospheric pressure ($P_o$) is 1.01325 bar. Hence, for locations other than sea level, the atmospheric pressure ($P_s$) varies thereby corrected air-mass ratio ($m_r$) is estimated as (Wald 2021)

$$m_r = \frac{mP_s}{P_o} \tag{2.15}$$

A simplest way to estimate air-mass ratio at any time in any location is by relating a vertical pole's height ($h$) and its shadow length ($s$) and is given as (Thorpe 2018)

$$m = \sqrt{1 + \left(\frac{s}{h}\right)^2} \tag{2.16}$$

Extraterrestrial solar radiation intensity while passing through the earth's atmosphere under goes extinction as per Bouguer's law. Intensity of direct normal solar radiation reaching the surface kept perpendicular to the sun rays after passing through the earth's atmosphere ($I_{b,N}$) is given as (Kreith and Krumdieck 2015)

$$I_{b,N} = I_o e^{-km} \tag{2.17}$$

where "$k$" is the optical depth and "$m$" air-mass ratio.

For Indian conditions on a clear day, the direct normal solar radiation intensity reaching a surface kept perpendicular to the sun's rays is estimated as (Garg and Prakash 2009)

$$I_{b,N} = \frac{1246}{1 + (0.3135)m} \tag{2.18}$$

**Table 2.1** Classification of day types based on sunshine duration and ratio of daily diffuse to daily total solar radiation intensity (Tiwari et al. 2016)

| Type of day | Sunshine hours | Ratio of daily diffuse to daily total solar radiation intensity |
|---|---|---|
| Clear day | Greater than 9.0 h | Less than 0.25 |
| Hazy day | Between 7.0 and 9.0 h | Between 0.25 and 0.50 |
| Hazy and cloudy day | Between 5.0 and 7.0 h | Between 0.50 and 0.75 |
| Cloudy day | Less than 5.0 h | Greater than 0.75 |

Clear day indicates the day with no clouds. On a clear day, some diffuse radiation will also be reaching the earth's surface. On a cloudy day, all the radiation reaching the earth's surface will be nearly diffuse. Different types of days based on sunshine duration and ratio of daily diffuse to daily total solar radiation intensity are listed in Table 2.1. According to World Meteorological Organization (WMO), sunshine duration refers to the number of hours in a day with direct normal solar radiation intensity greater than 120 W/m$^2$.

Instantaneous total or global solar radiation intensity reaching the horizontal surface ($I_h$) on the earth is schematically shown in Fig. 2.5, and it includes direct (or) beam and diffuse solar radiation components reaching the horizontal surface and is estimated by using clear sky model as (Kreith and Krumdieck 2015)

$$I_h = I_{b,h} + I_{d,h} \tag{2.19}$$

For a horizontal surface, the angle of incidence ($\theta_i$) is equal to the solar zenith angle ($\theta_z$). Instantaneous direct solar radiation intensity reaching the horizontal surface kept on the earth's surface ($I_{b,h}$) is estimated by (Kreith and Krumdieck 2015),

$$I_{b,h} = I_o e^{-km} \cos(\theta_z) \tag{2.20}$$

Instantaneous diffuse solar radiation intensity reaching the horizontal surface kept on the earth's surface ($I_{d,h}$) is estimated by (Kreith and Krumdieck 2015)

$$I_{d,h} = I_{b,N} \times \text{sky diffuse factor} \tag{2.21}$$

Sky diffuse factor varies between 0.057 and 0.136 for various months in sites of the United States (Kreith and Krumdieck 2015).

Instantaneous solar radiation intensity over a tilted surface on the earth will be different from the intensity over the horizontal surface due to the variation in angle of incidence. The schematic indicating various solar angles and tilted surface angles is shown in Fig. 2.6.

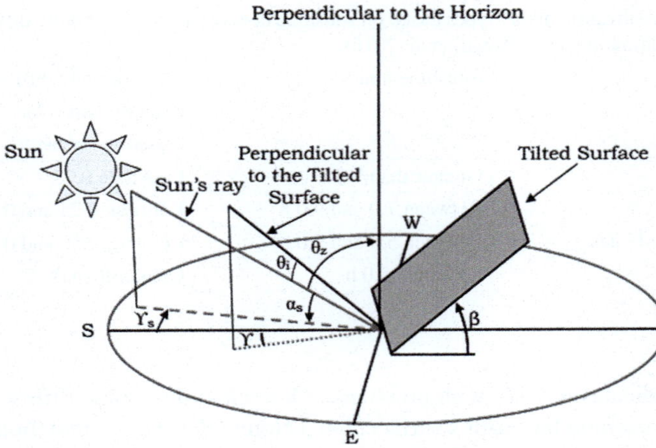


**Fig. 2.6** Schematic representing various solar angles and tilted surface angles

## 2.7.6 *Angle of Incidence* (θ$_i$)

It is the angle between the sun's ray to the normal to the tilted surface. At $\theta_i = 0°$, the sun's ray will be perpendicular to the surface, thereby the incident radiation does not spread out. For angle of incidence other than 0°, incident radiation gets reflected more and it also spreads out. Angle of incidence of sun's ray on any surface facing any direction and kept at any tilt angle can be estimated by

$$\cos(\theta_i) = \sin\delta \sin L \cos\beta - \sin\delta \sin L \sin\beta \cos\gamma + \cos\delta \cos L \cos\beta \cos\omega + \cos\delta \sin L \sin\beta \cos\gamma \cos\omega + \cos\delta \sin\beta \sin\gamma \sin\omega \quad (2.22)$$

## 2.7.7 *Tilt Angle* (β)

It is the angle between the horizontal and the collector surface. For a horizontal surface and vertical surface, $\beta = 0°$ and $\beta = 90°$, respectively.

## 2.7.8 *Surface Azimuth Angle* (ϒ)

It is the angle between the due south and the projection of line perpendicular to the surface over the horizon. For a surface facing due south, due north, due east, and due west, the surface azimuth angle is 0°, 180°, −90°, and +90°, respectively.

For any surface in northern hemisphere, facing due south ($\Upsilon = 0°$), and tilted at any angle, the angle of incidence of the sun's ray over the surface can be estimated by

$$\cos(\theta_i) = \sin(L - \beta)\sin(\delta) + \cos(L - \beta)\cos(\delta)\cos(\omega) \tag{2.23}$$

For any surface in southern hemisphere, facing due north ($\Upsilon = 180°$), and tilted at any angle, the angle of incidence of the sun's ray over the surface can be estimated by

$$\cos(\theta_i) = \sin(L + \beta)\sin(\delta) + \cos(L + \beta)\cos(\delta)\cos(\omega) \tag{2.24}$$

The orientation of the surface towards due south or due north to receive maximum radiation depends on the latitude and declination.

If $(L - \delta) > 0$, then tilt the surface towards due south ($\Upsilon = 0°$).
If $(L - \delta) < 0$, then tilt the surface towards due north ($\Upsilon = 180°$).

For southern latitudes, use $L = -L$ in above equations.

The optimum tilt angle of the surface to receive daily maximum solar radiation intensity over the surface kept in northern hemisphere facing due south when $(L - \delta) > 0$ has been estimated as

$\beta = (L - \delta)$ on positive declination days,
$\beta = (L + |\delta|)$ on negative declination days,
$\beta = L$ on zero declination days.

Total instantaneous solar radiation intensity reaching a tilted surface ($I_t$) kept on the earth has direct, diffuse, and reflected radiation components. Based on Liu and Jordan model, the total radiation over the tilted surface ($I_t$) is estimated by

$$I_t = I_{b,N}\cos(\theta_i) + I_{d,h}\left(\frac{1+\cos\beta}{2}\right) + \rho \underbrace{(I_{b,N}\cos(\theta_z) + I_{d,h})}_{\text{Total radiation reaching the horizontal surface}} \left(\frac{1-\cos\beta}{2}\right) \tag{2.25}$$

"$\rho$" is the albedo of the ground and is usually taken as 0.20.

## 2.8 Solar Radiation Spectrum

Any matter above 0 K having charged particles within it emits electromagnetic waves whose wavelength can range from 0 to infinity. The intensity of radiation emitted by the body per wavelength depends mainly on the temperature of the body and not on the type of the materials or atoms. The electromagnetic waves travel at a speed of light and the type of electromagnetic wave is described by its wavelength and frequency. Relationship between the wavelength ($\lambda$), frequency ($v$), and speed of electromagnetic wave ($C$) is given as

$$C = \lambda v \tag{2.26}$$

The monochromatic or spectral emissive power ($E_{b\lambda(\lambda,T)}$ in W/m$^2 \cdot \mu$m) of a black body at a temperature kept in space or vacuum can be estimated by Planck's law (Cengel 2002)

$$E_{b\lambda(\lambda,T)} = \frac{C_1}{\lambda^5 \left[\exp\left(\frac{C_2}{\lambda T}\right) - 1\right]} \tag{2.27}$$

where $C_1 = 3.742 \times 10^8$ W·$\mu$m$^4$/m$^2$ and $C_2 = 1.439 \times 10^4$ μm·K.

By varying the wavelength from 0 to infinity in Eq. (2.27) for a fixed black body temperature different spectral emissive power can be estimated and a plot can be made between the spectral emissive power of black body and the wavelength. The area under the obtained curve represents the total emissive power of the black body ($E_{b(\lambda)}$) at a given temperature ($T$).

Area under the curve can be obtained by integrating Eq. (2.27) between 0 and infinite wavelength and this result is famous Stefan-Boltzmann's Equation (Cengel 2002):

$$E_{b(\lambda)} = \int_0^\infty E_{b(\lambda,T)} d\lambda = \sigma T^4 \tag{2.28}$$

where $\sigma$ (Stefan-Boltzmann's constant)—$5.67 \times 10^{-8}$ W/m$^2$K$^4$.

A number of plots can be made for a black body at various temperatures and it is interestingly noticed by Wien that at higher temperature the maximum spectral emissive power of the black body occurs at shorter wavelength. As the temperature drops, the spectral emissive power drops and maximum value is noticed at longer wavelengths. The relation between the wavelength at which maximum spectral emissive power occurs and the black body temperature is called **Wien's displacement law** and is given as (Cengel 2002),

$$\lambda_{\max} T = 2898(\mu m.K) \tag{2.29}$$

The spectrum of the sun is shown in Fig. 2.7. The electromagnetic wave emission spectrum of the sun matches the emission spectrum of a perfect black body at a temperature of about 5780 K. Hence, the sun is considered to be a perfect black body whose surface temperature is 5780 K. Electromagnetic emission spectrum of the sun includes wide range of electromagnetic waves from shorter wavelength gamma rays to radio waves of longer wavelength. Electromagnetic spectrum is the term used to

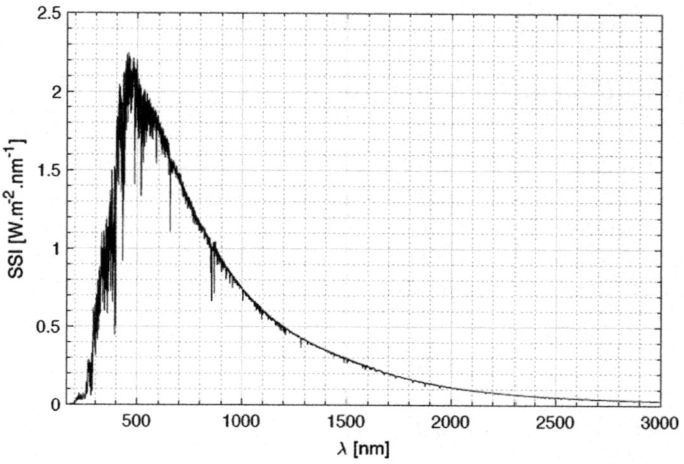


**Fig. 2.7** Solar spectrum (European Space Agency 2022)

describe the entire range of light that exists. **Light** is nothing but a wave of alternating electric and magnetic fields. Wavelength less than 4.0 μm falls under short-wave radiation and above 4.0 μm falls under long wave radiation. Electromagnetic waves between 0.4 and 0.7 μm, 0.7 and 4.0 μm, and 4.0 and 1000 μm wavelengths fall under visible radiation, near-infrared radiation, and infrared radiation, respectively. Electromagnetic waves in the range of $10^3$ to 105 μm and above 105 μm to few hundred km fall under micro-wave and radio wave radiation, respectively. Gamma rays, X-rays, and ultraviolet rays have wavelength in the range of about $10^{-6}$ to $10^{-2}$ nm, $10^{-2}$ to 10 nm, and 10 to 400 nm, respectively.

The maximum spectral emissive power of the sun falls within the visible radiation at a wavelength between 0.48 and 0.50 μm. Around 94% of the total radiant energy emitted by the sun is between 0.20 μm (ultraviolet) and 2.0 μm (near-infrared radiation). Similarly, 47.1% of the total radiant energy emitted by the sun falls within the visible radiation spectrum (0.40 to 0.70 μm). Ozone in the earth's atmosphere absorbs most of the radiation below 0.30 μm and some radiation in the range from 0.30 to 0.40 μm and some of the visible region of the spectrum. $CO_2$ and $H_2O$ vapor in the atmosphere absorbs infrared region of the spectrum. Apart from this, the dust particles and other components in the atmosphere also absorb radiation at different wavelengths of the spectrum. As a result, the spectral emissive power of the sun observed at the earth's surface will not be smooth as in the case of extraterrestrial spectrum. In general, the wavelength of solar radiation reaching the earth's surface will fall within the wavelength range between 0.30 and 2.50 μm. Radiation in the wavelength range between 0.10 μm and 100 μm is called thermal radiation because it is capable of affecting the thermal state of the matter.

## 2.9 Solar Radiation Intensity Measuring Instruments

### *2.9.1 Global Solar Radiation Intensity*

Global solar radiation intensity refers to the summation of the direct solar radiation intensity, diffuse solar radiation intensity, and reflected solar radiation intensity reaching any surface on the earth. It is measured by solar pyranometer whose field of view is 180°. Global solar radiation intensity over the horizontal surface and tilted surface is called **global horizontal irradiance** (GHI) and **global tilted irradiance** (GTI).

### *2.9.2 Diffuse Solar Radiation Intensity*

Scattered sky solar radiation reaching the earth's surface is called diffuse solar radiation intensity. It can be measured by solar pyranometer incorporated with a shade ring. The shade ring prevents direct solar radiation from reaching the pyranometer's sensor.

### *2.9.3 Direct Normal Solar Radiation Intensity*

Direct normal solar radiation intensity refers to the solar radiation received by a surface kept always perpendicular to the incident sun rays. It is measured by pyrheliometer which needs to be tracked towards the sun. The acceptance angle or field of view of pyrheliometer is about 5.0°.

### *2.9.4 Sunshine Duration*

Sunshine duration can be measured with pyrheliometer and Campbell-Stokes sunshine recorder. Number of hours in a day with direct normal solar radiation intensity greater than 120 W/m$^2$ refers to the **sunshine duration** and it can be obtained from pyrheliometer measurements. Campbell-Stokes sunshine recorder consists of a spherical solid glass ball which refracts the sun light to a graduated cardboard paper (sunshine sheet) kept beneath it at its focus. The refracted sunlight causes burning marks over the graduated cardboard paper and from the length of the burned marks the sunshine duration is estimated. The Campbell-Stokes sunshine recorder must be kept oriented at an angle corresponding to the local latitude.

The daily solar radiation potential over any surface at any location is generally given in MJ/m$^2$d or in kWh/m$^2$-d. If the radiation intensity ($I_h$ or $I_t$) is measured at every "$t$"-minute time interval throughout the day, then the daily solar radiation potential can be estimated by daily global horizontal solar radiation potential (GHI) in J/m$^2$d which is given as

$$GHI = \left( \sum_{sunrise}^{sunset} (I_h) \right) \times (t \times 60) \tag{2.30}$$

Daily global horizontal solar radiation potential in Wh/m$^2$d is given as

$$GHI = \frac{\left( \sum_{sunrise}^{sunset} (I_h) \right) \times (t \times 60)}{3600} \tag{2.31}$$

Daily global tilted solar radiation potential can be obtained by replacing "$I_h$" with "$I_t$" in Eqs. (2.30) and (2.31).

## 2.10 Capturing Solar Thermal Radiation Using Collectors

Solar radiation can be directly converted into electricity with the help of solar cells to power homes and industries. Solar radiation can also be utilized for water heating, space heating, steam generation, and various industrial processes with the aid of solar thermal collectors. Solar thermal collectors and photovoltaic modules are important components in solar energy-powered thermal and membrane desalination systems, respectively. Solar thermal collectors can be broadly classified as non-concentrating type and concentrating type as shown in Fig. 2.8, (Evangelisti et al. 2019).

### *2.10.1 Non-concentrating Type*

In non-concentrating-type collectors, no solar radiation concentrators/reflectors are used. The absorber and aperture area of the collector is nearly equal. Flat plate collector and evacuated tube collectors are well-known types of non-concentrating-type collectors. Flat plate collectors can heat working fluid to a temperature of about 85.0 °C at a thermal efficiency of about 34.0%. Evacuated tube collectors can heat working fluid to a temperature of about 125.0 °C at a thermal efficiency of about 30 to 45%. Evacuated tube collectors are more suitable in cold climates/winter as its performance will be superior to flat plate collector due to reduced heat loss from the collector to the ambient. Non-concentrating collectors do not require any sun tracking arrangement and are generally oriented at optimum tilt angle corresponding to the local latitude. Moreover, these collectors can be operated in thermo-siphon mode and forced circulation mode.

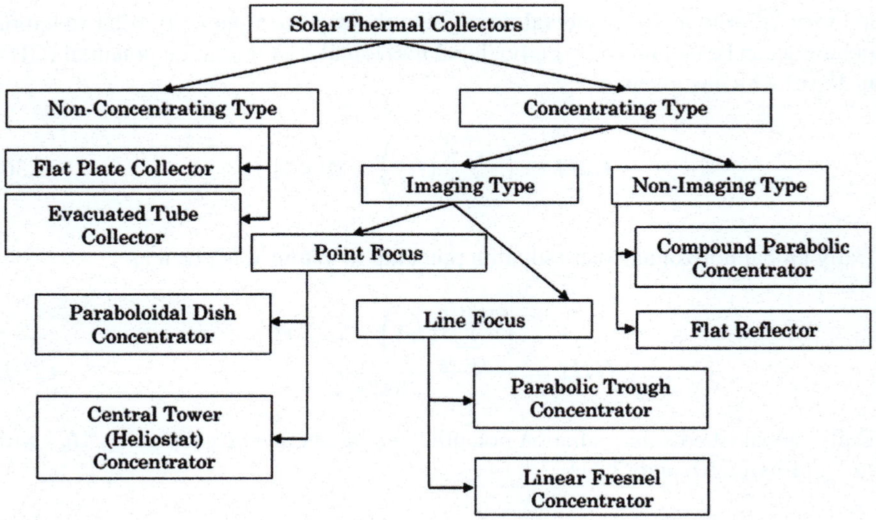


**Fig. 2.8** Classification of solar thermal collectors

### *2.10.2 Concentrating Type*

In concentrating type, reflectors are used to concentrate the solar radiation over the absorber/receiver. The aperture area will be higher than the receiver area. The ratio between the aperture area to the receiver area is called **concentration ratio**. In concentrating collectors, high temperature can be obtained and at the same time heat loss from the receiver to the ambient is also minimized due to its lower surface area. Concentrating type can also be classified as non-imaging-type and imaging-type concentrators.

### *2.10.3 Non-imaging Concentrator*

In non-imaging concentrator collectors, global solar radiation is concentrated over the receiver/absorber with the help of reflectors and it does not produce any image of the sun. In non-imaging concentrator, both direct and diffuse solar radiations are concentrated over the absorber. Compound parabolic concentrator is a non-imaging concentrator and is suitable for applications requiring temperature in the range of 100 to 250 °C. Compound parabolic concentrator solar collector has a concentration ratio less than 5.0 and a thermal efficiency of about 62.0 to 65.0%. Moreover, it has wide acceptance angle, hence tracking is not required. For temperatures higher than 200 °C, non-imaging concentrator collectors have significant heat losses to the ambient, thereby imaging concentrator collectors are highly preferred for temperatures higher than 200 °C.

### 2.10.4 Imaging Concentrator

In imaging concentrator collector, direct solar radiation is concentrated over the receiver/absorber with the help of reflectors and it forms sun's image over the receiver. The imaging concentrator can be line focusing type or point focusing type.

### 2.10.5 Line Focusing Type

Linear Fresnel reflector consists of a number of long flat/curved single axis tracking mirrors installed closer to the ground level which focuses the direct solar radiation over the line cavity receiver kept at its focus. Linear Fresnel reflector is capable of generating fluid temperature of about 100 to 350 °C at an efficiency of about 60 to 65%. Parabolic trough collector consists of parabolic trough-shaped single axis tracking mirror which focuses the direct solar radiation over the line receiver kept at its focus. Parabolic trough collector is capable of generating fluid temperature of about 150 to 400 °C at an efficiency of about 67 to 79%. Concentration ratio of linear Fresnel reflector and parabolic trough collector is about 30 to 60 and 30 to 100, respectively. Construction cost of linear Fresnel reflector is lower than the parabolic trough collector.

### 2.10.6 Point Focusing Type

In point focusing type, the direct solar radiation incident over the reflectors is reflected towards the receiver kept at its focus to produce a circular image. Paraboloidal dish concentrator and central tower are point focusing-type solar concentrators. Paraboloidal dish concentrator uses modified cavity receiver and has a concentration ratio of about 300 to 3000. It can generate heating fluid temperature of about 500 to 1200 °C at a thermal efficiency of 67 to 85%. In central tower, a number of heliostats are used to focus the incident direct solar radiation over the central volumetric receiver to generate heating fluid temperature of about 400 to 800 °C at a thermal efficiency of 58 to 70%. The concentration ratio of central tower unit is about 1500. Point focusing-type concentrators require double axis tracking.

Non-concentrating-type and non-imaging-type concentrators are highly suitable for small-scale solar thermal desalination systems of capacity less than 1.0 $m^3$/d and low-temperature desalination systems requiring feed water temperature around 60 to 80 °C. For large-scale solar thermal-based multi-stage flash and multi-effect distillation process, imaging concentrators like parabolic trough collectors and linear Fresnel reflectors are highly preferred to generate required steam of at least 120 °C to achieve top brine temperature of about 90 °C for the operation of desalination system.

## References

Cengel, Y.A. (2002). *Heat transfer: A practical approach* (2nd edn). McGraw Hill

European Space Agency. (2022). https://www.esa.int/ESA_Multimedia/Images/2017/12/Solar_spectrum.

Evangelisti, L., Vollaro, R. D. L., & Asdrubali, F. (2019). Latest advances on solar thermal collectors: A comprehensive review. *Renewable and Sustainable Energy Reviews,114*, 109318.

Foster, R., Ghassemi, M., & Cota, A. (2010). *Solar energy: Renewable energy and the environment*. CRC Press.

Garg, H. P., & Prakash, J. (2009). *Solar energy: Fundamentals and applications*. Tata McGraw-Hill.

Kreith, F., & Krumdieck, S. (2015). *Principles of sustainable energy systems*. CRC Press.

Thorpe, D. (2018). *Solar energy pocket reference*. Routledge.

Tiwari, G. N., Tiwari, A., & Shyam. (2016). *Handbook of solar energy: Theory*. Springer, Singapore: Analysis and Applications.

Wald, L. (2021). *Fundamentals of solar radiation*. CRC Press.

# Chapter 3
# Basin-Type Solar Thermal Desalination Systems

## 3.1 Basin-Type Solar Thermal Desalination System—Working Principle

The schematic of basin-type solar thermal desalination system is shown in Fig. 3.1. Its main components are insulated blackened basin, condensate collection trough, and glass cover. In general, the basin section is fabricated using galvanized iron sheet or fiber reinforced plastic sheet or mild steel or stainless-steel sheet. Inner surface of the basin section is coated with red-lead primer followed by matt-type black paint to increase solar radiation absorptivity. Provisions for feed water inlet, brine drain, and condensate collection are also provided in the basin. The top opening of the basin is covered by tempered glass cover and perfectly sealed against the basin using silicon sealant. Simple basin-type solar thermal desalination system is also called conventional solar still or basin solar still.

Heat loss from the blackened basin to the ambient is minimized by insulating the bottom and lateral sides of the basin with glass wool/rock wool/foam/thermocol (polystyrene). During sunshine hours, solar radiation passes through the glass cover, feed water, and reaches the basin liner where part of it gets absorbed. Basin liner heated by the absorbed solar radiation transfers majority of its heat energy to the basin water by convection and losses some amount to the ambient by conduction through insulation. The heated basin water then evaporates to form vapors which move towards the glass cover and gets condensed over its inner surface by releasing latent heat of condensation. Glass cover continuously losses heat energy by convection and radiation to the ambient. Hence, it will be relatively at a lower temperature than the basin liner. The condensed vapors reach the condensate collection trough with the aid of gravity. Basin solar thermal desalination system can be of single slope type or double slope type. Single slope type gives higher distillate production during winter season while double slope type is recommended for summer season (Fath 1998).

 
H. Sharon and K. Srinivas Reddy, *Solar Thermal Desalination Technologies for Potable Water*, https://doi.org/10.1007/978-3-031-49155-9_3

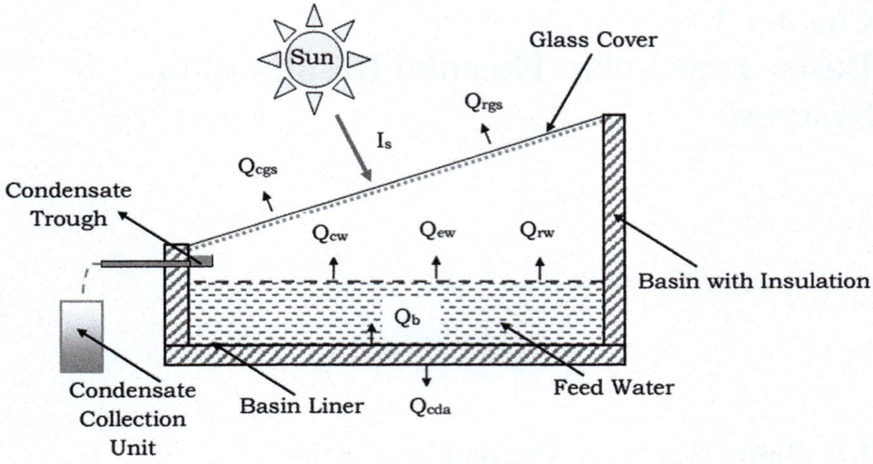


**Fig. 3.1** Schematic of basin-type solar thermal desalination system along with its associated energy transport processes

## 3.2 Thermodynamic Modeling of Basin-Type Solar Thermal Desalination System

The assumptions considered for thermodynamic modeling of basin-type solar thermal desalination system are listed below:

(a) The system is air tight and leakage free.
(b) Water level in the basin is always constant.
(c) Film condensation occurs over the inner surface of the glass cover.
(d) No thermal gradient across the thickness of glass cover and basin water depth.
(e) Produced distillate is free from salinity.

Energy balance in the glass cover,

$$\begin{pmatrix}\text{Rate of change of glass}\\ \text{cover temperature}\end{pmatrix} = \begin{pmatrix}\text{solar radiation absorbed}\\ \text{by glass cover}\end{pmatrix} + \begin{pmatrix}\text{convection, radiation and evaporation heat}\\ \text{transfer from basin water to glass cover}\end{pmatrix} - \begin{pmatrix}\text{convection \& radiation heat loss}\\ \text{cover to ambient}\end{pmatrix}$$

$$(mc)_{gs}\frac{dT_{gs}}{dt} = Q_{sgs} + (Q_c + Q_r + Q_e)_{(w-g)} - (Q_c + Q_r)_{(q-a)} \tag{3.1}$$

Energy balance in basin water,

$$\begin{pmatrix}\text{Rate of change of basin}\\ \text{water temperature}\end{pmatrix} = \begin{pmatrix}\text{solar radiation absorbed}\\ \text{by basin water}\end{pmatrix} + \begin{pmatrix}\text{convection heat transfer}\\ \text{from basin liner to basin water}\end{pmatrix} - \begin{pmatrix}\text{convection, radiation and evaporation heat}\\ \text{transfer from basin water to glass cover}\end{pmatrix}$$

$$(mc)_w \frac{dT_w}{dt} = Q_{sw} + Q_{cb} - (Q_c + Q_r + Q_e)_{(w-g)} \tag{3.2}$$

Energy balance in basin water,

$$\begin{pmatrix}\text{Rate of change of basin}\\ \text{liner temperature}\end{pmatrix} = \begin{pmatrix}\text{solar radiation absorbed}\\ \text{by basin liner}\end{pmatrix} - \begin{pmatrix}\text{convection heat transfer from}\\ \text{basin liner to basin water}\end{pmatrix} - \begin{pmatrix}\text{convection heat loss}\\ \text{from basin liner to ambient}\end{pmatrix}$$

$$(mc)_b \frac{dT_b}{dt} = Q_{sb} - Q_{cb} - Q_{cda} \tag{3.3}$$

Solar radiation absorbed by the glass cover ($Q_{sgs}$ in W) is given by

$$Q_{sgs} = I_s A_{gs} \alpha_{gs} \tag{3.4}$$

where

$I_s$—solar radiation intensity (W/m$^2$),
$A_{gs}$—surface area of the glass cover (m$^2$),
$\alpha_{gs}$—absorptivity of glass cover.

Area of glass cover surface ($A_{gs}$) in relation to its inclination is given as (Tiwari et al. 1994)

$$A_{gs} = A_b \sec\beta \tag{3.5}$$

where

$A_b$—area of the basin (m$^2$),
$\beta$—inclination angle of the glass cover in relation to the basin surface.

Solar radiation absorbed by the basin water ($Q_{sw}$ in W) is given by

$$Q_{sw} = I_s A_w \tau_{gs} \tau_w \alpha_w \tag{3.6}$$

where

$A_w$—surface area of basin water ($m^2$),
$\alpha_w$—absorptivity of basin water,
$\tau_{gs}$—transmissivity of glass cover,
$\tau_w$—transmissivity of basin water.

Solar radiation absorbed by the basin liner ($Q_{sb}$ in W) is given by

$$Q_{sb} = I_s A_b \tau_w \tau_{gs} \tau_w \alpha_b \tag{3.7}$$

where $\alpha_b$ is absorptivity of basin liner.

In both Eqs. (3.6) and (3.7), transmissivity of condensing water film over the inner surface of the glass cover is also considered.

Convection heat transfer from the basin liner to the basin water ($Q_{cb}$ in W) is given by

$$Q_{cb} = h_{cb} A_b (T_b - T_w) \tag{3.8}$$

Convection heat transfer coefficient between the basin liner and basin water ($h_{cb}$ in $W/m^2K$) is estimated by

$$h_{cb} = \frac{0.54(Gr \cdot Pr)^{1/4} K_w}{L} \tag{3.9}$$

where

Gr—Grashof number,
Pr—Prandtl number,
$K_w$—thermal conductivity of water (W/mK),
$L$—length of the basin (m).

Convection heat transfer from the water surface to the glass cover ($Q_{cw}$ in W) is given by

$$Q_{cw} = h_{cw} A_w (T_w - T_{gs}) \tag{3.10}$$

Convection heat transfer coefficient between the basin water and glass cover ($h_{cw}$ in $W/m^2K$) is given by (Singh and Tiwari 2019),

$$h_{cw} = 0.884 \left[ (T_w - T_{gs}) + \frac{(P_w - P_{gs}(T_w + 273.15)}{268900 - P_w} \right]^{1/3} \tag{3.11}$$

where

$P_w$—partial vapor pressure of water vapor at basin water temperature (Pa),
$P_{gs}$—partial vapor pressure of water vapor at glass cover temperature (Pa).

Radiation heat transfer from the basin water surface to the glass cover ($Q_{rw}$ in W) is given by

$$Q_{rw} = h_{rw} A_w (T_w - T_{gs}) \tag{3.12}$$

Radiation heat transfer coefficient between the basin water surface and the glass cover ($h_{rw}$ in W/m$^2$K) is given by

$$h_{rw} = \sigma \frac{1}{\left(\frac{1}{\varepsilon_w} + \frac{1}{\varepsilon_{gs}} - 1\right)} [(T_w + 273.15)^2 + (T_{gs} + 273.15)^2] \times [(T_w + 273.15) + (T_{gs} + 273.15)] \tag{3.13}$$

where

$\sigma$—Stefan-Boltzmann constant ($5.67 \times 10^{-8}$ W/m$^2$K$^4$),
$\varepsilon_w$—emissivity of basin water,
$\varepsilon_{gs}$—emissivity of glass cover.

Evaporation heat transfer from the basin water surface to the glass cover ($Q_{ew}$ in W) is given by

$$Q_{ew} = h_{ew} A_w (T_w - T_{gs}) \tag{3.14}$$

Evaporation heat transfer coefficient between the basin water surface and glass cover ($h_{ew}$ in W/m$^2$K) is given by

$$h_{ew} = 0.016273 h_{cw} \frac{(P_w - P_{gs})}{(T_w - T_{gs})} \tag{3.15}$$

Convection heat transfer from the glass cover to the ambient ($Q_{cgs}$ in W) is given by

$$Q_{cgs} = h_{cgs} A_{gs} (T_{gs} - T_a) \tag{3.16}$$

Convection heat transfer coefficient between the glass cover to the ambient ($h_{cgs}$ in W/m$^2$K) is given by

$$h_{cgs} = \begin{cases} 2.8 + 3.0(V), & \text{for } V \le 5.0\,\text{m/s} \\ 6.15(V)^{0.8}, & \text{for } V > 5.0\,\text{m/s} \end{cases} \tag{3.17}$$

where $V$ is wind speed in m/s.

Radiation heat transfer from the glass cover to the ambient ($Q_{rgs}$ in W) is given by

$$Q_{rgs} = h_{rgs} A_{gs} (T_{gs} - T_a) \tag{3.18}$$

Radiation heat transfer coefficient between the glass cover and the ambient ($h_{rgs}$ in $W/m^2K$) is given by

$$h_{rgs} = \sigma \varepsilon_{gs} \frac{[(T_{gs} + 273.15)^4 - (T_a + 273.15)^4]}{(T_{gs} - T_a)} \tag{3.19}$$

Heat loss from the basin liner to the ambient ($Q_{cds}$ in W) through insulation is given by

$$Q_{cda} = h_{cd} A_b (T_b - T_a) \tag{3.20}$$

Conduction heat transfer coefficient between the basin liner and the ambient is given by (Johnson et al. 2019)

$$h_{cd} = \frac{1}{\left(\frac{W_i}{K_i} + \frac{1}{h_{tba}}\right)} \tag{3.21}$$

where

$W_i$—thickness of insulation (m),
$K_i$—thermal conductivity of insulation ($W/m^2K$)

$$h_{tba} = 5.7 + (3.8(V)) \tag{3.22}$$

## 3.3 Parameters Influencing the Performance of Basin-Type Solar Thermal Desalination Systems

### *3.3.1 Surface Area of Basin Liner*

Basin liner's surface area can be increased by corrugation or by addition of fins. Increasing surface area favors enhanced basin water preheating effect and evaporation rate. Performance of finned basin liner depends on number of fins, fin thickness, fin height, and spacing between each fin. Increasing the fin height increases the solar radiation absorption area, thereby convection heat transfer between the fin and basin water increases leading to more distillate production. Increasing fin thickness and

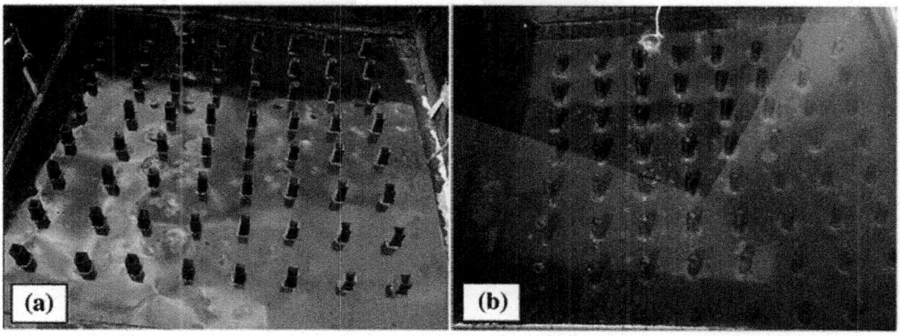


**Fig. 3.2** Photograph of basin liner with **a** hollow square fins and **b** hollow circular fins (Rajaseenivasan et al. 2016)

number of fins reduces distillate production due to reduction in effective absorber area and increase in shadow cast by the fins over the basin liner, respectively (Sebaii et al. 2015). Investigations on application of aluminum, iron, copper, glass, stainless steel, mica, and brass fins over basin liner revealed that the type of fin material has only negligible effect on desalination system's distillate productivity. However, black painted glass and mica-based fins are highly recommended as they don't corrode/deteriorate during operation (El-Sebaii and El-Naggar 2017).

Finned basin system produces more distillate (40.0% higher than conventional basin still) than the system with corrugated basin (21.0% higher than conventional basin still) (Omara et al. 2011). Panchal et al. (2020a) obtained about 26.77% and 24.19% higher distillate productivity by employing inclined and vertical fins over the basin liner. Wrapping fins with wicks have also been practiced by researchers to increase distillate productivity. Alaian et al. (2016) obtained nearly 23.0% higher productivity by utilizing number of wick wrapped vertical wires over basin liner. Rajaseenivasan et al. (2016) obtained around 36.0% and 45.8% higher distillate productivity by incorporating wick wrapped hollow circular and hollow square fins over the basin liner, respectively. Photograph of basin liner with circular and square fins is shown in Fig. 3.2a, b.

Abdullah et al. (2020) modified conventional solar still into trays solar still by incorporating trays on its back wall. Addition of trays increased evaporation area and effective solar radiation exposure area. Distillate yield of trays solar still with no reflectors and tray height of 0.5 cm and basin water depth of 1.0 cm is about 1.5 times higher than distillate yield of conventional basin still. Addition of internal reflectors over the back wall of trays solar still has resulted in 58.0% more distillate yield than conventional basin still due to prevention of heat loss to the ambient via back wall. The trays solar still with internal reflectors, and top and bottom external reflectors shown in Fig. 3.3 has 95.0% higher yield than conventional basin still. The thermal efficiencies of conventional basin still, trays solar still, and trays solar still with internal and external reflectors are about 34.0%, 41.0%, and 50.0%, respectively.

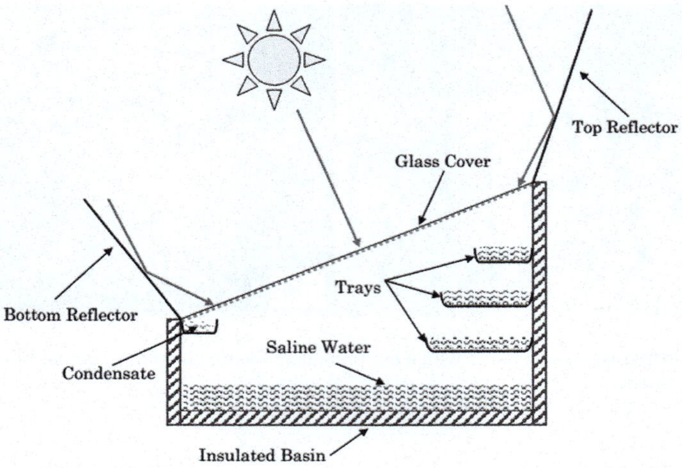


**Fig. 3.3** Schematic of trays solar still with internal and external reflectors (Abdullah et al. 2020)

### *3.3.2 Absorptivity of Basin Liner*

Basin liner must have high absorptivity and low emissivity. Absorptivity of basin liner can be increased by coating it with ordinary black paint or matt black polymethyl methacrylate (Boytorun acrylic) (Kalita et al. 2016). Distillate yield of basin solar thermal desalination system increased from 1.5 $kg/m^2d$ to 2.75 $kg/m^2d$ with increase in basin liner absorptivity from 0.50 to 0.80. Efficiencies of the system at 0.50 and 0.80 absorptivities are about 21 and 38%, respectively. Salem et al. (2020) covered the basin water with top surface black painted floating sponge to increase solar radiation absorptivity. Basin thermal desalination system's distillate productivity dropped with increase in sponge density from 16.0 $kg/m^3$ to 35.0 $kg/m^3$. This observation can be attributed to the drop in porosity with increase in density, thereby the generated vapors are blocked and throttled. The optimum sponge thickness is 30.0 mm and the productivity of basin solar thermal desalination system with floating sponge is about 58.1% higher than the basin solar thermal desalination system without sponge.

Balachandran et al. (2020) coated the basin liner with black enamel paint blended with nano-$Fe_2O_3$ particles at 10.0 wt/wt% and obtained 53.5% higher distillate productivity than conventional basin still without nano-particle-based black paint coating. Kabeel et al. (2017) obtained about 16.0% and 25.0% higher distillate productivities by coating the basin liner with black paint doped with 10.0 and 40.0% wt/wt% of CuO nano-particles. Similarly, yield enhancement of about 6.1% is obtained by coating the basin liner with black paint doped with 0.1 vol/vol% of $TiO_2$ nano-particles (Kabeel et al. 2019a). Increased distillate productivity of desalination systems with nano-particle-coated basin liner is mainly due to the ability of the nano-particles to effectively absorb the incident solar radiation and facilitate enhanced heat transfer between the basin liner and basin water.

### 3.3.3 *Depth of Basin Water*

Water depth represents the thickness of water layer inside the basin and it significantly affects the distillate productivity. Increased water depth indicates increased heat capacity, thereby the time taken for warming up the basin water above the ambient temperature to initiate evaporation process increases for a given amount of solar radiation input. Increased water depth also reduces the amount of solar radiation reaching the basin liner as transmissivity of water decreases with increase in water depth.

Correlation relating the basin water depth ($d_w$ in cm) and distillate productivity ($M_d$ in kg/m$^2$d) of basin-type solar thermal desalination system is given as (Khalifa and Hamood 2009a)

$$M_d = 3.884e^{-0.0458(d_w)} \quad (R^2 = 0.832) \tag{3.23}$$

Reduced water depth enhances evaporation rate due to its low heat capacity and better solar radiation transmission to the basin liner. However, increasing water depth increases nocturnal productivity but decreases overall daily productivity. This observation is due to the utilization of most of the incident solar radiation energy to increase the temperature of huge mass of basin water rather than causing its immediate evaporation. Depth of basin water must be selected in such a manner to prevent dry out of the basin to avoid salt deposition over the basin liner.

### 3.3.4 *Color of Basin Water*

Operating basin solar still at low water depth can cause salt deposition and increase maintenance cost. Moreover, maintaining constant level of low water depth in basin requires precision controls to prevent dry out. However, these problems can be avoided by adding dyes to the basin water, thereby the efficient operation of basin still even at increased water depth is made possible. Addition of dye increases the solar radiation absorptivity of the basin water, thereby salt deposition at the basin liner will not affect its distillate productivity. In addition, the addition of dyes also facilitates the use of deep basins instead of shallow basins, thereby precision controls can be avoided. The absorption spectra of black dye, red dye, and green dye are shown in Fig. 3.4.

Dye concentration above 400 ppm in basin water has shown no significant enhancement in productivity irrespective of the type of dye. Performance of solar desalination system with black dye is better than the system operating with green dye and red dye for dye concentration below 500 ppm. However, at a dye concentration of 500 ppm the yield of the desalination system tends to reach the same magnitude for all the dyes (Rajvanshi 1981). Moreover, for the same concentration of dyes in basin water, increased basin water depth has favored increased distillate productivity

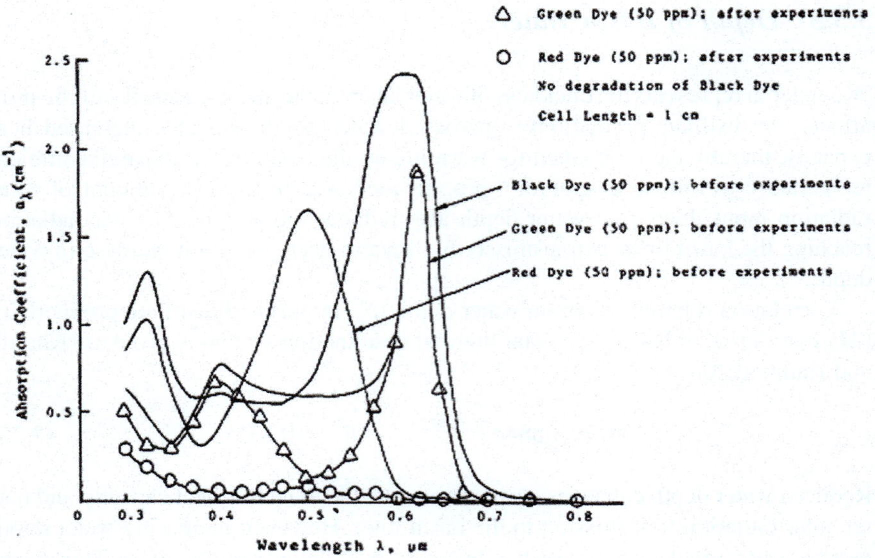


**Fig. 3.4** Absorption spectra of various dyes (Rajvanshi 1981)

because the basin liner temperature remains lower due to reduced heat losses to the ambient via conduction through the insulation. However, (Tamim 1987) suggested that the utilization of dyes in solar desalination systems may cause environment pollution if the reject brine is disposed improperly.

### *3.3.5 Thermal Conductivity of Basin Water*

Distillate productivity can be increased by modifying the thermo-physical properties of basin water. Addition of nano-particles to the basin water can enhance basin water thermal conductivity significantly which in turn can lead to enhanced water evaporation. Thermal conductivity of basin water incorporated with nano-particles depends on nano-particle type, aspect ratio, and its concentration. The thermo-physical properties of various nano-particles are tabulated in Table 3.1. $Al_2O_3$, ZnO, and $SnO_2$ (20.0 nm diameter) concentration of 0.1 wt% in basin water have improved the thermal conductivity of basin water by 10.34%, 5.25%, and 7.16%, respectively. Distillate productivity of solar desalination system with 0.1 wt% concentration of $Al_2O_3$, ZnO, and $SnO_2$ nano-particle is about 29.95%, 12.67%, and 18.63% higher than the system without nano-particle, respectively (Elango et al. 2015). Nano-particles can also absorb and scatter solar radiation, thereby the path length of solar radiation within the basin water increases which can lead to increased radiation absorption by

**Table 3.1** Thermo-physical properties of various nano-particles (Bait and Ameur 2018)

| Nano-particles | Density (kg/m$^3$) | Heat capacity (kJ/kg-K) | Thermal Conductivity (W/mK) |
|---|---|---|---|
| $Al_2O_3$ | 3960 | 0.773 | 40 |
| Cu | 8940 | 0.385 | 401 |
| CuO | 6000 | 0.551 | 33 |
| Graphite | 2160 | 0.701 | 120 |
| SiC | 3370 | 1.34 | 150 |
| $SiO_2$ | 3970 | 0.765 | 3970 |
| $TiO_2$ | 4230 | 0.692 | 8.40 |
| Carbon nano-tubes | 2100 | 9.124 | 30007.4 |
| $Cu_2O$ | 6320 | 42.36 J/mol K | 76.5 |
| Graphene oxide | 1910 | 0.710 | 1000 |
| $Fe_2O_3$ | 5250 | 0.650 | 20 |
| Ag | 104990 | 0.235 | 429 |
| $Fe_3O_4$ | 5810 | 0.670 | 161 |
| MgO | 3580 | 0.874 | 55 |
| Single-walled carbon nano-tubes | 2100 | 0.841 | 6000 |
| Multi-walled carbon nano-tubes | 2100 | 0.711 | 1500 |
| $Fe_3O_4$ + Multi-walled carbon nano-tubes (nano-composites) | 4845.4 | 0.681 | 509.14 |
| Ag + MgO (nano-composites) | 7035 | 0.555 | 242 |

basin water. Moreover, the optical properties of nano-particles can be adjusted by modifying their shape and size.

Nano-fluid refers to the mixture of nano-particles and base fluid. Base fluid refers to the fluid in which the nano-particles are dispersed. Stability of nano-fluid is evaluated by its absolute zeta potential. Nano-fluids having absolute zeta potential below 20 mV tend to have poor stability. Highly stable nano-fluids have absolute zeta potential above 30 mV. Absolute zeta potential of silver nano-fluid (0.04 wt% of Ag in water) is about 33 mV to 46 mV in the temperature range of 300 to 353 K, respectively (Parsa et al. 2020a). The stability periods for 0.1 wt% concentration of $Al_2O_3$, ZnO, $SnO_2$, and $Fe_2O_3$ in basin water are about 48, 72, 48, and 8 hours, respectively (Elango et al. 2015). Nano-fluid prepared in a single step using electrical explosion of wire method has stability period more than 12 months (Parsa et al. 2020a). $Al_2O_3$-loaded solar desalination system produces more distillate than $TiO_2$- and CuO-loaded systems (Sahota and Tiwari 2016). Temperature difference between the nano-particle-loaded basin water and bare basin water rises with increase in nano-particle concentration up to 0.25 wt%. However, further addition of nano-particle beyond this limit reduces temperature difference and this observation can be attributed to the increased heat capacity of nano-particles.

Nano-particles enhance distillate yield of solar stills significantly. However, it has few drawbacks like high production cost, poor stability, requirement of periodic cleaning, corrosion due to direct contact with saline water, and issues with recovery and reusability. Apart from these, discharge of spent nano-particles into the environment or water bodies can affect the environment and human health significantly. Hence, recovery of nano-particle from the base fluid is necessary before discharge. Separation of nano-particles from the nano-fluids using centrifugation and filtration is energy intensive and requires high operation cost. However, magnetic nano-particles can be separated easily from the nano-fluid using an external magnetic field. Researches have also been conducted in the areas of synthesis of bio-based nano-particles from agricultural residues and wastes to replace inorganic and metallic nano-particles (Iqbal et al. 2021). Coffee-based organic colloids have been used as an organic alternative to conventional inorganic nano-particles by (Essa et al. 2020) to enhance distillate yield of solar still. The distillate production enhancement of basin solar desalination system with coffee-based organic colloids is noticed to be about 35.14%.

### 3.3.6 *Salinity of Basin Water*

Salinity of basin water has negative effect on solar thermal desalination system's distillate productivity. The distillate productivity drops linearly with increase in salinity because as salinity increases, the vapor pressure decreases at the basin water surface, thereby the evaporation rate drops significantly. Distillate yield drops by 29.0% with increase in basin water salinity from 3000 ppm to 20000 ppm (Hoque et al. 2019). Moreover, chances of basin material corrosion also enhance with increase in salinity (Muftah 2014). Specific heat capacity of basin water drops from 4250 J/kgK to 3240 J/kgK with increase in water salinity from 0 ppm to 30000 ppm. However, surface tension of water increases from 68.4 mN/m to 69.5 mN/m with increase in salinity from 0 ppm to 30000 ppm. Surface tension enhancement with salinity may be the reason for low evaporation rates in solar thermal desalination systems (Dumka and Mishra 2019). Hence, it is recommended to drain off the un-evaporated brine every day and feed the system with new feed water to prevent distillate production loss and metal corrosion (Layek 2017).

### 3.3.7 *Condensing Cover Inclination*

The optimum inclination of condensing cover depends on location, seasons, and cover material. Inclination of cover must be at least $10^\circ$ to prevent back fall of condensing droplets into the basin water. The inclination angle of cover must be larger than the local latitude during winter season for obtaining maximum distillate productivity. This is because as inclination angle increases, amount of solar radiation incident over the cover increases, thereby a large fraction will be refracted towards the basin water. Moreover, increasing inclination angle also increases the surface area of glass cover causing increased cooling rate of the glass cover leading to enhanced condensation phenomena (Tiwari et al. 1994).

Distillate yield increases by 225% with increase in glass cover tilt angle from 10 degree to 60 degree during winter season. For summer season, maximum productivity is possible only at low inclinations as the sun will be nearly overhead, thereby higher inclinations can cause reflection of sun rays leading to low productivity. Inclination of glass cover also has an impact on average gap between the basin water surface and the cover. If the inclination is more from the basin water surface, then the volume of air space between the basin water and glass cover increases which leads to increased time requirement for saturation of air to start condensation phenomena. Temperature and water mass fraction are higher within the still with glass cover inclination of 20° which can be seen in Fig. 3.5a, c, respectively. Moreover, two vortices are seen inside the still for glass cover inclination of 20° in Fig. 3.5b. However, due to high-temperature difference between the basin water and glass cover distillate yield is higher for basin desalination system with glass cover inclination angle of 30° which is closer to the local latitude (Keshtkar et al. 2020). Hence, for optimum year-round performance, cover inclination angle closer to local latitude is highly recommended (Khalifa 2011).

Correlation relating cover inclination angle ($\beta$ in degree) and distillate productivity ($M_d$ in kg/m$^2$d) is given by (Khalifa and Hamood 2009a)

$$M_d = -0.0025(\beta)^2 + 0.1562(\beta) + 0.843 \quad R^2 = 0.734 \tag{3.24}$$

### 3.3.8 *Condensing Cover Material and Surface Properties*

Condensing cover material type, roughness, shape, transmittance, vibration, and wiping also affects the distillate productivity of solar thermal desalination system. Thinner glass cover absorbs less solar radiation and also gets cooled easily by the external environment. Reducing glass cover thickness from 0.004 m to 0.002 m has increased distillate productivity by about 3.50% (Keshtkar et al. 2020). On the other hand, increasing glass cover thickness from 0.002 m to 0.006 m has dropped productivity by about 70.0%. Wettability of condensing cover is one of the important parameters which influence the condensation rate of water vapor over it. Zanganeh et al. 2019 investigated the effect of drop-wise condensation on distillate production enhancement of solar thermal desalination system. Drop-wise condensation can be achieved by nano-coating of condensing covers with nano-silicon-based dip coating technique.

The mean surface roughness value and contact angle of condensing cover with and without nano-coating are shown in Table 3.2. Surface roughness of metallic surfaces is nearly ten times higher than that of glass and plexiglass. Glass and aluminum cover has high wettability (low contact angle) than other materials. Moreover, it can be seen that the contact angle of surfaces increases after nano-coating. Influence of contact

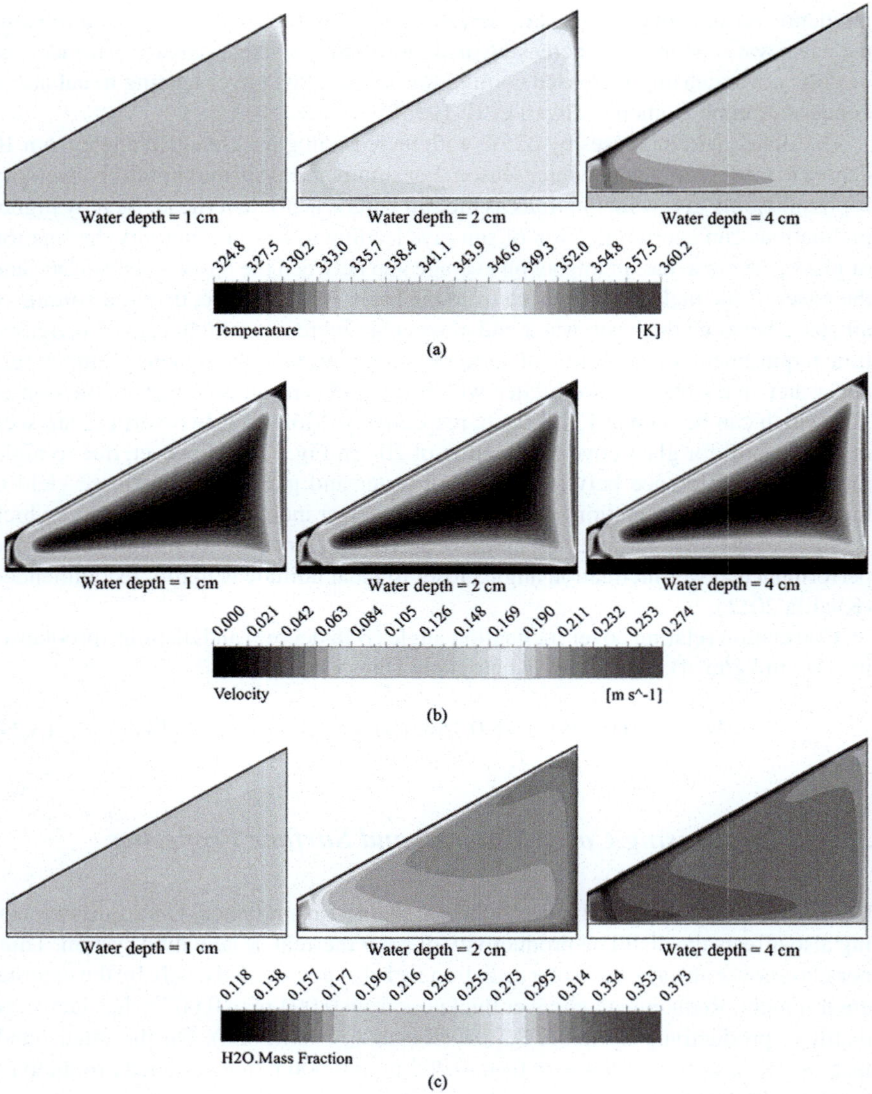


**Fig. 3.5** Contours of **a** Temperature, **b** Velocity, and **c** Water mass fraction variation inside solar still for varying glass cover inclination for solar radiation intensity of 800 W/m$^2$ (Keshtkar et al. 2020)

**Table 3.2** Effect of nano-coating on mean roughness and contact angle of various condensing cover materials (Zanganeh et al. 2019, 2020)

| | Mean roughness (nm) | | Contact angle (degree) | |
|---|---|---|---|---|
| Materials | Without nano-coating | With nano-coating | Without nano-coating | With nano-coating |
| Glass | 177.30 | 1.21 | 31 | 61 |
| Plexiglass (PMMA) | 315.20 | 1.69 | 75 | 109 |
| Aluminum | 9.85 | 10.66 | 20 | 57 |
| Copper | 8.70 | 10.09 | 56 | 83 |
| Stainless steel | 9.13 | 10.32 | 60 | 98 |
| Galvanized iron | 9.93 | 11.56 | 66 | 93 |
| Brass | 8.89 | 9.74 | 49 | 78 |
| Iron | 10.17 | 11.31 | 69 | 87 |
| Polycarbonate | – | – | 81 | 117 |
| Polyvinyl chloride | – | – | 85 | 123 |
| Polyethylene terephthalate | – | – | 67 | 94 |

angle and cover inclination angle on distillate productivity of stills with different cover materials without nano-coating and with nano-coating is shown in Figs. 3.6 and 3.7, respectively. Generally, there is an inverse relationship between the contact angle and distillate production rate. Reducing wettability (increased contact angle) reduces distillate collection rate.

Nano-coating increases distillate collection rate of solar still with high cover inclination (30 ° and 50 °) while it has negative effect on system with low cover inclination (10 °). This observation is attributed to the increased dripping effect at low inclination angle. Incorporation of additional condensate collection troughs is highly recommended for stills with nano-coated cover. For the same solar radiation input, distillate productivity of basin still with acrylic cover is about 30.0% lower than the distillate productivity of basin still with glass cover. This observation is mainly due to the large surface tension associated with acrylic cover in comparison to the glass cover (Tiwari et al. 1994). Condensation phenomena over glass cover and polyethylene terephthalate (PET) cover is shown in Fig. 3.8(a,b,c). Hydrophilic surfaces with low contact angle are found to be the suitable condensing covers for solar distillers (Bhardwaj et al. 2013). For high-temperature basin solar stills, highly conductive metallic condensing cover is recommended rather than glass and polyvinyl chloride cover (Dimri et al. 2008). Film-wise condensation is observed over the glass cover without nano-coating while drop-wise condensation is observed over the glass cover with nano-coating.

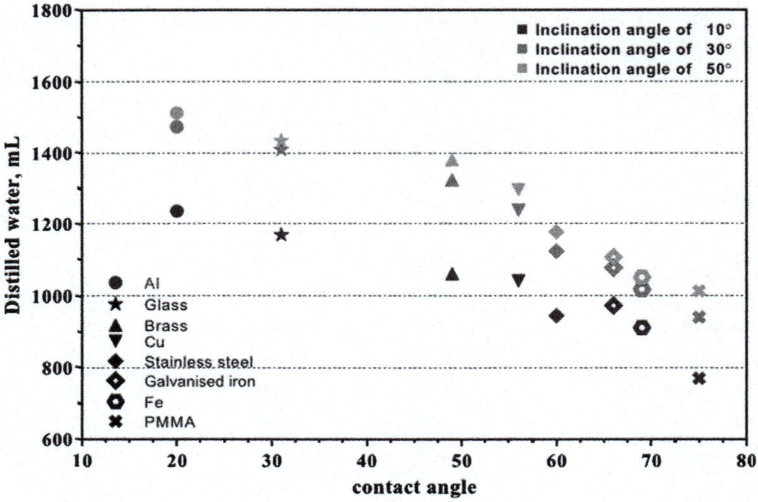


**Fig. 3.6** Effect of contact angle and inclination angle of various condensing covers without nano-coating on solar still's productivity (Zanganeh et al. 2019)

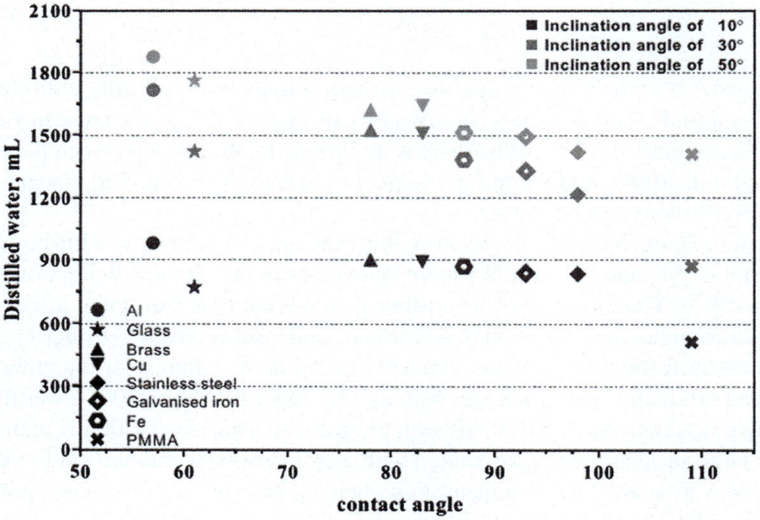


**Fig. 3.7** Effect of contact angle and inclination angle of various condensing covers with nano-coating on solar still's distillate productivity (Zanganeh et al. 2019)

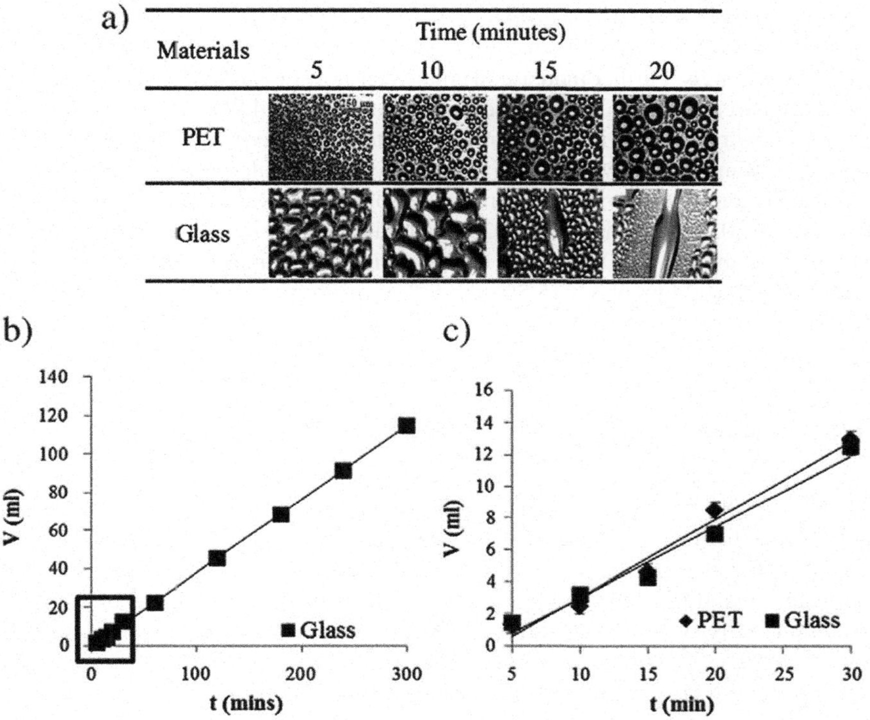


**Fig. 3.8** **a** Condensation phenomena over PET and glass cover, **b** Volume of condensate collected over glass cover up to 5 h, **c** Comparison of volume of condensate collected over PET and glass cover in 30 m (Bhardwaj et al. 2013)

### *3.3.9 Condensing Cover Temperature*

Reducing condensing cover temperature can increase temperature difference between the basin water and the condensing cover, thereby increased evaporation and condensation rates are facilitated. Condensing cover cooling can be achieved by flowing water or by blowing air over the condensing cover (Omara et al. 2017). Cooling of condensing cover by flowing water is mainly of two types: (a) feedback flow type and (b) counter-flow type. In **feedback flow type**, cooling water flowing over the glass cover is sent back to the basin to utilize its gained sensible heat energy. In case of **counter-flow type**, the water used for cooling the glass cover is left unused. Feedback flow type is highly recommended as it gives increased distillate productivity Fath (1998). Moreover, flowing water at wet bulb temperature over the glass cover is more effective in increasing the distillate productivity of the still in comparison to flowing water at ambient temperature. In addition, increased mass flow rate of flowing water at wet bulb temperature increased productivity while the productivity dropped with increase in mass flow rate of flowing water at ambient temperature. Distillate yield increased from 5.93 to 6.30 kg/m$^2$d with increase in cold water flow rate from 0.010 kg/s to 0.075 kg/s. Further increase in cold water flow rate has not shown significant yield improvement. Distillate yield dropped from 5.28 to 4.96 kg/m$^2$d with increase in mass flow rate of flowing water at ambient temperature from 0.001 kg/s to 0.025 kg/s (Somwanshi and Tiwari 2014). Thermoelectric modules have also been used to reduce the condensing cover temperature of solar stills (Rahbar and Esfahani 2012; Parsa et al. 2020a; Panchal et al. 2020b). Cooling of glass cover can reduce its temperature by about 6 to 20 °C which can result in enhancement in thermal efficiency and distillate yield by about 15.5% and 20% maximum, respectively (Omara et al. 2017).

### 3.3.10 *Basin Insulation Thickness*

Minimizing heat losses from solar still's basin liner to the ambient can increase solar still's distillate productivity. This task can be achieved by insulating the basin effectively. Insulation used for solar still applications must have resistance to high temperature and must have required strength to withstand the weight of the basin. Addition of Styrofoam insulation to the basin increases distillate productivity of still by about 80.0%. The optimum insulation thickness for basin solar stills is suggested to be 0.06 m. Further addition of insulation beyond 0.06 m has not resulted in significant distillate production enhancement. The energy efficiencies of solar still with and without insulation are about 28.5% and 26.17%, respectively (Manokar 2019).

The relationship between Styrofoam insulation thickness ($W_i$ in m) and distillate production rate of basin solar still ($M_d$ in kg/m$^2$d) is given by (Khalifa and Hamood 2009b)

$$M_d = 1023(W_i^3) - 408.8(W_i^2) + 45.34(W_i) + 1.81 \quad (R^2 = 1) \tag{3.25}$$

Conduction and convection heat losses also occur from the solar still's basin to the ambient. During wind flow, convection heat transfer coefficient dominates conduction heat transfer coefficient, thereby heat loss to the ambient increases. Convection heat loss from the basin to the ambient can be avoided by placing the insulated still directly over the ground (Abu-Hijleh 2003).

### 3.3.11 *Geometrical Dimensions of Basin Solar Still*

Performance of basin solar still is also affected by its geometrical dimensions like height of front wall, length of the basin, and width of the basin. The height of both front and back wall can be altered by an amount to have same glass cover tilt angle. (Using the height of front wall and glass cover tilt angle, height of back wall can be estimated.). In this case, the yield drops with increase in front wall height because more solar radiation will be intercepted by the walls and more shade will be casted over the basin water. Increasing height of front wall diminishes efficiency of the still by about 55.0% and 73.0% in summer and winter seasons, respectively. Higher drop in winter seasons is due to low solar altitude angle during winter. Height of the front wall cannot be reduced to very low value, i.e., below 0.10 m, because of required provisions for holding basin water of 2.0 cm depth and safe collection of distillate (Feilizadeh et al. 2017). Yield is lower at lower length of the still because of the increased effect of shade caused by the walls and this effect diminishes for increased still length. Increasing ratio of width to length of the still increases yield up to an optimum value and further increase in ratio beyond optimum value results in reduced distillate yield. Optimum width to length ratios for summer, winter, and year-round condition are around 0.15, 0.60, and 0.40, respectively. Reduced yield beyond optimum ratio of width to length is due to the increased interception of incident solar radiation by the walls of the still (Feilizadeh et al. 2017).

### 3.3.12 Thermal Energy Storage Materials

Purpose of using thermal energy storage materials in solar still is to store the incident excess solar radiation energy during day time and discharge it during night/non-sunshine hours to increase cumulative distillate productivity. Thermal energy storage materials are of two types, namely, sensible and latent heat energy storage material. Sensible heat energy storage materials do not undergo phase change while latent heat storage materials undergo phase change (Shukla et al. 2017).

Sensible heat energy storage materials like cement-concrete pieces, red brick pieces, quartzite rock, and iron scraps have been utilized in solar stills to improve the distillate productivity (Murugavel et al. 2010). Incorporation of pebbles and spherical clay balls as sensible heat energy storage materials in solar still has increased distillate productivity by about 47.3% and 37.9%, respectively (Arunkumar et al. 2020). Bilal et al. (2019) obtained negative results by incorporating pumice stones as sensible energy storage material in solar still. Addition of pumice stones increased night distillate productivity but decreased the daytime productivity resulting in reduced overall distillate productivity. This observation was observed in still with increased mass of pumice stone. Deshmukh and Thombre (2017) also observed similar results while using sand and servotherm oil as sensible heat energy storage medium. Kabeel et al. (2018a) obtained around 18.0% increase in distillate productivity by incorporating jute knitted sand as sensible energy storage material in basin solar still. (Mohamed et al. 2019) observed around 19.09 to 33.37% increased distillate production by incorporating black basalt stone as sensible energy storage material. Kabeel et al. (2018b) obtained around 75.28% higher productivity by incorporating high conductive graphite as sensible energy storage material in basin solar still.

Sensible heat storage materials have less energy storage capacity and for storing large amount of energy it occupies large volume/mass. A better way to store large amount of energy is to utilize latent heat energy-based storage materials. Latent heat energy storage materials are also called phase change materials (PCMs). PCM helps to store thermal energy and prevents temperature swing within a specific range (Javadi et al. 2020). PCM must have high specific heat capacity, high density, and high latent heat of fusion to improve energy storage density. The melting point of PCM should be close to the required operation temperature range (Javadi et al. 2020). Paraffin wax is a widely used PCM for solar still applications. Kabeel et al. (2016) obtained 67.18% higher yield in basin solar still by incorporating 3.0-cm-thick paraffin wax layer below the absorber plate. Kabeel et al. (2018c) found that the thickness of PCM has no significant impact on solar still's productivity and suggested the use of low PCM thickness for achieving low distillate production cost. Inorganic PCM-like capric-palmatic and organic PCM A48 has been recommended as best PCMs for better and low-cost distillate production in solar still (Kabeel et al. 2018c).

Increasing the mass ratio of PCM to basin water from 10 to 100 has reduced the still's distillate productivity by about 30% (Arabi et al. 2018). Shalaby et al. (2016) observed 7.4% drop in daytime productivity and about 72.7% enhancement in nocturnal productivity by incorporating PCM in basin solar still. Kateshia and Lakhera (2021) suggested incorporation of pin fins within PCM layer to enhance distillate productivity of solar still from 24% to 30%. Kabeel et al. (2019b) observed 37.55% higher distillate productivity for a still loaded with PCM + black gravel (composite PCM) in comparison to the still with sole PCM. Vigneswaran et al. (2021) recommended construction of basin with low thermal conductivity material to obtain better/positive results during incorporation of PCMs in basin solar still. Use of low conductive basin material caused earlier/fast melting of PCMs during sunshine hours and delayed its solidification rate to late hours leading to enhanced night time yield, thereby the overall distillate production of basin still increases significantly. The productivity of acrylic basin solar still with PCM is about 10.1% higher than the galvanized iron basin solar still with PCM. Use of multiple PCMs in solar still helps to enhance nocturnal productivity but the selection of PCMs must be done very carefully by considering appropriate phase change temperature range such that one of the PCMs should start releasing the stored heat energy when solar radiation starts diminishing and the other one must start releasing the stored heat energy when the former PCM has almost discharged its entire stored energy.

### *3.3.13 Reflectors*

Reflectors are employed in solar still to increase the amount of solar radiation incident over its basin liner. Reflectors can be integrated to the exterior or interior or on both to collect solar radiation from one direction or from more than one direction. Reflectors generally used are ordinary mirrors, polished metal sheets, and acrylic sheets. Distillate productivity can be increased by about 2.5 times by incorporating vertical external reflector to solar still (Khalifa and Ibrahim 2010). Distillate productivity of basin solar still can be enhanced by about 70.0 to 100.0% by incorporating internal

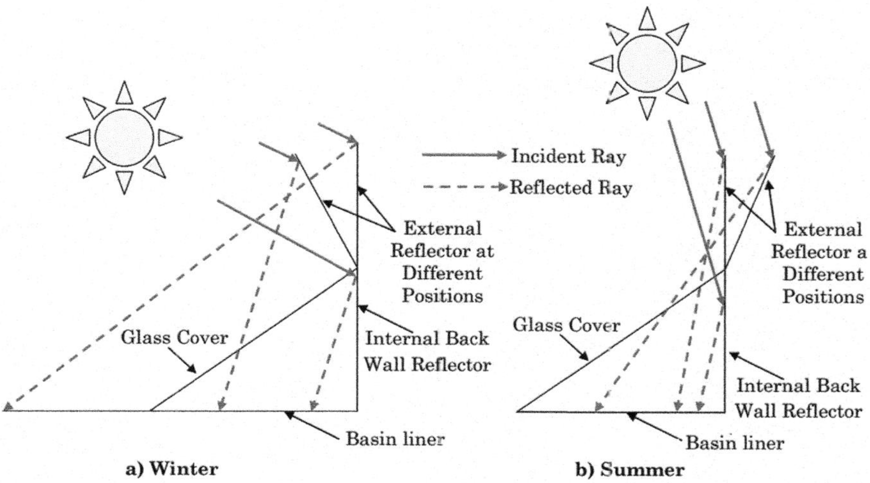


**Fig. 3.9** Path of incident solar radiation reflected from vertical and internal reflectors of solar still during **a** winter and **b** summer seasons

and external reflectors. External reflector must be inclined slightly backward from vertical during summer seasons to effectively reflect the incident sun rays towards the basin liner. Similarly, during winter season, external reflector must be inclined slightly forward from vertical to take care of low solar altitude angle during this period (Tanaka 2009). Path of reflected solar radiation from the top vertical and inclined external reflector during summer and winter seasons is shown in Fig. 3.9. Optimum length of top external reflector is about half of the solar still's length (Tanaka 2010).

The schematic of solar still with flat plate bottom reflector and internal reflector is shown in Fig. 3.10. The still with external bottom reflector and internal reflectors produced 41.0%, 25.0%, and 62.0% higher distillate than conventional solar still in spring equinox, summer, and winter solstice days, respectively. Optimum length of flat plate bottom external reflector is equal to the length of the basin liner (Tanaka 2011). Internal reflectors are more useful for increasing the productivity during winter season rather than at summer season. For completely cloudy conditions, internal reflectors are more suitable and they can increase the yearly distillate productivity of solar still by about 18.0%. Enhancement in yield due to incorporation of internal reflectors on all walls of the still is about 65% and 22% at winter and summer, respectively (Estahbanati et al. 2016). Bataineh and Abbas (2020) recommended the use of internal reflectors over the inner sidewalls of the still in combination with basin fins to achieve significant distillate yield during winter season.

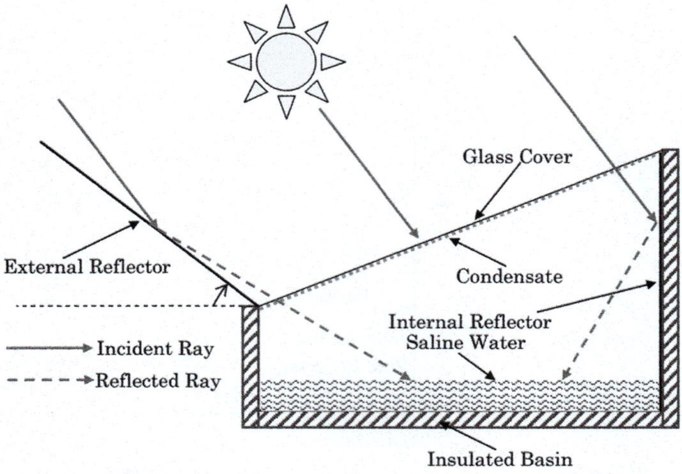


**Fig. 3.10** Schematic of solar still with bottom external reflector and internal back wall reflector

### *3.3.14 Condensing Surface Area*

Increasing heat transfer area of condensing surface can make its temperature relatively closer to ambient temperature, thereby it can facilitate more vapor condensation. Increasing condensing surface area by 6.5 times can enhance yield of the basin still by about 5 times (Bhardwaj et al. 2015). Bhardwaj et al. (2016) employed three vertical external plastic channels to an inflatable solar still to increase condensing surface area. Qudais et al. (1996) obtained about 47.0% increased distillate productivity by integrating external water-cooled condenser unit with basin solar still. External condensers are generally employed with vacuum fans to suck water vapor from the basin still to the condenser unit as shown in Fig. 3.11. Kabeel et al. (2014) obtained about 116.0% and 76.0% higher distillate productivity for nano-fluid-loaded basin still with external condenser with and without vacuum fan operation, respectively.

Elfadl et al. (2020) obtained about 41.6%, 31.9%, 52.7%, and 24.7% higher distillate productivity by replacing the transparent back wall of basin still with corrugated aluminum sheet, aluminum sheet with rectangular fins at its outer surface, aluminum sheet with pin fins at its outer surface, and aluminum sheet with pin fins at its both outer and inner surfaces, respectively. Mohaisen et al. (2020) modified back wall of solar still as condenser surface by shading it to prevent incidence of solar radiation over it. Around, 86.0% and 92.3% higher distillate productivity was noticed in the modified system for back wall without and with external plate fins, respectively. Rahmani et al. (2021) observed negative effect in distillate productivity of basin solar still incorporated with air-cooled external condenser (shown in Fig. 3.12) during cold and hot weather conditions. During cold weather condition, high loss of thermal energy from the basin to the ambient via the external condenser was observed. Similarly,

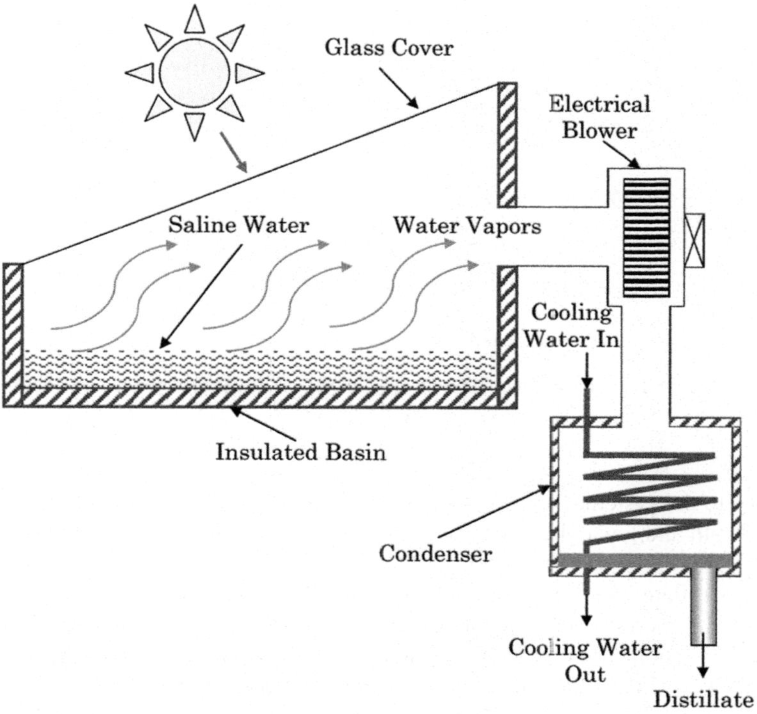


**Fig. 3.11** Schematic of basin solar still with active external condenser (supported by blower) (Qudais et al. 1996)

during hot weather condition, wall temperature of external air-cooled condenser increased, thereby condensation phenomenon got affected significantly.

### *3.3.15 Feed Water Preheating*

Enhanced evaporation rate in basin solar still can be achieved by rising feed water temperature with the aid of solar collectors. Distillate yield and thermal efficiency of solar still coupled with solar collector are about 2.3 L/m$^2$d and 22.26%, respectively, while for the system without solar collector the distillate yield and thermal efficiency are about 1.5 L/m$^2$d and 28.56%, respectively (Xiao et al. 2013). Thermal efficiencies of solar collector-coupled solar still dropped from 40.0% to 28.0% with increase in collector-to-basin area ratio from 0.88 to 3.52. High solar collector area in combination with low basin water depth may cause boiling of basin water which is inappropriate as this reduces thermal efficiency of the system and may also cause

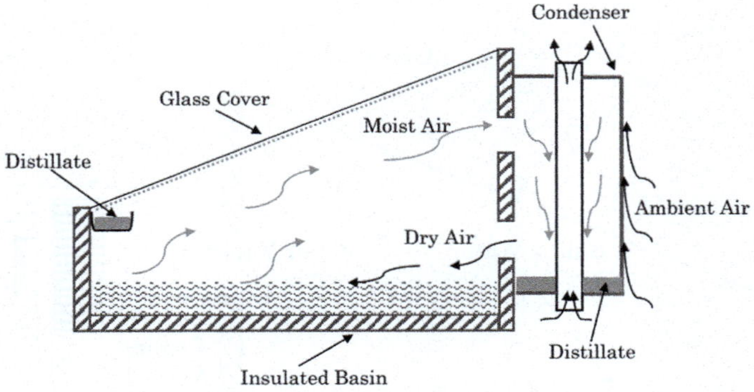


**Fig. 3.12** Schematic of basin solar still with air-cooled passive external condenser (Rahmani et al. 2021)

damage to the system. Hence, for better performance and safe operation, low basin water depth and low collector-to-basin area ratio are recommended (Taghvaei et al. 2015).

The distillate yield of PV/T solar collector (0.325 $m^2$ each)-coupled basin solar still (1.0 $m^2$ aperture area) increased from 10 kg/d to 24 kg/d with increase in number of PV/T collectors from two to ten at a feed water circulation rate of 0.055 kg/s. However, the efficiency of the system dropped from 22.0% to 13.0% with increase in number of PV/T collectors from two to ten (Gaur and Tiwari 2010). The reduced thermal efficiency of the system with increase in collector area is due to the increased heat losses associated with increased aperture area of the collector. The thermal efficiency of concentrator-assisted regenerative solar still is higher than non-concentrator-assisted regenerative solar still due to less thermal loss associated with concentrator in comparison to ordinary flat plate collector (Kumar and Sinha 1995). Compound parabolic concentrator (Prasad and Tiwari 1996), parabolic trough collector (Hassan et al. 2020), and paraboloidal dish collector (Prado et al. 2016)-assisted basin solar still have also been investigated. The photograph of parabolic trough collector-assisted basin solar still is shown in Fig. 3.13. The distillate yield of paraboloidal dish collector-assisted solar still is about 4.11 to 4.95 L/$m^2$d (Prado et al. 2016).

Circulation pumps are essential for solar collector- or concentrator-assisted solar stills, thereby initial and operation costs of these systems will be higher. This drawback has been eliminated by using lens instead of collectors. Johnson et al. (2019) concentrated solar radiation over the aperture of basin solar still (0.45 × 0.45 m) using Fresnel lens (0.508 × 0.559 × 0.003 m) as shown in Fig. 3.14, and obtained a distillate yield of about 8.32 L/$m^2$d which is about 6.38 times higher than the yield of the system without Fresnel lens. The yield enhancement is due to increment in solar radiation intensity over the aperture of solar still by a factor of the concentration ratio (area of Fresnel lens divided by area of the solar still aperture).

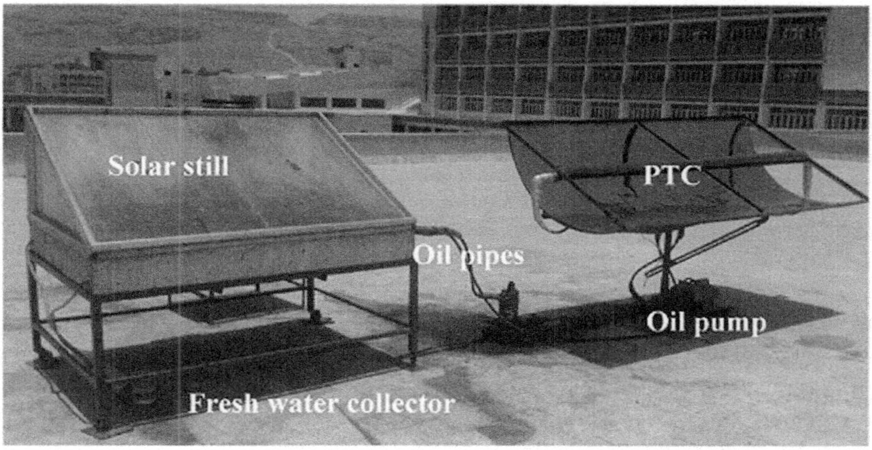


**Fig. 3.13** Photograph of parabolic trough collector-assisted basin solar still (Hassan et al. 2020)

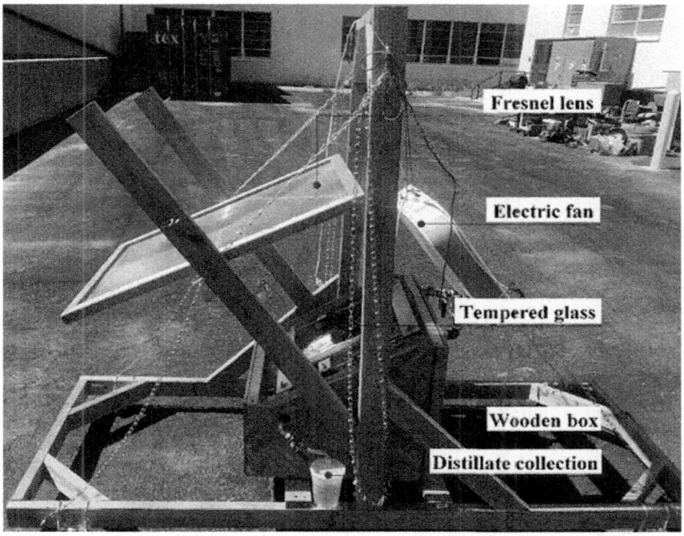


**Fig. 3.14** Photograph of Fresnel lens-assisted basin solar still (Johnson et al. 2019)

## 3.4 Features of Basin-Type Solar Thermal Desalination Systems

Simple in design and can be constructed with locally available materials in rural and remote regions:

(a) Requires less maintenance and no/minimal operation cost.
(b) Performance characteristics of the system is very well documented.
(c) Higher capacity units have already been in practical application in countries like India, Chile, and some African countries.
(d) Low distillate output per $m^2$ area.
(e) Large ground area occupancy for higher production rates.
(f) Productivity is highly affected by seasons.
(g) Scaling may be a problem if not maintained properly or if low-cost materials are used for construction.
(h) Impurities and salt deposition over the basin liner with time can reduce solar radiation absorption significantly.
(i) Algae growth within the distillation chamber has also been reported in many cases.

## References

Abdullah, A. S., Younes, M. M., Omara, Z. M., & Essa, F. A. (2020). New design of trays solar still with enhanced evaporation methods—Comprehensive study. *Solar Energy, 203*, 164–174.

Abo-Elfadl, S., Yousef, M.S., & Hassan, H. (2020). Energy, exergy, economic and environmental assessment of using Different passive condenser designs of solar distiller. *Process Safety and Environmental Protection*. https://doi.org/10.1016/j.psep.2020.10.022.

Abu-Arabi, M., Al-harahsheh, M., Mousa, H., & Alzghoul, Z. (2018). Theoretical investigation of solar desalination with solar still having phase change material and connected to a solar collector. *Desalination, 448*, 60–68.

Abu-Qudais, M., Abu-Hilleh, B. A/K., & Othman, O. N. (1996). Experimental study and numerical simulation of a solar still using an external condenser. *Energy, 21*(10), 851–855.

Abu-Hijleh, B. A. K. (2003). Effect of water emissivity on solar still efficiency. *International Journal of Sustainable Energy,23*(1–2), 13–19.

Ahsan, A., Imteaz, M., Thomas, U. A., Azmi, M., Rahman, A., & Nik Daud, N. N. (2013). Parameters affecting the performance of a low cost solar still. *Applied Energy*. https://doi.org/10.1016/j.apenergy.2013.08.066.

Alaian, W. M., Elnegiry, E. A., & Hamed, A. M. (2016). Experimental investigation on the performance of solar still augmented with pin- finned wick. *Desalination, 379*, 10–15.

Arunkumar, T., Wang, J., Rufus, D. D. W., Denkenberger, D., & Kabeel, A. E. (2020). Sensible desalting: Investigation of sensible thermal storage materials in solar stills. *Journal of Energy Storage, 32*, 101824.

Bait, O., & Si-Ameur, M. (2018). Enhanced heat and mass transfer in solar stills using nanofluids: A review. *Solar Energy, 170*, 694–722.

Balachandran, G. B., David, P. W., Mariappan, R. K., Kabeel, A. E., Athikesavan, M. M., & Sathyamurthy, R. (2020). Improvising the efficiency of single-sloped solar still using thermally conductive nano-ferric oxide. *Environmental Science and Pollution Research, 27*, 32191–32204.

Bataineh, K. M., & Abbas, M. A. (2020). Performance analysis of solar still integrated with internal reflectors and fins. *Solar Energy, 205*, 22–36.

Bhardwaj, R., Kortenaar, M. V. T., & Muddle, R. F. (2013). Influence of condensation surface on solar distillation. *Desalination, 326*, 37–45.

Bhardwaj, R., Kortenaar, M. V. T., & Mudde, R. F. (2015). Maximized production of water by increasing area of condensation surface for solar distillation. *Applied Energy, 154*, 480–490.

Bhardwaj, R., Kortenaar, M. V. T., & Mudde, R. F. (2016). Inflatable plastic solar still with passive condenser for single family use. *Desalination, 398*, 151–156.

Bilal, A., Jamil, B., Haque, N. U., & Ansari, M. A. (2019). Investigating the effect of pumice stones sensible heat storage on the performance of a solar still. *Groundwater for Sustainable Development, 9*, 100228.

Deshmukh, H. S., & Thombre, S. B. (2017). Solar distillation with single basin solar still using sensible heat storage materials. *Desalination, 410*, 91–98.

Dhiman, N. K. (1988). Transient analysis of a spherical solar still. *Desalination, 69*, 47–55.

Dimri, V., Sarkar, B., Singh, U., & Tiwari, G. N. (2008). Effect of condensing cover material on yield of an active solar still: An experimental validation. *Desalination, 227*, 178–189.

Dumka, P., & Mishra D. R. (2019). Influence of salt concentration on the performance characteristics of passive solar still. *International Journal of Ambient Energy*. https://doi.org/10.1080/01430750.2019.1611638.

Durkaieswaran, P., & Murugavel, K. K. (2015). Various special designs of single basin passive solar still—a review. *Renewable and Sustainable Energy Reviews, 49*, 1048–1060.

Elango, T., Kannan, A., & Murugavel, K. K. (2015). Performance study on single basin single slope solar still with different water nanofluids. *Desalination, 360*, 45–51.

El-Sebaii, A. A., Ramadan, M. R. I., Aboul-Enein, S., & El-Naggar, M. (2015). Effect of fin configuration parameters on single basin solar still performance. *Desalination, 365*, 15–24.

El-Sebaii, A. A., & El-Naggar, M. (2017). Year round performance and cost analysis of a finned single basin solar still. *Applied Thermal Engineering, 110*, 787–794.

Essa, F. A., Elsheikh, A. H., Algazzar, A. A., Sathyamurthy, R., Ali, M. K. A., Elaziz, M. A., & Salman, K. H. (2020). Eco-friendly coffee-based colloid for performance augmentation of solar stills. *Process Safety and Environmental Protection, 136*, 259–267.

Estahbanati, M. R. K., Ahsan, A., Feilizadeh, M., Jafarpur, K., Ashrafmansouri, S. -S., & Feilizadeh, M. (2016). Theoretical and experimental investigation on internal reflectors in a single-slope solar still. *Applied Energy, 165*, 537–547.

Fath, H. E. S. (1998). Solar distillation: a promising alternative for water provision with free energy, simple technology and a clean environment. *Desalination, 116*, 45–56.

Feilizadeh, M., Soltanieh, M., Estahbanati, M. R. K., Jafarpur, K., & Ashrafmansouri, S. -S. (2017). Optimization of geometrical dimensions of single-slope basin-type solar stills. *Desalination, 424*, 159–168.

Gaur, M. K., & Tiwari, G. N. (2010). Optimization of number of collectors for integrated PV/T hybrid active solar still. *Applied Energy, 87*, 1763–1772.

Hassan, H., Yousef, M. S., Fathy, M., & Ahmed, M. S. (2020). Assessment of parabolic trough solar collector assisted solar still at various saline water mediums via energy, exergy, exergoeconomic, and enviroeconomic approaches. *Renewable Energy, 155*, 604–616.

Hoque, A., Abir, A. H., & Shourov, K. P. (2019). Solar still for saline water desalination for low-income coastal areas. *Applied Water Science, 9*, 104. https://doi.org/10.1007/s13201-019-0986-9.

Iqbal, A., Mahmoud, M. S., Sayed, E. T., Elsaid, K., Abdelkareem, M. A., Alawadhi, H., & Olabi, A. G. (2021). Evaluation of the nanofluid-assisted desalination through solar stills in the last decade. *Journal of Environmental Management, 277*, 111415.

Ismail, B. I. (2009). Design and performance of a transportable hemispherical solar still. *Renewable Energy, 34*, 145–150.

Javadi, F. S., Metselaar, H. S. C., & Ganesan, P. (2020). Performance improvement of solar thermal systems integrated with phase change materials (PCM), a review. *Solar Energy, 206*, 330–352.

Johnson, A., Mu, L., Park, Y. H., Valles, D. J., Wang, H., Xu, P., Kota, K., & Kuravi, S. (2019). A thermal model for predicting the performance of a solar still with Fresnel lens. *Water, 11*, 1860.

Kabeel, A. E. (2009). Performance of solar still with a concave wick evaporation surface. *Energy, 34*, 1504–1509.

Kabeel, A. E., Omara, Z. M., & Essa, F. A. (2014). Enhancement of modified solar still integrated with external condenser using nanofluids: An experimental approach. *Energy Conversion and Management, 78*, 493–498.

Kabeel, A. K., & Abdelgaid, M. (2016). Improving the performance of solar still by using PCM as a thermal storage medium under Egyptian conditions. *Desalination, 383*, 22–28.

Kabeel, A. E., Omara, Z. M., Essa, F. A., Abdulla, A. S., Arunkumar, T., & Sathyamurthy, R. (2017). Augmentation of a solar still distillate yield via absorber plate coated with black nano particles. *Alexandria Engineering Journal, 56*, 433–438.

Kabeel, A. E., El-Agouz, S. A., Sathyamurthy, R., & Arunkumar, T. (2018). Augmenting the productivity of solar still using jute cloth knitted with sand heat energy storage. *Desalination, 443*, 122–129.

Kabeel, A. E., Abdelgaied, M., & Eisa, A. (2018). Enhancing the performance of single basin solar still using high thermal conductivity sensible storage materials. *Journal of Cleaner Production, 183*, 20–25.

Kabeel, A. E., El-Samadony, Y. A. F., & El-Maghlany, W. M. (2018). Comparative study on the solar still performance utilizing different PCM. *Desalination, 432*, 89–96.

Kabeel, A. E., Sathyamurthy, R., Sharshir, S. W., Muthumanokar, A., Panchal, H., Prakash, N., Prasad, C., Nandakumar, S., & ElKady, M. S. (2019). Effect of water depth on a novel absorber plate of pyramid solar still coated with $TiO_2$ nano black paint. *Journal of Cleaner Production, 213*, 185–191.

Kabeel, A. E., Abdelaziz, G. B., & El-Said, E. M. S. (2019). Experimental investigation of a solar still with composite material heat storage: Energy, exergy and economic analysis. *Journal of Cleaner Production, 231*, 21–34.

Kalita, P., Dewan, A., & Borah, S. (2016). A review on recent developments in solar distillation units. *Sadhana 41*(2), 203–223.

Kateshia, J., & Lakhera, V. J. (2021). Analysis of solar still integrated with phase change material and pin fins as absorbing material. *Journal of Energy Storage, 35*, 102292.

Keshtkar, M., Eslami, M., & Jafarpur, K. (2020). Effect of design parameters on performance of passive basin solar stills considering instantaneous ambient conditions: A transient CFD modeling. *Solar Energy, 201*, 884–907.

Khalifa, A. J. N., & Hamood, A. M. (2009). Performance correlations for basin type solar stills. *Desalination, 249*, 24–28.

Khalifa, A. J. N., & Hamood, A. M. (2009). Effect of insulation thickness on the productivity of basin type solar stills: An experimental verification under local climate. *Energy Conversion and Management, 50*, 2457–2461.

Khalifa, A. J. N., & Ibrahim, H. A. (2010). Effect of inclination of the external reflector of simple solar still in winter: An experimental investigation for different cover angles. *Desalination, 264*, 129–133.

Khalifa, A. J. N. (2011). On the effect of cover tilt angle of the simple solar still on its productivity in different seasons and latitudes. *Energy Conversion and Management, 52*, 431–436.

Kumar, S., & Sinha, S. (1995). Transient model and comparative study of concentrator coupled regenerative solar still in forced circulation mode.

Layek, A. (2017). Effect of depth and salinity of basin water on performance of solar still. In *Proc. of the 2017 IEEE Region 10 Conference (TENCON), Malaysia, November 5–8, 2017*.

Manokar, A. M., Taamneh, Y., Kabeel, A. E., Winston, D. P., Vijayabalan, P., Balaji, D., Sathyamurthy, R., Sundar, S. P. & Mageshbabu, D. (2019). Effect of water depth and insulation on the productivity of an acrylic pyramid solar still—an experimental study. *Ground Water for Sustainable Development*. https://doi.org/10.1016/j.gsd.2019.100319

Muftah, A. F., Alghoul, M. A., Fudholi, A., Abdul-Majeed, M. M., & Sopian, K. (2014). Factors affecting basin type solar still productivity: A detailed review. *Renewable and Sustainable Energy Reviews, 32*, 430–447.
Mohaisen, H. S., Esfahani, J. A., & Ayani, M. B. (2020). Improvement in the performance and cost of passive solar stills using a finned-wall/built-in condenser: An experimental study. *Renewable Energy*. https://doi.org/10.1016/j.renene.2020.12.056.
Mohamed, A. F., Hegazi, A. A., Sultan, G. I., & El-Said, E. M. S. (2019). Enhancement of a solar still performance by inclusion the basalt stones as a porous sensible absorber: Experimental study and thermo-economic analysis. *Solar Energy Materials and Solar Cells, 200*, 109958.
Munoz, F., Barrera, E., Ruiz, A., Martinez, E. M., & Chargoy, N. (2020). Long-term experimental theoretical study on several single-basin solar stills. *Desalination, 476*, 114241.
Murugavel, K. K., Sivakumar, S., Ahamed, J. R., Chockalingam, K. K., & Srithar, K. (2010). Single basin double slope solar still with minimum basin depth and energy storing materials. *Applied Energy, 87*, 514–523.
Nayi, K. H., & Modi, K. V. (2018). Pyramid solar still: A comprehensive review. *Renewable and Sustainable Energy Reviews, 81*, 136–148.
Omara, Z. M., Hamed, M. H., & Kabeel, A. E. (2011). Performance of finned and corrugated absorbers solar stills under Egyptian conditions. *Desalination, 277*, 281–287.
Omara, Z. M., Abdullah, A. S., Kabeel, A. E., & Essa, F. A. (2017). The cooling techniques of the solar stills' glass covers—a review. *Renewable and Sustainable Energy Reviews, 78*, 176–193.
Panchal, H., Mevada, D., Sadasivuni, K. K., Essa, F. A., Shanmugan, S., & Khalid, M. (2020). Experimental and water quality analysis of solar stills with vertical and inclined fins. *Groundwater for Sustainable Development, 11*, 100410.
Panchal, H., Sadasivuni, K. K., Prajapati, C., Khalid, M., Essa, F. A., Shanmugan, S., Pandya, N., Suresh, M., Israr, M., Dharaskar, S., & Khechekhouche, A. (2020). Productivity enhancement of solar still with thermoelectric modules from groundwater to produce potable water: A review. *Groundwater for Sustainable Development*.
Parsa, S. M., Rahbar, A., Koleini, M. H., Javadi, Y. D., Afrand, M., Rostami, S., & Amidpour, M. (2020). First approach on nanofluid-based solar still in high altitude for water desalination and solar water disinfection (SODIS). *Desalination, 491*, 114592.
Parsa, S. M., Rahbar, A., Koleini, M. H., Aberoumand, S., Afrand, M., & Amidpour, M. (2020b). A renewable energy-driven thermoelectric-utilized solar still with external condenser loaded by silver/nanofluid for simultaneously water disinfection and desalination. *Desalination*. https://doi.org/10.1016/j.desal.2020.114354.
Prado, G. O., Vieira, L. G. M., & Damasceno, J. J. R. (2016). Solar dish concentrator for desalting water. *Solar Energy, 136*, 659–667.
Prasad, B., & Tiwari, G. N. (1996). Effect of glass cover inclination and solar distillation system parametric studies of concentrator-assisted solar distillation system. *International journal of Energy Research, 20*, 495–505.
Rahbar, N., & Esfahani, J. A. (2012). Experimental study of a novel portable solar still by utilizing the heat pipe and thermoelectric module. *Desalination, 284*, 55–61.
Rahmani, A., Khemmar, F., & Saadi, Z. (2021). Experimental investigation on the negative effect of the external condenser on the conventional solar still performance. *Desalination, 501*, 114914.
Rajvanshi, A. K. (1981). Effect of various dyes on solar distillation. *Solar Energy, 27*(1), 51–65.
Rajaseenivasan, T., & Srithar, K. (2016). Performance investigation on solar still with circular and square fins in basin with $CO_2$ mitigation and economic analysis. *Desalination, 380*, 66–74.
Rajaseenivasan, T., Prakash, R., Vijayakumar, K., & Srithar, K. (2017). Mathematical and experimental investigation on the influence of basin height variation and stirring of water by solar PV panels in solar still. *Desalination, 415*, 67–75.
Salem, M. R., Salem, M. R., Higazy, M. G., & Abdrabbo, M. F. (2020). Performance enhancement of a solar still distillation unit: A field investigation. *Solar Energy, 202*, 326–341.
Selvaraj, K., & Natarajan, A. (2017). Factors influencing the performance and productivity of solar stills—a review. *Desalination*. https://doi.org/10.1016/j.desal.2017.09.031.

Singh, R. G., & Tiwari, G. N. (2019). Simulation performance of single slope solar still by using iteration method for convective heat transfer coefficient. *Ground Water for Sustainable Development*. https://doi.org/10.1016/j.gsd.2019.100287.

Somwanshi, A., & Tiwari, A. K. (2014). Performance enhancement of a single basin solar still with flow of water from an air cooler on the cover. *Desalination, 352*, 92–102.

Shalaby, S. M., El-Bialy, E., & El-Sebaii, A. A. (2016). An experimental investigation of a v-corrugated absorber single-basin solar still using PCM. *Desalination, 398*, 247–255.

Sharon, H., Vivar, M., & Fuentes, M. (2022). A review on role of solar photovoltaic (PV) modules in enhancing sustainable water production capacity of solar distillation units. *Journal of Environmental Management, 320*, 115781.

Sahota, L., & Tiwari, G. N. (2016). Effect of nano fluids on the performance of passive double slope solar still: A comparative study using characteristic curve. *Desalination, 388*, 9–21.

Shukla, A., Kant, K., & Sharma, A. (2017). Solar still with latent heat energy storage: A review. *Innovative Food Science and Emerging Technologies, 41*, 34–46.

Taghvaei, H., Taghvaei, H., Jafarpur, K., Feilizadeh, M., & Estahbanati, M. R. K. (2015). Experimental investigation of the effect of solar collecting area on the performance of active solar stills with different brine depths. *Desalination, 358*, 76–83.

Tamim, A. (1987). Performance of a solar still with reflectors and black dye. *Sol. Wind. Technol., 4*(4), 443–446.

Tanaka, H. (2009). Experimental study of a basin type solar still with internal and external reflectors in winter. *Desalination, 249*, 130–134.

Tanaka, H. (2010). Monthly optimum inclination of glass cover and external reflector of a basin type solar still with internal and external reflector. *Solar Energy, 84*, 1959–1966.

Tanaka, H. (2011). A theoretical analysis of basin type solar still with flat plate external bottom reflector. *Desalination, 279*, 243–251.

Tiwari, G. N., Thomas, J. M., & Khan, E. (1994). Optimization of glass cover inclination for maximum yield in a solar still. *Heat Recovery Systems & CHP, 14*(4), 447–455.

Vigneswaran, V. S., Kumar, P. G., Sakthivadivel, D., Balaji, K., Meikandan, M., Dinakar, B. V., Kamal, K. K., & Kumaresan, G. (2021). Energy, exergy, and economic analysis of low thermal conductivity basin solar still integrated with phase change material for energy storage. *Journal of Energy Storage, 34*, 102194.

Xiao, G., Wang, X., Ni, M., Wang, F., Zhu, W., Luo, Z., & Cen, K. (2013). A review on solar stills for brine desalination. *Applied Energy, 103*, 642–652.

Younis, O., Hussein, A. K., Attia, M. E. H., Rashid, F. L., Kolsi, L., Biswal, U., Abderrahmane, A., Mourad, A., & Alazzam, A. (2022). Hemispherical solar still: Recent advances and development. *Energy Reports, 8*, 8236–8258.

Zanganeh, P., Goharrizi, A. S., Ayatollahi, S., & Feilizadeh, M. (2019). Productivity enhancement of solar stills by nano-coating of condensing surface. *Desalination, 454*, 1–9.

Zanganeh, P., Goharrizi, A. S., Ayatollahi, S., & Feilizadeh, M. (2020). Nano-coated condensation surfaces enhanced the productivity of the single-slope solar still by changing the condensation mechanism. *Journal of Cleaner Production, 265*, 121758.

# Chapter 4
# Tilted Solar Thermal Desalination Systems

## 4.1 System Description and Working

The schematic of tilted solar thermal desalination system along with its associated energy transport process is shown in Fig. 4.1. Major components of tilted solar thermal desalination system are insulated distillation chamber, blackened wick, glass cover, feed water trough, condensate collection trough, and provisions for feed water inlet and brine drain. The distillation chamber is fabricated by placing smaller metallic rectangular tray within relatively larger metallic tray and filling the gap between them with insulation. The exposed area of smaller tray is then black painted to facilitate effective absorption of incident solar radiation. Feed water trough is fabricated by maintaining its length and height closer to the width and height of the distillation chamber, respectively. Breadth of the feed water trough is kept 1/10th of the distillation chamber's length. Blackened wick is lined over the inner surface of the distillation chamber such that its upper end is within feed water trough and its lower end is left free. Condensate collection trough is attached along the width of the lower edge of the distillation chamber by maintaining the required gap from the inner surface of the distillation chamber. Provisions for supplying feed water and draining excess brine are made in the distillation chamber. The top opening of the distillation chamber is covered by tempered glass cover and is sealed perfectly by silicon sealant to prevent vapor leakage to the ambient. This system is generally mounted at an angle to the horizontal surface, hence it is termed as tilted solar thermal desalination system. It is also called as tilted solar still or inclined solar still. The optimum length-to-height ratio for tilted solar still is about 16 (Alvarado-Juárez et al., 2013). Height refers to the gap between the evaporating and condensing surface.

Blackened wick becomes wet by absorbing feed water from the feed water trough. Solar radiation incident over the wetted wick heats up the wick to generate vapor which in turn condenses over the inner surface of the glass cover by releasing its latent heat. Condensed vapors then roll down to the lower end of the glass cover by gravity and get diverted into condensate collection trough by a glass strip fixed along the width of the lower end of glass cover. The excess un-evaporated brine leaves the



H. Sharon and K. Srinivas Reddy, *Solar Thermal Desalination Technologies for Potable Water*, https://doi.org/10.1007/978-3-031-49155-9_4

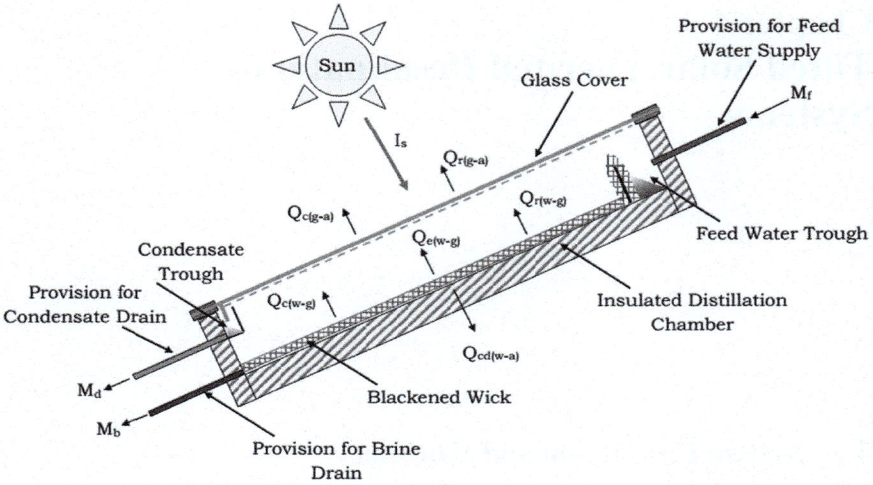


**Fig. 4.1** Schematic of tilted solar thermal desalination system showing its associated energy transport processes

system through a drain located at the lower end of the distillation chamber. Feed water is supplied to the feed water through every 15 min using external overhead tank arrangement. Feed water flow rate is generally maintained two to three times higher than the distillation rate to facilitate continuous evaporation and to ensure prevention of salt deposition and drying of wicks. Distillate productivity of wick-based systems is about two to three times higher than bare plate systems (Aybar et al., 2005). Commercial tilted solar thermal desalination systems have been developed and marketed successfully throughout the globe by an Australian company F Cubed Limited. The commercial system is made of coated polycarbonate plastic (Mashaly & Alazba, 2017). The tilted solar still can have wick-lined absorber surface or partitioned absorber or weir-type absorber as shown in Fig. 4.2.

## 4.2 Thermodynamic Modeling of Tilted Solar Still

Major assumptions considered for thermodynamic modeling of tilted solar still are listed below (Shukla & Sorayan, 2005):

(a) Still is vapor tight and free from vapor leakage.
(b) Glass cover is thin and there is no conductance resistance across its thickness.
(c) No thermal gradient across the thickness of wetted wick.
(d) Feed water flow rate/capillary effect of wick is always constant and in streamline.
(e) Reject brine leaves the still at wetted wick temperature.

**Fig. 4.2** Photograph of tilted solar still along with its partitioned basin and felt sheet (Sharon et al., 2017)

Energy balance in blackened wetted wick,

$$\begin{pmatrix}\text{Rate of change of wetted}\\ \text{wick temperature}\end{pmatrix} = \begin{pmatrix}\text{solar radiation absorbed}\\ \text{by wetted wick}\end{pmatrix}$$
$$- \begin{pmatrix}\text{convection, radiation \& evaporation heat}\\ \text{loss from wetted wick to glass cover}\end{pmatrix}$$
$$+ \begin{pmatrix}\text{Heat energy content}\\ \text{of feed water}\end{pmatrix} - \begin{pmatrix}\text{Heat energy content}\\ \text{of reject brine}\end{pmatrix}$$
$$- \begin{pmatrix}\text{convection, radiation \& evaporation heat}\\ \text{loss from wetted wick to glass cover}\end{pmatrix}$$
$$- \begin{pmatrix}\text{convection heat loss from wetted}\\ \text{wick to ambient via insulation}\end{pmatrix}$$

$$(mc)_w \frac{dT_w}{dt} = \tau_g \alpha_w A_g I_s - (Q_s + Q_r + Q_e)_{(w-g)} + \dot{m}_f c_f (T_{f,in}) - \dot{m}_b c_b (T_{b,out}) - Q_{cd(w-a)} \tag{4.1}$$

Energy balance in glass cover,

$$\begin{pmatrix}\text{Rate of change of glass}\\ \text{cover temperature}\end{pmatrix} = \begin{pmatrix}\text{solar radiation absorbed}\\ \text{by glass cover}\end{pmatrix} + \begin{pmatrix}\text{convection \& radiation heat}\\ \text{loss from wetted wick to glass cover}\end{pmatrix} + \begin{pmatrix}\text{Heat content of}\\ \text{condensing vapors}\end{pmatrix} - \begin{pmatrix}\text{convection \& radiation heat loss}\\ \text{from glass cover to ambient}\end{pmatrix}$$

$$(mc)_g \frac{dT_g}{dt} = \tau_g \alpha_w A_g I_s - (Q_c + Q_r)_{(w-g)} + Q_{cond} - (Q_c + Q_r)_{(g-a)} \tag{4.2}$$

Feed water can also be allowed to flow over the bare absorber plate of the tilted still in the absence of wick. The thickness of the water film ($\delta_w$) and its velocity ($U$) in this case are estimated by (El-Agouz et al., 2015)

$$\delta_w = \sqrt[3]{\frac{3\mu_w \dot{V}_w}{Br \rho_w g \sin\alpha}} \tag{4.3}$$

$$U = \frac{\dot{V}_w}{\delta_w Br} \tag{4.4}$$

where

$\alpha$—tilt angle of the still from horizon,
$\mu_w$—dynamic viscosity of water (kg/ms),
$\dot{V}_w$—volumetric flow rate of water (m$^3$/s),
Br—breadth of the tilted still (m)
$g$—acceleration due to gravity (m/s$^2$), and
$\rho_w$—density of water (kg/m$^3$).

For simplicity, the temperature of water film can be taken as the temperature of the absorber plate. For more accurate estimation, separate energy balance equations are needed for absorber plate and the water film. In that case, convective heat transfer between the absorber plate and the water film needs to be estimated (El-Agouz et al., 2015).

The convection heat transfer between the absorber plate and flowing water film ($Q_{c(ab-w)}$ in W) can be estimated by (Oliveria de Paula & Ismail, 2020)

$$Q_{c(ab-w)} = h_{cw} A_{ab} (T_{ab} - T_w) \tag{4.5}$$

where

$h_{cw}$—convective heat transfer coefficient between absorber plate and flowing water ($W/m^2K$),
$A_{ab}$—surface area of absorber plate ($m^2$),

$$h_{cw} = \begin{cases} \frac{K_w}{L} 0.664 Re^{0.5} Pr^{0.33}, & \text{if } Re_L \le 5 \times 10^5 \\ \frac{K_w}{L} (0.037 Re^{3/4} - 871) Pr^{0.33}, & \text{if } Re_L > 5 \times 10^5 \end{cases}$$

$K_w$—thermal conductivity of water (W/mK),
$L$—length of the still (m),
Re—Reynolds number, and
Pr—Prandtl number.

$$Re = \frac{4\dot{m}_f}{Br\mu_w} \tag{4.6}$$

where $\dot{m}_f$ is mass flow rate of feed water (kg/s).

Heat energy content of the brine leaving the desalination system ($Q_{bl}$ in W) is estimated by (Shukla & Sorayan, 2005)

$$Q_{bl} = \frac{\dot{m}_f c_f}{U_L} A_w (\tau_w I_t - U_L (T_{f,in} - T_a)) \tag{4.7}$$

where

$c_f$—specific heat capacity of water (J/kgK),
$U_L$—overall heat loss coefficient ($W/m^2K$),
$A_w$—surface area of the evaporating surface of the still ($m^2$),
$I_t$—solar radiation intensity over the still ($W/m^2$),
$T_{f,in}$—feed water inlet temperature (°C), and
$T_a$—ambient temperature (°C).

Heat energy loss to the ambient through the wetted wick/absorber surface via insulation ($Q_{cd(w-a)}$ in W) is given as

$$Q_{cd(w-a)} = \frac{(T_w - T_a)}{\left(\frac{W_i}{K_i A_i}\right) + \left(\frac{1}{h_{ca} A_i}\right)} \tag{4.8}$$

where

$W_i$—thickness of insulation (m),
$K_i$—thermal conductivity of insulation (W/m$^2$K),
$A_i$—surface area of insulation (m$^2$), and
$h_{ca}$—convective heat transfer coefficient between the insulation and ambient (W/m$^2$K).

Radiation heat transfer between the wetted wick/absorber surface and glass cover ($Q_{r(w-g)}$ in W) is estimated as

$$Q_{r(w-g)} = \frac{\sigma A_w (T_w + 273.15)^4 - (T_a + 273.15)^4}{\left(\frac{1}{\varepsilon_w} + \frac{1}{\varepsilon_g} - 1\right)} \tag{4.9}$$

where

$\sigma$—Stefan–Boltzmann constant ($5.67 \times 10^{-8}$ W/m$^2$K$^4$),
$\varepsilon_w$—emissivity of water, and
$\varepsilon_g$—emissivity of glass cover.

Evaporation heat transfer ($Q_{e(w-m)}$ in W) and condensation heat transfer ($Q_{cond}$ in W) between the wetted wick/absorber surface and glass cover are estimated by (Aybar, 2006)

$$Q_{e(w-m)} = 0.027(\Delta T)^{1/3} P_{sat}(1-\phi)A_w \tag{4.10}$$

$$Q_{cond} = 85.0(T_m - T_g)\phi A_g \tag{4.11}$$

where

$\Delta T$—temperature difference between the evaporating and condensing surface (K),
$\phi$—relative humidity,
$P_{sat}$—saturation vapor pressure at evaporating surface temperature (Pa),
$A_w$—surface area of the evaporating surface of the still (m$^2$), and
$A_g$—surface area of the condensing surface of the still (m$^2$).

Relative humidity ($\phi$) inside the desalination chamber is estimated by (Oliveria de Paula & Ismail, 2020)

$$\phi = \frac{P_v}{P_{sat}} = \frac{\frac{m_v R_v (T_m + 273.15)}{V}}{\exp\left(25.317 - \frac{5144}{(T_m + 273.15)}\right)} \tag{4.12}$$

where

$m_v$—vapor mass inside the distillation chamber (kg),
$V$—volume of the distillation chamber (m$^3$),
$R_v$—gas constant of water vapor (J/kgK), and
$T_m$—air-vapor mixture temperature (°C).

From earlier experiments, it has been found that the air-vapor mixture temperature is about 2.0 °C higher than the water/wetted wick temperature (Aybar, 2006). Hence, $\Delta T(\Delta T = T_w - T_m)$ has been taken as 2.0 °C in some of the earlier studies for simulation (El-Agouz et al., 2015; Oliveira de Paula and Ismail, 2020).

Mass of vapor ($m_v$) inside the still is estimated by (Aybar, 2006)

$$\frac{dm_v}{dt} = A_w \dot{m}_e - A_g \dot{m}_{cond} \tag{4.13}$$

where

$\dot{m}_e$—evaporation rate (kg/s) and
$\dot{m}_{cond}$—condensation rate (kg/s).

Convection heat transfer between the glass cover and the ambient ($Q_{c(g-a)}$ in W) is given as

$$Q_{c(g-a)} = h_{ca} A_g (T_g - T_a) \tag{4.14}$$

$$h_{ca} = \begin{cases} 2.8 + 3.0V_{wind} & \text{if } V_{wind} \le 5.0\,\text{m/s} \\ 6.15V_{wind}^{0.8} & \text{if } V_{wind} > 5.0\,\text{m/s} \end{cases}$$

Radiation heat loss from the glass cover ($Q_{r(g-sky)}$ in W) is estimated by considering its heat exchange with the ambient. In some cases, it has been estimated by considering glass cover's heat exchange with the sky (El-Agouz et al., 2015).

$$Q_{r(g-sky)} = \sigma A_g \varepsilon_g ((T_g + 273.15)^4 - (T_{sky} + 273.15)^4) \tag{4.15}$$

$$T_{sky} = T_a - 6.0 \tag{4.16}$$

where

$T_{sky}$—sky temperature (°C) and
$T_a$—ambient temperature (°C).

## 4.3 Parameters Influencing the Performance of Tilted Solar Still

In general, the performance of solar stills can be improved by (Sodha et al., 1981)

(a) Placing evaporating surface/water layer at optimum inclination to receive maximum solar radiation.

(b) Reducing reflection losses from the glass cover by placing it parallel to the water layer.
(c) Lowering thermal capacity of water by reducing film thickness or by incorporating wicks.

These modifications are highly possible in tilted solar still (Maheswaran et al., 2021). Hence, tilted solar stills have better yield per $m^2$ of solar energy collection area in comparison to the conventional basin solar still. Some of the well-known modifications made in tilted solar still to improve its performance significantly are discussed in this section.

### *4.3.1 Incorporation of Wicks*

Incorporation of wicks helps to minimize the heat capacity of evaporating liquid by reducing its film thickness. Single wick or multiple wicks can be incorporated inside the distillation chamber. The problem observed by researchers with single wick solar still is dry out of the wick during operation which leads to reduced daily yield. Multi-wicks prevent dry out problem and make capillary effect to be carried out for a long duration. The schematic of double slope multi-wick tilted solar still is shown in Fig. 4.3. In multi-wick systems, wicks of different length are placed above the absorber plate and each wick is separated by thin polythene sheet. Upper end of the wicks is dipped inside a trough containing feed water while the lower end is left free. Capillary action of the wicks helps to maintain the portion of the wicks exposed to solar radiation to be wet all the time (Tiwari & Singh, 1991).

In addition, portion of wick covered by polythene sheet is not subjected to evaporation, thereby the exposed portion is maintained always wet and subjected to evapora-

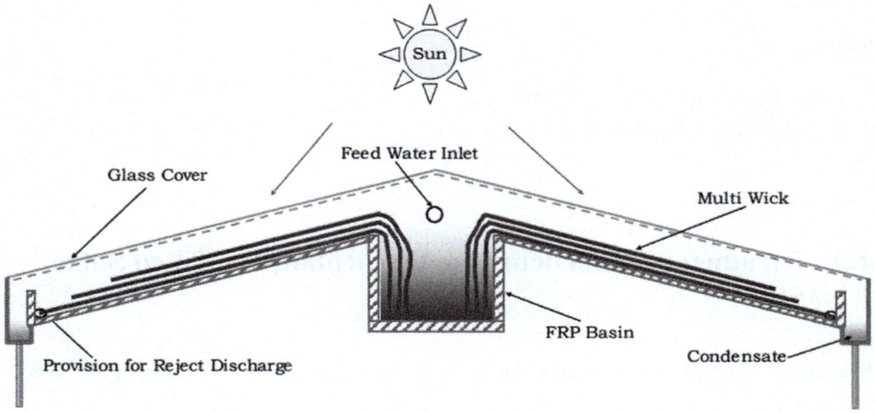


**Fig. 4.3** Schematic of double slope multi-wick tilted solar still

**Table 4.1** Wicking characteristics of wick materials (Hansen et al., 2015)

| Wick materials | Wicking characteristics | | | | |
|---|---|---|---|---|---|
| | Porosity (%) | Absorbency (s) | Repellent | Capillary rise (mm/h) | Evaporative heat transfer coefficient (W/m$^2$K) |
| Cotton | 28.50 | 1.00 | 0 | 120.00 | 36.00 |
| Wool | 27.00 | 150.00 | 0 | 110.00 | 45.80 |
| Nylon | 14.50 | 1.00 | 0 | 160.00 | 28.00 |
| Waste cotton | 28.23 | 10.00 | 0 | 90.00 | 41.04 |
| Jute cloth | 16.70 | 128.00 | 0 | 10.00 | 15.40 |
| Coir mate | 34.26 | 2.00 | 0 | 60.00 | 18.20 |
| Charcoal cloth | 16.20 | 2.00 | 0 | 180.00 | 58.40 |
| Wood pulp paper | 17.00 | 2.00 | 0 | 65.00 | 37.30 |
| Polystyrene sponge | 52.06 | 300.00 | 0 | 0.00 | 29.05 |
| Water coral fleece | 69.67 | 2.00 | 0 | 10.00 | 34.21 |

tion. Distillate yield and thermal efficiency of single-slope multi-wick still are about 2.50 L/m$^2$d and 34.0%, respectively, even under cloudy weather conditions (Sodha et al., 1981). Thermal efficiency of double slope tilted wick solar still is lower than single-slope tilted wick solar still in winter due to high reflection losses (Shukla & Sorayan, 2005). Multi-wick tilted solar still of area 28.0 m$^2$ and capacity 70.0 L/d has been successfully operated for 1 year to meet the distilled water requirements of a small chemical lab at New Delhi, India. Little maintenance is recommended to sustain the continuous successful operation of the system for long years (Tiwari, 1984). Incorporation of multiple wicks also prevents bottom heat loss from the system to the ambient. Type of wicks employed in tilted solar stills along with their characteristics and evaporative heat transfer coefficient between the wetted wick surface and the glass cover is tabulated in Table 4.1.

### 4.3.2 *Feed Water Flow Rate, Salt Concentration, and Temperature*

Increasing the feed water flow rate reduces the residence time of feed water in the wick and increases wetted wick heat capacity, thereby it leads to low evaporation rates. Drop in efficiency with increase in salt concentration can be attributed to the increased surface tension and reduced vapor pressure over the evaporating surface. Increased absorber/wetted wick temperature favors increased distillate production rate due to enhancement in evaporation process. The relation between thermal efficiency ($\eta_{th}$)

of tilted solar still with wick and feed water flow rate ($\dot{m}$ in kg/m$^2$.h) is given by (Mahdi et al., 2013)

$$\eta_{th}(\%) = 54.83 - 8.36(\dot{m}) + 0.58(\dot{m})^2 - 0.014(\dot{m})^3 \tag{4.17}$$

Valid for $2 \leq \dot{m} \leq 14$ kg/m$^2$.h.

The relation between thermal efficiency of tilted solar still with wick ($\eta_{th}$) and feed water salt concentration ($X_{\mathrm{NaCl}}$ in %) is given by (Mahdi et al., 2013)

$$\eta_{th}(\%) = 37.7 - 1.7(X_{\mathrm{NaCl}}) \tag{4.18}$$

Valid for $0\% < X_{\mathrm{Nacl}} < 11\%$.

The relation between hourly distillate productivity ($\dot{m}_h$ in kg/m$^2$.h) and wetted wick temperature ($T_w$ in °C) of tilted solar still with wick is given by (Mahdi et al., 2011)

$$\dot{m}_h = 0.33 + 0.011(T_w) \tag{4.19}$$

Valid for $30\,°\mathrm{C} < T_w < 60\,°\mathrm{C}$.

The above correlations are subjected to the following conditions of 420 W/m$^2$ solar radiation intensity, wind speed of 0.7 m/s, and ambient temperature of 22.1 to 22.5 °C (Mahdi et al., 2011).

### 4.3.3 Absorber Plate Modifications

The absorber plate of tilted solar still can be of weir type and partitioned basin type. The schematic of weir-type tilted solar still is shown in Fig. 4.4a. In weir-type tilted solar still, absorber plate is modified to have a number of weirs along the feed water flow direction in addition to the top basin and bottom basin which can be seen from Fig. 4.4b. This type of design combines the advantages of both basin and tilted solar still. The height of weirs is limited to 1.60 mm to reduce water thickness while the top and bottom basins can hold feed water that is to be treated at least for 1 day. Feed water flows from the top basin to the bottom basin via weirs and these weirs increase the residence time of feed water to facilitate increased evaporation.

One of the major limitations of this arrangement is the requirement of circulation pump to recirculate feed water between the bottom basin and top basin. The feed water flow rate from the top basin to bottom basin is about 5.0 g/s. Distillate production rate of this still is about 5.50 L/m$^2$d which is about 23.0% higher than the distillate production rate of conventional basin still under similar climatic conditions. Single glass cover is recommended for this system as double glass cover reduces the solar radiation intensity reaching the absorber plate due to the reduced transmissivity value of dual glass cover in comparison to single glass cover (Sadineni et al., 2008). Pump and wick are some of the essential components of weir-type solar still. This

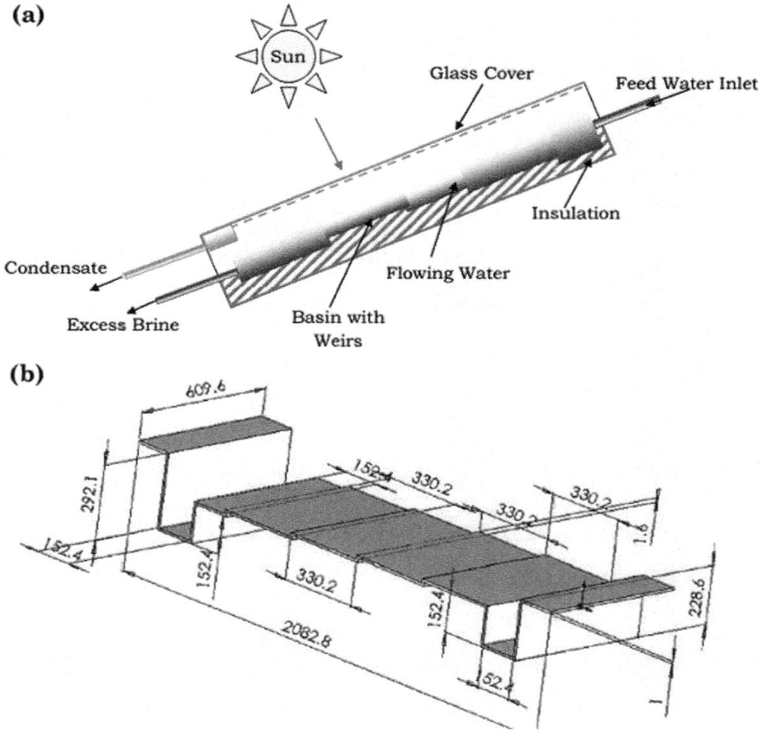


**Fig. 4.4** Schematic of **a** weir-type tilted solar still, **b** weir-type basin (Sadineni et al., 2008)

limitation has been addressed in tilted solar still with partitioned basin, thereby both pump and wicks are avoided which makes partitioned basin system more economical than weir-type tilted still.

The schematic of tilted solar still with partitioned basin is shown in Fig. 4.5. Partitioned basin holds the feed water throughout the day, thereby the need for feed water recirculation is prevented. The daily yield of tilted solar still with copper partitions over its absorber is about 4.38 L/m$^2$ (Abujazar et al., 2018). Nocturnal productivity of tilted solar still with partitioned basin is about 17.75% of the total distillate production rate of the system in cloudy days. Moreover, its optimum tilt angle helps to capture more solar radiation for the given location. The tilted solar still with partitioned basin has a thermal efficiency of about 43.0% (Sharon et al., 2017). Moreover, its performance can outsmart most of the solar stills including weir-type tilted solar still which can be seen from the plot shown in Fig. 4.6.

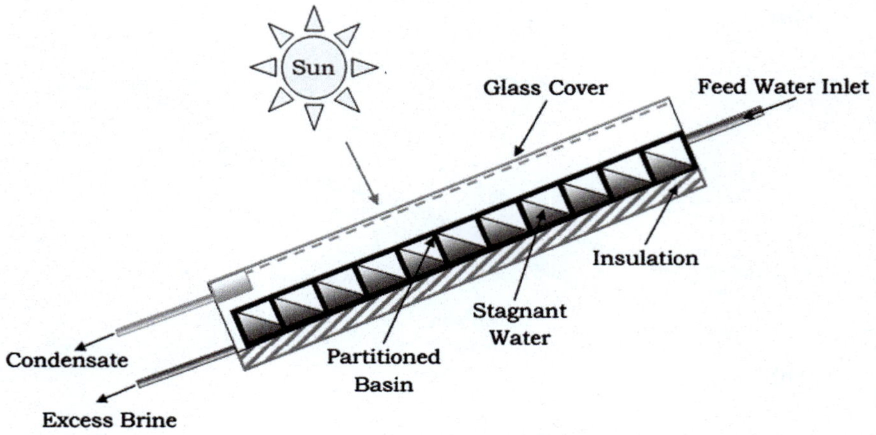


**Fig. 4.5** Schematic of tilted solar still with partitioned basin (Sharon et al., 2017)

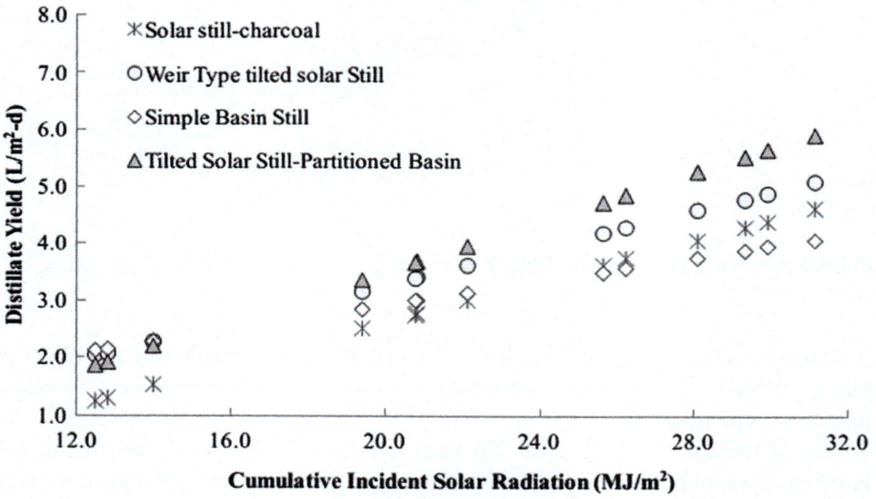


**Fig. 4.6** Comparison of distillate yield of tilted solar still with partitioned basin with other stills (Sharon et al., 2017)

### *4.3.4 Incorporation of Vertical Reflector*

Distillate yield of tilted wick solar still can be enhanced by increasing the intensity of solar radiation incident over its aperture. One of the simple and least expensive methods to enhance incident solar radiation intensity is by incorporating reflectors.

The schematic of tilted solar still with flat plate reflector extending from its back wall is shown in Fig. 4.7. Fixed vertical external reflectors can enhance the distillate

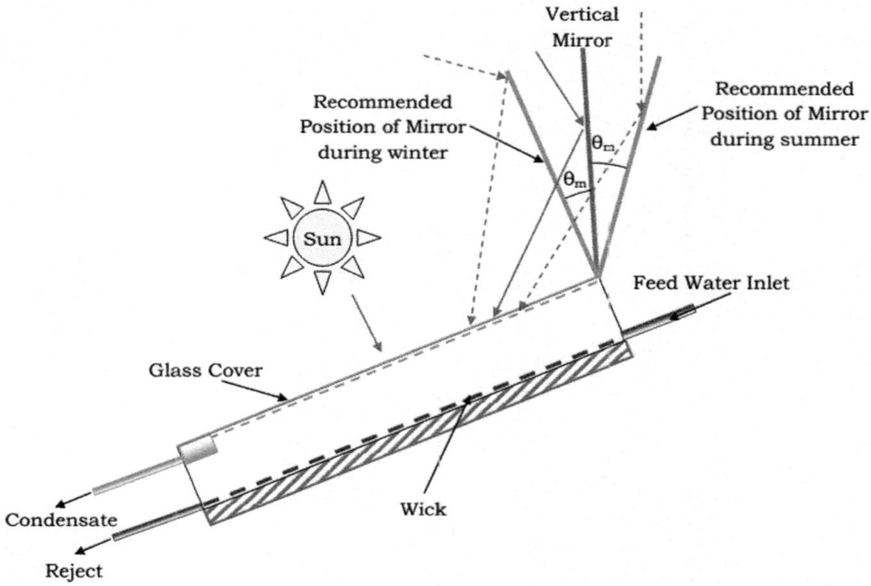


**Fig. 4.7** Schematic of tilted wick solar still with top reflector showing position of mirror in summer and winter seasons

productivity of tilted wick still during spring and autumn seasons to a great extent rather than in summer and winter seasons. Increasing tilt angle of still in winter beyond local latitude diminishes the effect of fixed vertical reflector because during winter the altitude angle of the sun is minimum, thereby most of the radiation reflected by the reflector falls on the ground rather than over the still's aperture. Hence, it becomes necessary to adjust the tilt angle of reflector during winter and summer seasons to improve reflector's utilization and still's yield. In winter season, it is recommended to incline the reflector slightly forward to facilitate the reflected ray to incident over still's aperture even at low solar altitude angle.

The impact of reflector's inclination angle from vertical and tilted solar still's tilt angle from ground on distillate yield of tilted solar still with external reflector above its back wall on winter solstice day is shown in Fig. 4.8a. Higher distillate yield of about 4.80 L/m$^2$d is observed for solar still's tilt angle of 40° and reflector's inclination angle of 15°. The impact of reflector length on distillate yield during winter solstice day is shown in Fig. 4.8b. For fixed vertical orientation of the reflector ($\theta_m = 0°$) and length of reflector greater or equal to half of the still length, varying tilt angle of still has no significant impact on distillate yield. For optimum reflector inclination angle of 15°, increasing reflector length shows significant improvement in yield of the still. From Fig. 4.8c, it can be noticed that the yield of the system is higher by about 100.0% in all seasons at the corresponding optimum tilt angle of still and optimum inclination angle of the reflector in comparison to tilted solar still

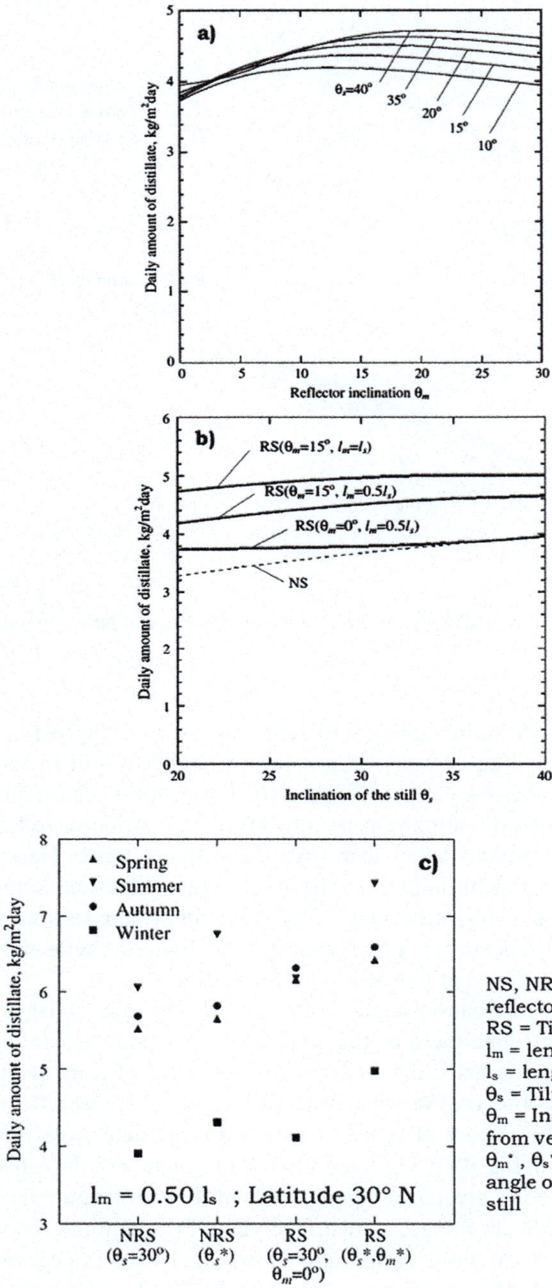


NS, NRS = Tilted solar still without reflector
RS = Tilted solar still with reflector
$l_m$ = length of reflector
$l_s$ = length of tilted solar still
$\theta_s$ = Tilt angle of still from ground
$\theta_m$ = Inclination angle of reflector from vertical
$\theta_m^*$ , $\theta_s^*$ = Optimum Inclination angle of reflector and tilt angle of still

**Fig. 4.8** **a** Effect of tilted solar still's tilt angle and reflector's inclination angle on distillate yield during winter solstice day. **b** Effect of tilted solar still's tilt angle and reflector's length on distillate yield during winter solstice day. **c** Comparison of distillate yield of tilted solar still under various conditions (Tanaka & Nakatake, 2009a; Tanaka, 2009)

**Table 4.2** Optimum tilt angle of tilted wick solar still and inclination angle of top reflector for various months in 30 °N latitude (Tanaka, 2009)

| Month | Tilted solar still with no reflector | Tilted solar still with reflector | |
|---|---|---|---|
| | Tilt angle of still ($\theta_s$) | Tilt angle of still ($\theta_s$) | Inclination angle of reflector from vertical ($\theta_m$) |
| January | 50° | 45° | 20° |
| February | 45° | 35° | 10° |
| March | 30° | 25° | 0° |
| April | 15° | 10° | 15° |
| May | 10° | 10° | 20° |
| June | 10° | 10° | 20° |
| July | 10° | 10° | 20° |
| August | 15° | 10° | 15° |
| September | 30° | 25° | 0° |
| October | 45° | 35° | 10° |
| November | 50° | 45° | 20° |
| December | 50 | 50 | 25 |

without top reflector. The optimum tilt angle of still and inclination of reflector for various months for 30 °N latitude is tabulated in Table 4.2. Reflector adjustment is highly recommended every month to have high distillate yield because with fixed vertical reflector, yield enhancement is just 9.0% in comparison to the system without reflector (Tanaka & Nakatake, 2007).

The schematic representing one-step azimuth tracking of tilted wick solar stills with fixed vertical reflector half the still's length is shown in Fig. 4.9a. The loci of the sun in the sky for various seasons are shown in Fig. 4.9b. In case of one-step azimuth tracking, the still with vertical reflector will be tracked manually once in a day (near noon) to change its azimuth position. Azimuth angles of still facing sole south, east, and west directions are 0°, −90°, and +90°, respectively. The optimum azimuth angle and tilt angle of the still with fixed vertical reflector in 30 °N latitude are tabulated in Table 4.3. One-step manual azimuth tracking of tilted wick solar still with fixed vertical reflector has increased its distillate production rate significantly which can be seen clearly in Table 4.3. Yield of the system due to one-step manual azimuth tracking is about 12.50–42.90% higher than the fixed system without tracking.

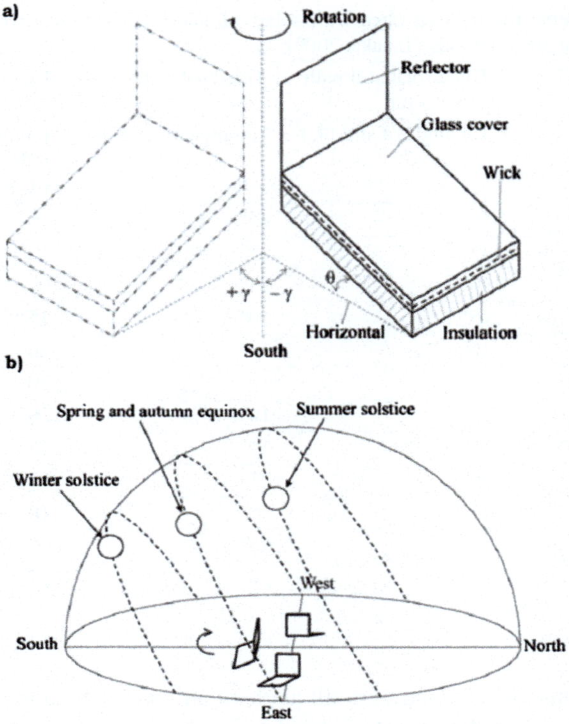


**Fig. 4.9** **a** Schematic representing one-step manual azimuth tracking of tilted wick solar still with fixed vertical reflector, **b** Loci of the sun in winter solstice, summer solstice, spring equinox, and autumn equinox at 30 °N latitude (Tanaka & Nakatake, 2009b)

**Table 4.3** Optimum azimuth angle and tilt angle of tilted wick solar with fixed vertical reflector (Tanaka & Nakatake, 2009b)

| Season | Rotated still with reflector | | | Fixed still with reflector | | | Yield enhancement due to manual azimuth tracking (%) |
|---|---|---|---|---|---|---|---|
| | $\theta°$ | $y°$ | Yield (kg/m$^2$d) | $\theta°$ | $y°$ | Yield (kg/m$^2$d) | |
| Spring | 30 | ±50 | 7.80 | 30 | – | 6.50 | 20.0% |
| Summer | 30 | ±80 | 10.00 | 10 | – | 7.00 | 42.90% |
| Autumn | 30 | ±50 | 8.20 | 30 | – | 6.80 | 20.60% |
| Winter | 60 | ±30 | 4.50 | 60 | – | 4.00 | 12.50% |

$\theta°$ = Tilt angle of still
$y°$ = Azimuth angle of still; + = south west (Afternoon hours); − = south east (morning hours)
For fixed still, $y = 0°$

### 4.3.5 *Incorporation of Bottom Reflector*

The schematic of tilted wick solar still with bottom reflector is shown in Fig. 4.10. The declination angle of the sun varies day to day, hence the entire ray reflected from the bottom reflector may not reach the wick of the still, as some may get reflected towards the ground or reflected back to the sky. Hence, it is necessary to identify the optimum orientation of the bottom reflector for various seasons to have enhanced distillate production from the system. The variation of optimum inclination angle of the bottom reflector from horizontal along with still's distillate yield for various seasons is shown in Table 4.4. Distillate yield of the still with bottom reflector at optimum inclination is about 10% (in winter) to 25% (in summer) higher than the still without bottom reflector. For yearly average bottom reflector inclination, yield of the still is about 13.0% higher the still without bottom reflector (Tanaka, 2011).

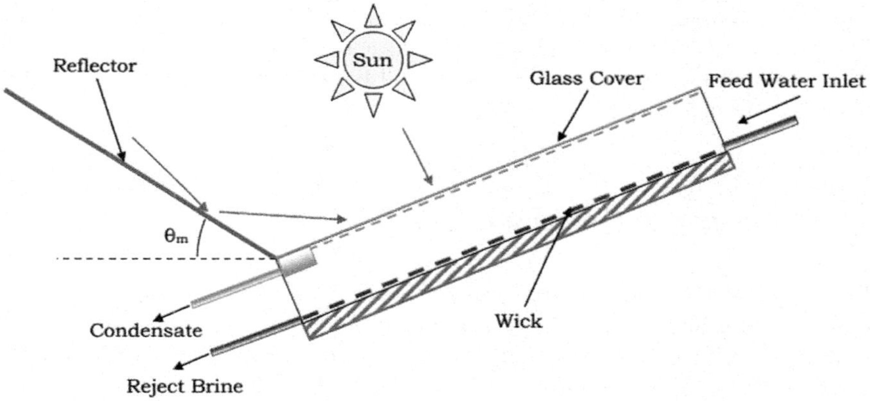


**Fig. 4.10** Tilted wick solar still with bottom reflector

**Table 4.4** Yield and optimum inclination of bottom reflector of the tilted wick still for various seasons in 30 °N latitude (Tanaka, 2011)

| Seasons | Spring | Summer | Autumn | Winter | Average |
|---|---|---|---|---|---|
| Global solar radiation intensity (MJ/m$^2$d) | 23.3 | 30.3 | 23.5 | 12.6 | 22.4 |
| Yield of still without bottom reflector (kg/m$^2$d) | 5.66 | 6.01 | 5.83 | 3.66 | 5.29 |
| Optimum reflector inclination from horizontal ($\theta_m$) | 30° | 50° | 30° | 20° | 35° |
| Yield of still with bottom reflector oriented at optimum inclination (kg/m$^2$d) | 6.83 | 7.50 | 6.99 | 4.02 | 6.03 |

Length of reflector is equal to the length of the still

### 4.3.6 Glass Cover Cooling

Cooling the glass cover of tilted multi-wick solar still by flowing thin film of water over it reduces the glass cover temperature and increases condensate productivity of the still by about 10.0% (Dhiman & Tiwari, 1990). Optimum flow velocity of cooling water over the glass cover is about 1.5 m/s (Janarthanan et al., 2006). The convective heat transfer coefficient between the glass cover and the flowing water is at least 135.0 $W/m^2K$. Distillate yield of single-slope tilted multi-wick solar still with glass cover cooling is about 3.50% higher than conventional basin still with glass cover cooling. Thermal efficiency of single-slope tilted multi-wick solar still with glass cover cooling is about 47.8%. For larger capacity plants, evaporation of flowing cooling water results in loss of precious water resources. Water loss by evaporation from the flowing water over the glass cover of the still is more significant during summer with low humidity and under thin film flow conditions. Maximum area requirement of tilted wick solar still with glass cover cooling and conventional basin solar still to produce 300.0 L/d is about 107.0 $m^2$ and 177.0 $m^2$, respectively, under hot and dry climates of India (Tiwari & Somwanshi, 2018). Shading the uppermost 25% of the glass cover of the tilted wick still has enhanced the distillate yield by about 5% (Deniz, 2012).

### 4.3.7 Shading of Condensing Cover

Condensation of vapors over the inner surface of glass cover leads to reduced temperature difference between the absorber plate and glass cover. In addition, it also reflects a portion of solar radiation incident over the glass cover, thereby leads to reduced yield. Moreover, some droplets may fall back over the evaporating surface by gravity, thereby some amount of distillate is also lost. Apart from these, direct exposure of the wick to the sunlight reduces the life of the wick and diminishes the color of the wick rapidly (Tiwari & Gupta, 1988). Hence, in order to avoid these drawbacks, tilted inverted trickle solar still has been proposed. The schematic of tilted inverted trickle solar still is shown in Fig. 4.11. In this still, the feed water is trickled down the back of the blackened absorber plate and the wick is kept attached to the plate by means of wire mesh welded to the back of the absorber plate. The absorber plate is covered by glass cover above and below by maintaining required gap. Insulated back plate is provided between the absorber plate and bottom glass cover to prevent the mixing of accidentally dripped saline water from the wetted wick with the formed condensate.

Vapors generated move through the gap between insulated back plate and bottom glass cover by convection and get condensed over the bottom glass cover. The productivity of this system is about 1.70–3.83 $L/m^2d$. If feed water salinity of about 32360 ppm is distilled, the salinity of water collected at intermediate header and condenser header is noticed to be about 9327 ppm and 356 ppm, respectively (Bad-

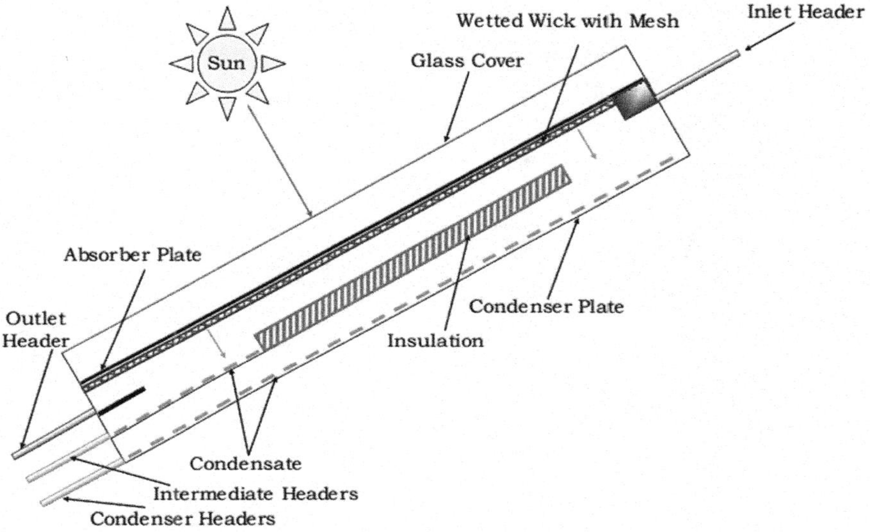


**Fig. 4.11** Schematic of tilted inverted trickle-type solar still (Badran & Hamdan, 1995)

ran & Hamdan, 1995). The high salinity of water observed at intermediate header indicates dripping of saline water from the wick to the back plate by gravity. Badran (2001) claimed that the condensate and saline water mixture obtained in intermediate header can be used for agricultural activities. Recovering latent heat of condensation for preheating feed water has resulted in 12.0% increase in distillate productivity of the tilted inverted trickle-type solar still (Badran, 2001). Reduction in salinity of feed water from 32360 to 6000 ppm has resulted in 18.0% increase in distillate production. The optimum feed water flow rate for this still configuration is recommended to be about 0.70 g/s (Badran et al., 2004).

### *4.3.8 Provision of Separate Condenser Unit*

Vapors generated inside the distillation chamber of solar still condense over the inner surface of glass cover, thereby glass cover temperature rises. Especially during peak sunshine hours the temperature of glass cover becomes higher than the dew point temperature of air-vapor mixture inside the still. Hence, the still will not be able to generate more condensate/distillate. In order to overcome this drawback, separate condensers are employed in still, thereby most of the vapors condense in the condenser leading to lower glass cover temperature and reduced heat loss from the glass cover to the ambient. Condenser can be installed within the still or external to the still and the condensation process can be carried out in passive or active mode. In passive mode, flow of vapors towards the condensing surface is driven by buoyancy,

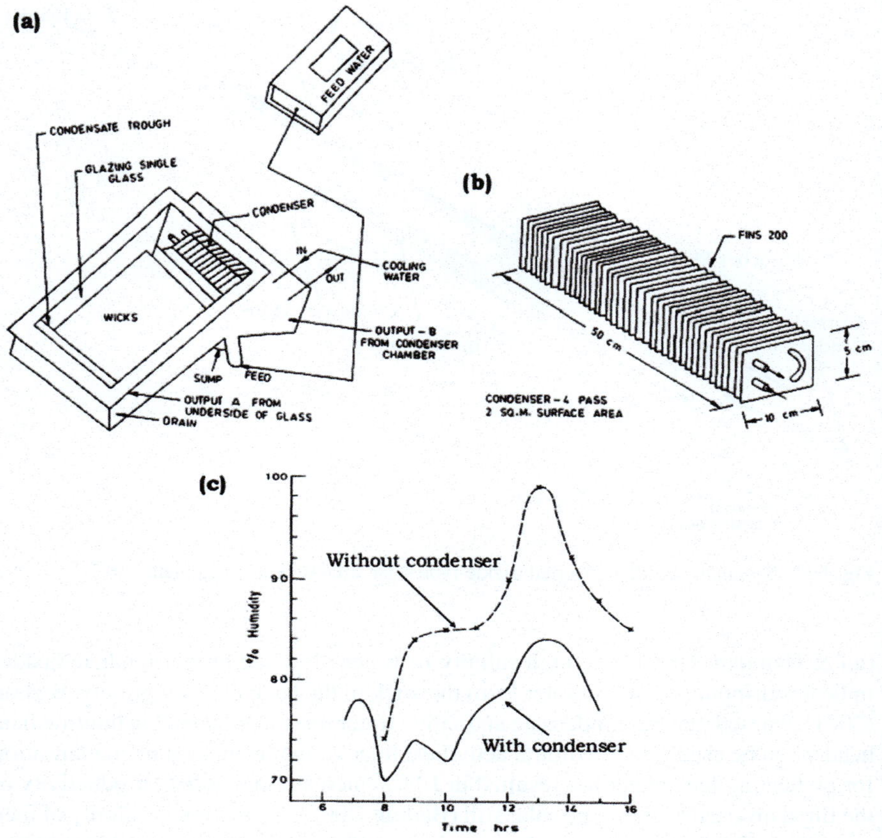


**Fig. 4.12** **a** Tilted wick solar still with internal condenser, **b** Finned condenser, **c** Effect of condenser on relative humidity within the tilted solar still (Reddy et al., 1983)

while in active mode, fans are used to suck the vapors and move them towards the condensing surface.

**Provision of Internal Condenser** The schematic of tilted wick solar still (0.5 $m^2$) with internal condenser operating in passive mode is shown in Fig. 4.12a. Four pass condenser with 200 fins and surface area of 2.0 $m^2$ installed inside the still is shown in Fig. 4.12b. Cooling water is circulated through the pipes of the condenser. From Fig. 4.12c, it can be seen that for the still without condenser, relative humidity of air within the distillation chamber is high and sometimes around 100% which clearly indicates that at these times neither condensation nor evaporation occurs. For the system with condenser, the relative humidity inside the still is less and the maximum observed value is around 84.0% which indicates the occurrence of continuous evaporation and condensation inside the system. Distillate yield of the system with

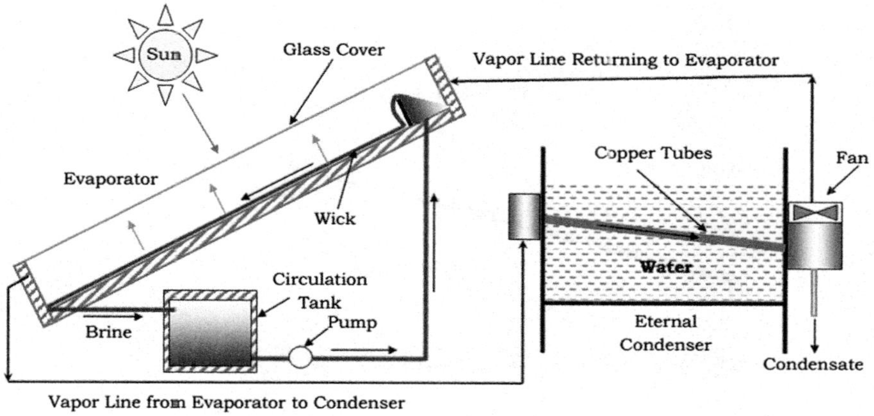


**Fig. 4.13** Tilted wick solar still with external condenser (Rahim, 1995)

condenser is around 4.45 L/m$^2$d for a solar radiation intensity of 22.0 MJ/m$^2$d which is about 15.0–25.0% higher than the system without condenser.

**Provision of External Condenser** The schematic of tilted solar still with forced external condenser is shown in Fig. 4.13. In this system, vapor generated inside the still is forced to condense inside the copper tube of the external condenser by sucking the vapor using an 80 W capacity fan. Condenser unit consists of copper tubes dipped inside the water bath. The optimum vapor suction rate and volumetric circulation rate of brine are found to be 2.0 m/s and 0.0004 m$^3$/s, respectively. Increasing the vapor suction rate beyond the optimum value leads to re-evaporation of condensed water. Daily average temperature difference of about 30.0 °C is noticed between the absorber surface and glass cover in this system. The nocturnal productivity of this system is zero while the diurnal productivity is about 4.20 L/m$^2$d for a cumulative solar radiation intensity of 15.0 MJ/m$^2$d which is about three times higher than conventional basin solar still's productivity under similar climatic conditions (Rahim, 1995).

### *4.3.9 Recirculation of Brine*

Temperature of un-evaporated brine leaving the tilted wick solar still will be generally higher than the ambient temperature. Hence, simply dumping it outside without utilizing its heat energy content indicates loss of useful energy potential. The rejected hot un-evaporated brine can be recirculated back to the distillation chamber using small pump for further evaporation, thereby enhanced yield and efficiency can be achieved. The schematic of tilted wick solar still with brine recirculation is shown in Fig. 4.14. Distillate productivity of the tilted wick solar still with brine recirculation is

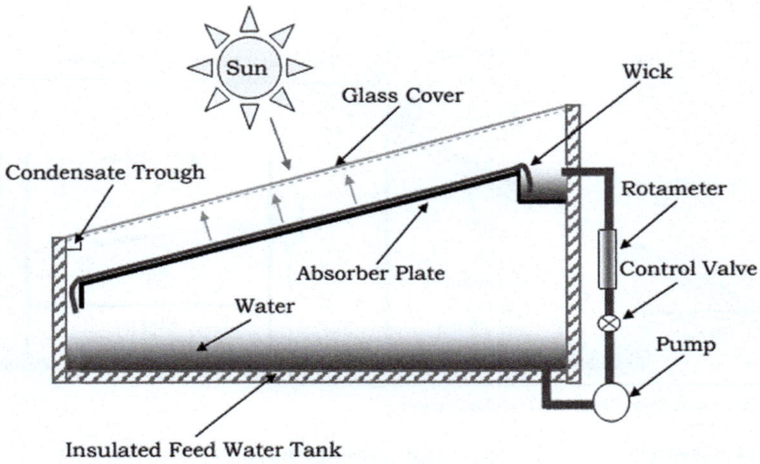


**Fig. 4.14** Tilted wick solar still with continuous brine recirculation (El-Agouz et al., 2015)

about 41.2% higher than conventional basin solar still. Along with brine recirculation, addition of makeup water to maintain constant water level in bottom basin has been effective in enhancing the distillate productivity further by 16.0%. This observation can be attributed to the dilution of basin water salinity due to addition of makeup water (El-Agouz et al., 2015).

### *4.3.10 Thermal Energy Storage Material*

In general, nocturnal distillate productivity of tilted wick solar still is nearly zero. In addition, the absorber and glass cover temperature of the still is also high which favors increased heat losses to the ambient during sunshine hours. Hence, in order to overcome these drawbacks and to improve distillate productivity, incorporation of phase change material (PCM) below the absorber plate of tilted wick solar still is recommended. Phase change material stores excess heat energy during high solar radiation intensity hours and releases it during low sunshine or after sunshine hours (afternoon and evening hours). The released heat energy from the PCM causes evaporation of feed water during late evening hours. Ambient temperature during the late evening hours are low, thereby enhanced distillate production can be expected from the PCM-integrated tilted wick solar still.

The schematic of tilted wick solar still with half barrel absorber plate incorporated with PCM and tilted wick solar still with corrugated absorber plate incorporated with PCM is shown in Figs. 4.15, and 4.16, respectively. The distillate productivities of tilted wick solar still, tilted wick solar still with corrugated absorber plate, and tilted wick solar still with half barrel absorber plate are about 3.85 $L/m^2d$, 4.40 $L/m^2d$,

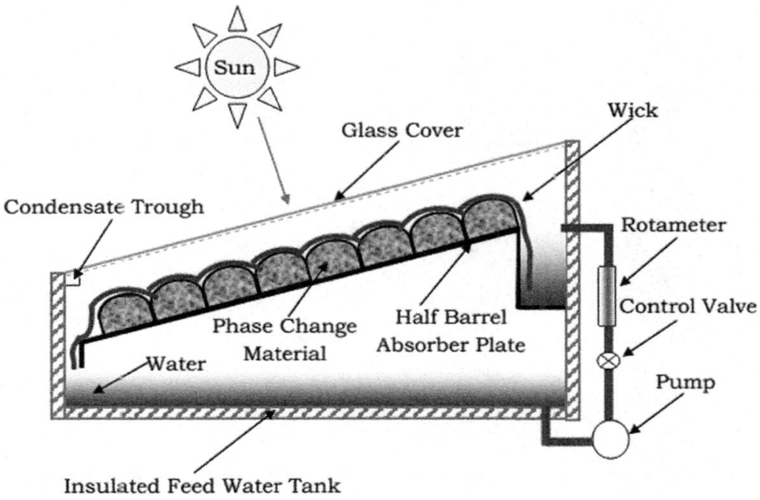


**Fig. 4.15** Tilted wick solar still with half barrel absorber plate and phase change material (no continuous recirculation) (Younes et al., 2021)

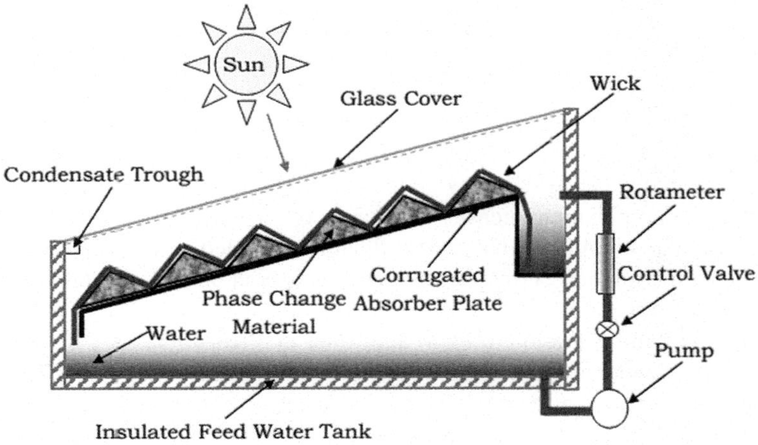


**Fig. 4.16** Tilted wick solar still with corrugated absorber plate and phase change material (no continuous recirculation) (Younes et al., 2021)

and 4.25 $L/m^2d$, respectively. Surface area of corrugated and half barrel absorber is relatively higher than flat absorber, thereby more solar radiation can be collected by the still.

In addition, addition of 2.5 wt/wt% of CuO nano-particle to paraffin wax (PCM) has improved its thermal conductivity from 0.21 to 0.28 W/mK. Incorporation of PCM with nano-particles to the tilted wick solar still with corrugated absorber plate has increased its distillate yield from 4.40 to 5.60 $L/m^2d$. The distillate productivity

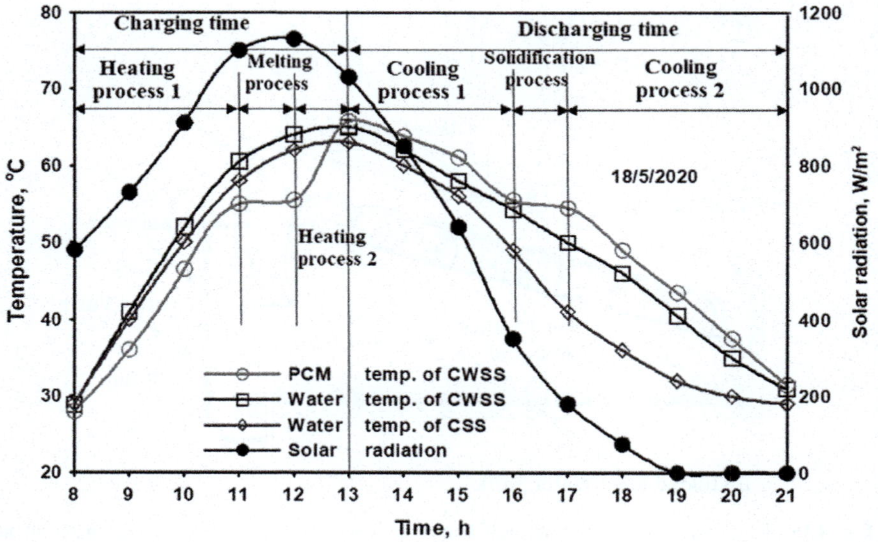


**Fig. 4.17** Variation of solar radiation intensity, PCM temperature, and water temperature of tilted wick solar still with corrugated absorber and PCM (CSS—conventional basin still; CWSS—tilted wick solar still with corrugated absorber and CuO-doped PCM) (Younes et al., 2021)

of tilted wick solar still with half barrel absorber plate and nano-particle-based PCM has increased by about 124.0% in comparison to conventional solar still. The mass flow rate of feed water by capillary effect will be varying throughout the day due to variation in solar radiation intensity with time. The average feed water flow rates in tilted wick solar still, tilted wick solar still with corrugated absorber plate, and tilted wick solar still with half barrel absorber plate are about 0.0112, 0.0065, and 0.0075 kg/min.$m^2$ of still area.

The temperature variation of PCM and water in tilted solar still with corrugated basin and conventional basin solar still with time is shown in Fig. 4.17. This figure indicates the charging and discharging process happening in PCM. Up to 11:00 A.M., sensible heating of PCM is observed, above 11:00 A.M. melting of PCM occurs at a temperature of about 55.0 °C and it lasts up to 12:00 P.M. After phase change, again sensible heating occurs from 12:00 P.M. to 13:00 P.M. shooting up the temperature to 65.0 °C. It can be noticed that there is a shift/time lag between the occurrence of peak solar radiation and peak temperature and this is due to the high thermal capacity of PCM. Discharging of heat energy from PCM is observed after 13:00 P.M. during which solar radiation intensity declines. Sensible heat energy discharge, latent heat energy discharge, and again sensible heat energy discharge are observed during PCM energy discharge process. It can be noticed that the temperature of PCM and water of tilted wick solar still with corrugated absorber plate and PCM are closer to each other and are higher than the temperature of water in conventional basin still even after sunset (around 17:00 P.M.). This clearly indicates the high nocturnal distillate production capability of PCM-integrated stills.

## 4.4 Features of Tilted Wick Solar Thermal Desalination Systems

(a) Less weight, portable and orientation can be adjusted without major modifications (Sodha et al., 1981).
(b) For same water production capacity cost of tilted wick still is nearly half the basin solar still cost (Sodha et al., 1981).
(c) High distillate productivity than basin still (Manikandan et al., 2013).
(d) Tilt angle can be easily adjusted as per seasons and locations (Manikandan et al., 2013).
(e) Can produce distillate even in cloudy days (Sodha et al., 1981).
(f) Shadowing effect is nearly nil as the side walls are small (Sodha et al., 1981).
(g) High temperatures are achieved which indicates effective removal of bacteria and other pathogens (Sodha et al., 1981).
(h) Color of wick may fade away quickly, thereby black dye needs to be added (Tiwari et al., 1988).
(i) Life of natural wicks is limited and requires frequent replacement (Tiwari et al., 1988).
(j) Maintaining constant level of water in feed water troughs to prevent dry out of wicks is difficult (Tiwari, 1984).
(k) Algae deposits in pipe lines and tubes can block flow of feed water in long-term operation. Hence, colored pipelines and copper sulfate are needed to avoid this problem (Tiwari, 1984).

## References

Abujazar, M. S. S., Fatihah, S., Lotfy, E. R., Kabeel, A. E., & Sharil, S. (2018). Performance evaluation of inclined copper-stepped solar still in a wet tropical climate. *Desalination, 425*, 94–103.

Alvarado-Juárez, R., Alvarez, G., Xaman, J., & Hernandez-Lopez, I. (2013). Numerical study of conjugate heat and mass transfer in a solar still device. *Desalination, 325*, 84–94.

Aybar, H. S., Egelioglu, F., & Atikol, U. (2005). An experimental study on an inclined solar water distillation system. *Desalination, 180*, 285–289.

Aybar, H. S. (2006). Mathematical modeling of an inclined solar water distillation system. *Desalination, 190*, 63–70.

Badran, A. A. (2001). Inverted trickle solar still: Effect of heat recovery. *Desalination, 133*, 167–173.

Badran, A. A., & Hamdan, M. A. (1995). Inverted trickle solar still. *International Journal of Solar Energy, 17*(1), 51–60.

Badran, A. A., Assaf, L. M., Kayed, K. S., Ghaith, F. A., & Hammash, M. I. (2004). Simulation and experimental study for an inverted trickle solar still. *Desalination, 164*, 77–85.
Deniz, E. (2012). An investigation of some of the parameters involved in inclined solar distillation systems. Environmental Progress & Sustainable Energy.
Dhiman, N. K., & Tiwari, G. N. (1990). Effect of water flowing over the glass cover of a multi-wick solar still. *Energy Conversion and Management, 30*(3), 245–250.
El-Agouz, S. A., El-Samadony, Y. A. F., & Kabeel, A. E. (2015). Performance evaluation of a continuous flow inclined solar still desalination system. *Energy Conversion and Management, 101*, 606–615.
Hansen, R. S., Narayanan, C. S., & Murugavel, K. K. (2015). Performance analysis on inclined solar still with different new wick materials and wire mesh. *Desalination, 358*, 1–8.
Janarthanan, B., Chandrasekaran, J., & Kumar, S. (2006). Performance of floating cum tilted-wick type solar still with the effect of water flowing over the glass cover. *Desalination, 190*, 51–62.
Mahaly, A. F., & Alazba, A. A. (2017). Thermal performance analysis of an inclined passive solar still using agricultural drainage water and artificial neural network in arid climate. *Solar Energy, 153*, 383–395.
Maheswaran, G., Ravi, R., Maheswar, R., Shanmugasundaram, K., Padmanathan, K., Subramaniyan, U., Alavandar, S., Ganesh, R., & Manikandan, V. (2021). Conceptual context of tilted wick type solar stills: A productivity study on augmentation of parameters. *Solar Energy, 221*, 348–374.
Manikandan, V., Shanmugasundaram, K., Shanmugan, S., Janarthanan, B., & Chandrasekaran, J. (2013). Wick type solar stills: A review. *Renewable and Sustainable Energy Reviews, 20*, 322–335.
Mahdi, J. T., Smith, B. E., & Sharif, A. O. (2011). An experimental wick-type solar still system: Design and construction. *Desalination, 267*, 233–238.
Oliveri de Paula, A. C., Ismail, K. A. R. (2020). Comprehensive investigation of water film thickness effects on the heat and mass transfer of an inclined solar still. Desalination. https://doi.org/10.1016/j.desal.2020.114895
Reddy, M. S., Chandra, D. J. N., Sehgal, H. K., & Sabberwal, S. P. (1983). Performance of a multiple-wick solar still with condenser. *Applied Energy, 13*, 15–21.
Rahim, N. H. A. (1995). Utilization of a forced condensing technique in a moving film inclined solar desalination still. *Desalination, 101*, 255–262.
Sadineni, S. B., Hurt, R., Halford, C. K., & Boehm, R. F. (2008). Theory and experimental investigation of a weir-type inclined solar still. *Energy, 33*, 71–80.
Sharon, H., Reddy, K. S., Krithika, D., & Philip, L. (2017). Experimental performance investigation of tilted solar still with basin and wick for distillate quality and enviro-economic aspects. *Desalination, 410*, 30–54.
Shukla, S. K., & Sorayan, V. P. S. (2005). Thermal modeling of solar stills: An experimental validation. *Renewable Energy, 30*, 683–699.
Sodha, M. S., Kumar, A., Tiwari, G. N., & Tyagi, R. C. (1981). Simple multiple wick solar still: Analysis and performance. *Solar Energy, 26*, 127–131.
Tanaka, H., & Nakatake, Y. (2007). Improvement of the tilted wick solar still by using a flat plate reflector. *Desalination, 216*, 139–146.
Tanaka, H. (2009). Tilted wick solar still with external flat plate reflector: Optimum inclination of still and reflector. *Desalination, 249*, 411–415.
Tanaka, H., & Nakatake, Y. (2009a). Increase in distillate productivity by inclining the flat plate external reflector of a tilted-wick solar still in winter. *Solar Energy, 83*, 785–789.
Tanaka, H., & Nakatake, Y. (2009b). One step azimuth tracking tilted-wick solar still with a vertical flat plate reflector. *Desalination, 235*, 1–8.
Tanaka, H. (2011). Tilted wick solar still with flat plate bottom reflector. *Desalination, 273*, 405–413.
Tiwari, G. N. (1984). Demonstration plant of a multi wick solar still. *Energy Conversion and Management, 24*(4), 313–316.

Tiwari, G. N., & Gupta, S. P. (1988). Analytical model of inverted multiwick solar still. *International Journal of Solar Energy, 6*, 139–150.

Tiwari, G. N., & Singh, A. K. (1991). Thermal efficiency of double slope FRP solar distiller: An analytical and experimental studies. *Desalination, 82*, 223–232.

Tiwari, A. K., & Somwanshi, A. (2018). Techno-economic analysis of mini solar distillation plants integrated with reservoir of garden fountain for hot and dry climate of Jodhpur (India). *Solar Energy, 160*, 216–224.

Younes, M. M., Abdullah, A. S., Essa, F. A., Omara, Z. M., & Amro, M. I. (2021). Enhancing the wick solar still performance using half barrel and corrugated absorbers. *Process Safety and Environmental Protection, 150*, 440–452.

# Chapter 5
# Tubular Solar Thermal Desalination Systems

## 5.1 Working Concept

The schematic of tubular solar thermal desalination system is shown in Fig. 5.1. Tubular solar thermal desalination system consists of a tubular/cylindrical condensing cover/tube within which a blackened basin loaded with feed water is placed. The working principle of tubular solar thermal desalination system is similar to that of basin-type solar thermal desalination system discussed earlier in Chap. 3. Solar radiation passing through the cylindrical condensing cover heats up the blackened basin, which in turn heats up the basin water, and initiates the evaporation process. The vapors thus generated saturate the air inside the desalination chamber. This saturated air moves up by natural convection and comes in contact with the inner surface of the condensing cover which is relatively at a lower temperature than the basin water, thereby condensation of water vapor occurs. The condensate thus formed rolls down along the cylindrical wall by gravity and gets collected. Tubular solar thermal desalination system has more advantages than simple basin-type solar desalination system as it has large condensing surface area, no basin insulation, and improved condensate quality and quantity (Tiwari, 1988). Curvature of the upper half of the tubular glass cover increases the condensing area by about 50.0% in comparison to the flat condensing surface which is commonly seen in other solar thermal desalination systems (Kumar & Anand, 1992). Tubular solar thermal desalination system is also called tubular solar still.



H. Sharon and K. Srinivas Reddy, *Solar Thermal Desalination Technologies for Potable Water*, https://doi.org/10.1007/978-3-031-49155-9_5

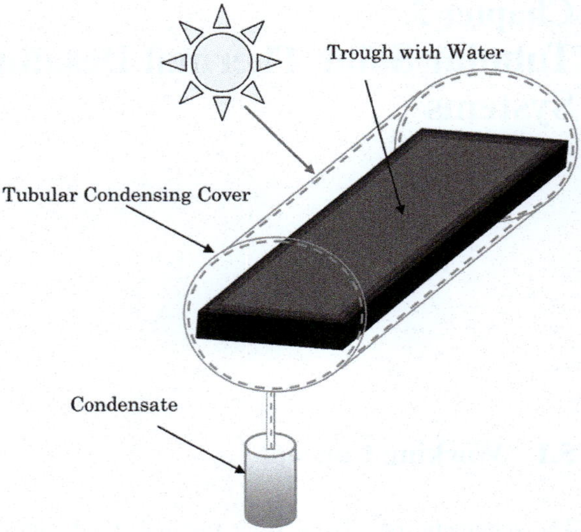


**Fig. 5.1** Schematic of tubular solar still (Ahsan et al., 2012)

## 5.2 Thermodynamic Modeling of Tubular Solar Thermal Desalination System

Assumptions considered for thermodynamic modeling of tubular solar still are (Tiwari, 1988)

(a) The system is air tight and free from vapor leakage.
(b) No thermal gradient across the thickness of water layer and condensing cover.
(c) Vapor condensation is uniform over the curved condensing surface, thereby condensing cover temperature is uniform throughout its surface.
(d) The bottom heat loss coefficient due to trapped air between the bottom of the tray and lower curved glass surface is equal to the air thermal conductance of half of this distance.

The schematic representing the energy transport processes in various components of the tubular solar still (Ahsan & Fuhuhara, 2010) is shown in Fig. 5.2.

Energy balance in water trough,

$$\begin{pmatrix}\text{Rate of change of water}\\ \text{trough temperature}\end{pmatrix} = \begin{pmatrix}\text{solar energy absorbed}\\ \text{by trough}\end{pmatrix} - \begin{pmatrix}\text{convective heat transfer}\\ \text{between trough and water}\end{pmatrix} - \begin{pmatrix}\text{convective heat transfer}\\ \text{between trough and humid air}\end{pmatrix}$$

$$(mc)_t \frac{dT_t}{dt} = Q_{st} - Q_{c(t-w)} - Q_{c(t-ha)} \tag{5.1}$$

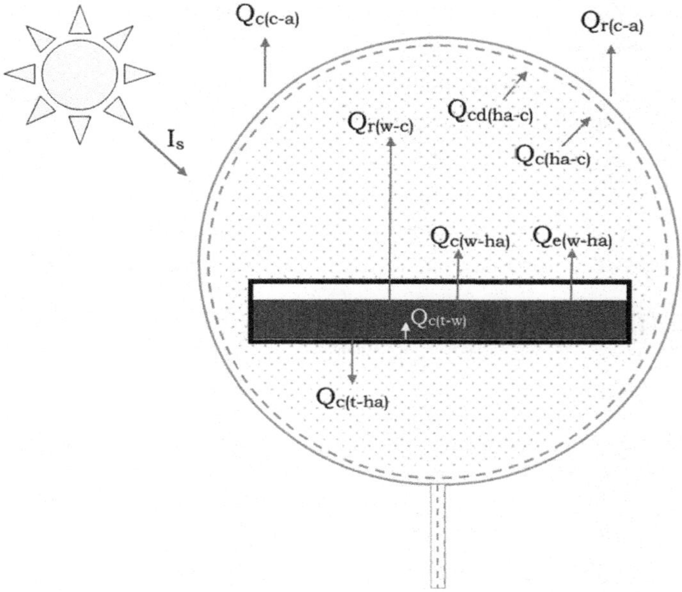


**Fig. 5.2** Energy transport process in tubular solar still (Ahsan & Fukuhara, 2010)

Energy balance in basin water,

$$\begin{pmatrix}\text{Rate of change of basin}\\ \text{water temperature}\end{pmatrix} = \begin{pmatrix}\text{solar energy absorbed}\\ \text{by water}\end{pmatrix}$$
$$+\begin{pmatrix}\text{convective heat transfer}\\ \text{between trough and water}\end{pmatrix} - \begin{pmatrix}\text{evaporation and convection heat}\\ \text{transfer between water and humid air}\end{pmatrix}$$
$$-\begin{pmatrix}\text{radiation heat transfer}\\ \text{between water and cover}\end{pmatrix}$$

$$(mc)_w \frac{dT_w}{dt} = Q_{sw} + Q_{c(t-w)} - (Q_e + Q_c)_{(w-ha)} - Q_{r(w-c)} \tag{5.2}$$

Energy balance in humid air,

$$\begin{pmatrix}\text{Rate of change of humid}\\ \text{air temperature}\end{pmatrix} = \begin{pmatrix}\text{solar energy absorbed}\\ \text{humid air}\end{pmatrix} + \begin{pmatrix}\text{evaporation and convection heat}\\ \text{transfer between water and humid air}\end{pmatrix} + \begin{pmatrix}\text{convective heat transfer}\\ \text{between trough and humid air}\end{pmatrix} - \begin{pmatrix}\text{convention heat transfer}\\ \text{between humid air and cover}\end{pmatrix} - \begin{pmatrix}\text{condensation heat transfer}\\ \text{between humid air and cover}\end{pmatrix}$$

$$(mc)_{ha}\frac{dT_{ha}}{dt} = Q_{sha} + (Q_e + Q_c)_{(w-ha)} + Q_{c(t-ha)} - Q_{c(ha-c)} - Q_{cd(ha-c)} \tag{5.3}$$

Energy balancing in condensing cover,

$$\begin{pmatrix}\text{Rate of change of cover}\\ \text{temperature}\end{pmatrix} = \begin{pmatrix}\text{solar energy absorbed}\\ \text{by cover}\end{pmatrix} + \begin{pmatrix}\text{convection and condensation heat}\\ \text{transfer between humid air and cover}\end{pmatrix} + \begin{pmatrix}\text{radiation heat transfer}\\ \text{between water and cover}\end{pmatrix} - \begin{pmatrix}\text{convention and radiation heat transfer}\\ \text{between cover and atmosphere}\end{pmatrix}$$

$$(mc)_c\frac{dT_c}{dt} = Q_{sc} + Q_{c(ha-c)} + Q_{cd(ha-c)} + Q_{r(w-c)} - (Q_{c(c-a)} + Q_{r(c-a)}) \tag{5.4}$$

Water vapor content in the humid air inside the desalination chamber is given as

$$\begin{pmatrix}\text{Rate of change of water}\\ \text{vapor in humid air}\end{pmatrix} = [\text{evaporated water}] - [\text{condensed water}]$$

$$V_{ha} = \frac{d\rho_{ha}}{dt} = A_w m_e - A_c m_c \tag{5.5}$$

Condensation heat transfer between the humid air and cover ($Q_{cd(ha-c)}$ in W) is given as

$$Q_{cd(ha-c)} = h_{fg} h_{cd(ha-c)} (\rho_{vha} - \rho_{vc}) A_c \tag{5.6}$$

where

$h_{fg}$—latent heat of vaporization (J/kg),
$\rho_{vha}$—water vapor density at humid air temperature (kg/m$^3$),
$\rho_{vc}$—water vapor density at condensing cover temperature (kg/m$^3$), and
$A_c$—surface area of condensing cover (m$^2$).

Condensation mass transfer coefficient for film-wise condensation ($h_{cd(ha-c)}$ in m/s) is given as

$$h_{cd(ha-c)} = 0.996\gamma_c^{(3/4)}\left[\frac{g\rho_l(\rho_l - \rho_{vha})\lambda_l^3 a^3(T_{ha} - T_a)^3}{\mu_l D h_{fg}^3(\rho_{vha} - \rho_{vc})^4}\right]^{(1/4)} \tag{5.7}$$

where

$\gamma_c$—condensation coefficient,
$\rho_l$—density of saturated liquid (kg/m$^3$),
$\gamma_l$—thermal conductivity of saturated liquid (W/mK),
$a$—temperature difference fraction,
$\mu_1$—dynamic viscosity of saturated liquid (kg/ms),
$D$—diameter of tubular solar still (m),
$g$—acceleration due to gravity (m/s$^2$),
$T_{ha}$—temperature of humid air (K), and
$T_a$—ambient temperature (K).

Condensation rate ($m_c$) inside the tubular solar desalination system is given as

$$m_c = h_{cd(ha-c)}(\rho_{vha} - \rho_{vc})A_c \tag{5.8}$$

Convection heat transfer between the humid air and condensing cover ($Q_{c(ha-c)}$ in W) is given as

$$Q_{c(ha-c)} = h_{c(ha-c)}(T_{ha} - T_c)A_c \tag{5.9}$$

Convection heat transfer coefficient between the humid air and condensing cover ($h_{c(ha-c)}$ in W/m$^2$K) is given as

$$h_{c(ha-c)} = \frac{h_{cd(ha-c)}(\rho_{vha} - \rho_{vc})\rho_a C_a}{\rho_{vha}} \tag{5.10}$$

where

$\rho_a$—density of ambient air (kg/m$^3$) and
$C_a$—specific heat capacity of ambient air (J/kgK).

Evaporative mass transfer coefficient ($h_{e(w-ha)}$ in m/s) inside the tubular solar desalination system is given as

$$h_{e(w-ha)} = \left(\frac{\gamma_e M_v R_v(T_w + T_{ha})}{2R_g T_{ha}}\right)\left(\frac{g\beta D_v^2(T_w - T_c)}{v}\right)^{(1/3)} \tag{5.11}$$

where

$\gamma_e$—evaporation coefficient,
$M_v$—molecular weight of water vapor (kg/kmol),
$R_v$—specific gas constant of water vapor (J/kgK),
$R_g$—universal gas constant (J/kmolK),
$D_v$—molecular diffusion coefficient of water vapor in air (m$^2$/s),
$\beta$—volumetric thermal expansion coefficient (1/K),
$\varepsilon$—kinematic viscosity of water vapor (m$^2$/s),
$T_{ha}$—temperature of humid air (K),
$T_a$—ambient temperature (K).

Evaporation heat transfer between trough water and humid air ($Q_{e(w-ha)}$ in W) is given as

$$Q_{e(w-ha)} = h_{fg} h_{e(w-ha)}(\rho_{vm} - \rho_{vha})A_w \tag{5.12}$$

Evaporation rate ($m_e$) inside the tubular solar desalination system is given as

$$m_e = h_{e(w-ha)}(\rho_{ew} - \rho_{vha})A_w \tag{5.13}$$

where

$A_w$—surface area of water in trough (m$^2$).

Convective heat transfer coefficient between trough water and humid air ($h_{c(w-ha)}$ in W/m$^2$K) is given as

$$h_{c(w-ha)} = \frac{h_{e(w-ha)}(\rho_{vw} - \rho_{vha})\rho_a c_a}{\rho_{vw}} \tag{5.14}$$

where

$\rho_{vha}$—water vapor density at humid air temperature (kg/m$^3$) and
$\rho_{vw}$—water vapor density at trough water temperature (kg/m$^3$).

Convective heat transfer between trough water and humid air ($Q_{c(w-ha)}$ in W) is given as

$$Q_{c(w-ha)} = h_{c(w-ha)}(T_w - T_{ha})A_w \tag{5.15}$$

Convective heat transfer between condensing cover and the atmosphere ($Q_{c(c-a)}$ in W) is given as

$$Q_{c(c-a)} = h_{c(c-a)}(T_c - T_a)A_c \tag{5.16}$$

Natural convection heat transfer coefficient between condensing cover and the atmosphere ($h_{c(c-a)}$ in W/m$^2$K) is given as

$$h_{c(c-a)} = \frac{\lambda_a}{D}(Nu_L^{3.3} + Nu_T^{3.3})^{(1/3.3)} \tag{5.17}$$

where

$\lambda_a$—thermal conductivity of ambient air (W/mK),
$D$—diameter of tubular still (m), and
Nu—Nusselt number.

$$Nu_L = \frac{2f}{\ln\left(1 + \frac{2f}{Nu_1}\right)} \tag{5.18}$$

where

$$F = 1 - \frac{0.13}{(Nu_1)^{0.16}} \tag{5.19}$$

$$Nu_1 = 0.772 C_1 Ra^{(1/4)} \tag{5.20}$$

$$Nu_T = C_3 Ra^{1/3} \tag{5.21}$$

where

$C_1$, $C_3$—constants and
Ra—Rayleigh number.

The values of $C_1$ and $C_3$ for various Prandtl numbers are listed in Table 5.1.

Forced convection heat transfer coefficient between condensing cover and the atmosphere ($h_{c(c-a)}$ in W/m$^2$K) is given as

$$h_{c(c-a)} = D_2 \frac{\lambda_a}{x}\left(\frac{\rho_a v_a x}{\mu_a}\right)^{n_1} \tag{5.22}$$

$$Nu = D_2 Re^{n_1} \tag{5.23}$$

where

$D_2$, $n_1$—constants,
$x$—characteristic dimension (in this case it is diameter of the still, $m$),

**Table 5.1** Values of $C_1$ and $C_3$ for various Prandtl numbers (Ahsan & Fuhuhara, 2010)

| Pr | 0.01 | 0.10 | 0.71 | 2.0 | 6.0 |
|---|---|---|---|---|---|
| $C_1$ | 0.242 | 0.387 | 0.515 | 0.568 | 0.608 |
| $C_3$ | 0.074 | 0.088 | 0.104 | 0.110 | 0.111 |

**Table 5.2** Values of $D_2$ and $n_1$ for various Reynolds numbers (Ahsan & Fuhuhara, 2010)

| Reynolds number, Re | $D_2$ | $n_1$ |
|---|---|---|
| 0.0001–0.004 | 0.437 | 0.0895 |
| 0.004–0.09 | 0.565 | 0.136 |
| 0.09–1.00 | 0.800 | 0.280 |
| 1.00–35.00 | 0.795 | 0.384 |
| 35.00–5000.00 | 0.583 | 0.471 |
| 5000.00–50000.00 | 0.148 | 0.633 |
| 50000.00–200000.00 | 0.0208 | 0.814 |

$v_a$—air velocity (m/s),
$\rho_a$—density of ambient air (kg/m$^3$), and
$\mu_a$—dynamic viscosity of ambient air (kg/ms).

The values of $D_2$ and $n_1$ for various range of Reynolds numbers are listed in Table 5.2.

Radiation heat transfer coefficient between the condensing cover and the atmosphere ($h_{r(c-a)}$ in W/m$^2$K) is given as

$$h_{r(c-a)} = \frac{\sigma(\varepsilon_c T_c^4 - \varepsilon_{sky} T_{sky}^4)}{(T_c - T_a)} \tag{5.24}$$

where

$\varepsilon_{sky}$—emissivity of the sky and
$\varepsilon_c$—emissivity of condensing cover.

Emissivity of the sky ($\varepsilon_{sky}$) is given as

$$\varepsilon_{sky} = 0.58\rho_{va}^{1/7} \tag{5.25}$$

where

$\rho_{va}$—vapor density in atmosphere (g/m$^3$).

The sky temperature (in Kelvin) is given as

$$T_{sky} = 0.5T_a + 137.6 \tag{5.26}$$

Radiation heat transfer between the condensing cover and the atmosphere ($Q_{r(c-a)}$ in W) is given as

$$Q_{r(c-a)} = h_{r(c-a)}(T_c - T_a)A_c \tag{5.27}$$

Radiation heat transfer between trough water and the condensing cover ($h_{r(c-a)}$ in W/m$^2$K) is given as

$$Q_{r(w-c)} = h_{r(w-c)}(T_w - T_c)A_w \tag{5.28}$$

Radiation heat transfer coefficient between trough water and the condensing cover ($h_{r(w-c)}$ in W/m$^2$K) is given as

$$h_{r(w-c)} = \frac{\sigma(T_w^2 + T_c^2)(T_w + T_c)}{(1/\varepsilon_w) + (A_w/A_c)[(1/\varepsilon_c) - 1]} \tag{5.29}$$

Convective heat transfer between the trough and humid air ($Q_{c(t-ha)}$ in W) is given as

$$Q_{c(t-ha)} = h_{c(t-ha)}(T_t - T_{ha})A_{tha} \tag{5.30}$$

where

$A_{tha}$—contact surface area between trough and humid air (m$^2$).

Convective heat transfer between the trough and trough water ($Q_{c(t-w)}$ in W) is given as

$$Q_{c(t-w)} = h_{c(t-w)}(T_t - T_w)A_{tw} \tag{5.31}$$

where

$A_{tw}$—contact surface area between trough and water (m$^2$).

Solar radiation absorbed by the trough ($Q_{st}$ in W) is given as

$$Q_{st} = \tau_{t1}I_cA_w + \tau_{t2}I_c(BL_t - A_w) \tag{5.32}$$

where

$$\tau_{t1} = (1 - a_c)(1 - \alpha_c)(1 - \alpha_{ha})(1 - a_w)(1 - \alpha_w)(1 - a_t)\alpha_t \tag{5.33}$$

$$\tau_{t2} = (1 - a_c)(1 - \alpha_c)(1 - \alpha_{ha})(1 - a_t)\alpha_t \tag{5.34}$$

Solar radiation absorbed by the condensing cover ($Q_{sc}$ in W) is given as

$$Q_{sc} = \tau_c I_c DL_c \tag{5.35}$$

where

$$\tau_c = (1 - a_c)\alpha_c \tag{5.36}$$

where

$\tau_c$—transmissivity of the cover.

Solar radiation absorbed by the humid air ($Q_{sha}$ in W) inside the desalination chamber is given as

$$Q_{sha} = \tau_{ha} I_c A_{ha} \tag{5.37}$$

where

$$\tau_{ha} = (1 - a_c)(1 - \alpha_c)\alpha_{ha} \tag{5.38}$$

Solar radiation absorbed by the trough water ($Q_{sw}$ in W) is given as

$$Q_{sw} = \tau_w I_c A_w \tag{5.39}$$

where

$$\tau_w = (1 - a_c)(1 - \alpha_c)(1 - \alpha_{ha})(1 - \alpha_w)\alpha_w \tag{5.40}$$

where

$B$—width of the trough (m),
$L_t$—length of the trough (m),
$D$—diameter of the tubular still (m),
$I_c$—solar radiation intensity incident over the still (W/m$^2$),
$a_c$—albedo of the cover,
$a_w$—albedo of water,
$a_t$—albedo of trough,
$\alpha_w$—absorptivity of water,
$\alpha_t$—absorptivity of trough,
$\alpha_{ha}$—absorptivity of humid air,
$\alpha_c$—absorptivity of cover,
$\tau_c$—transmissivity of the humid air, and
$\tau_w$—transmissivity of water.

The evaporative mass transfer coefficient ($h_m$ in m/s) of the tubular solar desalination system under low-pressure operation conditions can be estimated by (Xie et al., 2018)

$$h_m = 0.46\left(\frac{P_o}{P_t}\right)^{1/8}\left[\frac{L^3\rho_m g\beta_m}{\mu_m}\left(\Delta T + \frac{(P_w - P_c)T_w}{2.64P_t - P_w}\right)\right]^{1/4} \times \left[D_o\frac{P_o}{P_t}\left(\frac{T_w}{T_o}\right)^{(3/2)}\right]^{1/4}\frac{1}{L} \tag{5.41}$$

where

$\Delta T$—temperature difference between water surface and condensing surface (K),
$P_t$—operating pressure (Pa),

$P_o$—ambient pressure (101325 Pa),
$D_o$—mass diffusivity of vapor in air at ambient pressure and ambient temperature of 25 °C (m$^2$/s),
$T_w$—water temperature (K),
$T_o$—ambient temperature (K),
$P_w$, $P_c$—partial vapor pressure of water vapor at water, condensing cover temperature (Pa),
$L$—average gap between evaporating and condensing surface (m),
$\rho_m$—density of humid air (kg/m$^3$), and
$\mu_m$—dynamic viscosity of humid air (kg/m$^3$).

Mass expansion coefficient ($\beta_m$) is given as

$$\beta_m = \frac{M_a - M_v}{(M_a - M_v)\bar{c}_m + M_v} \tag{5.42}$$

where

$M_a$—molecular weight of air (28.96 kg/kmol),
$M_v$—molecular weight of water vapor (18 kg/kmol), and
$\bar{c}_m$—average water vapor concentration in humid air with in the still (kg of vapor per kg of humid air).

## 5.3 Performance Enhancement of Tubular Solar Thermal Desalination Systems

Performance of tubular solar still is affected by metrological parameters like solar radiation intensity, ambient temperature, air temperature, relative humidity, and cloud and dust cover (Sharshir et al., 2019).

Tubular solar still's performance can be enhanced by

(a) Modifying the absorber surface to enhance evaporation process.
(b) Incorporating reflectors/concentrators to increase solar radiation exposure.
(c) Incorporating energy storage materials to utilize excess solar radiation effectively.
(d) Reducing condensing cover temperature to improve vapor condensation process.
(e) Reducing air resistance within the desalination chamber.

### *5.3.1 Adoption of Wicks*

In this type, the simple basin of tubular solar thermal desalination system is replaced with a tray lined by multiple black wick clothes whose one end is dipped inside saline

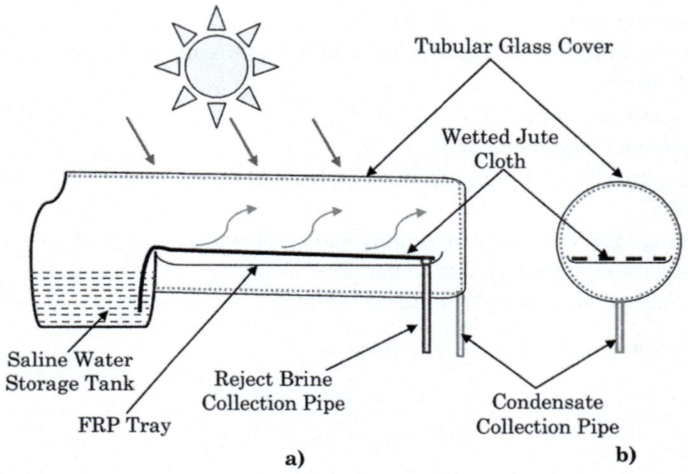


**Fig. 5.3** Schematic of tubular multi-wick solar still **a** Side view, **b** Front view (Kumar & Anand, 1992)

water reservoir as shown in Fig. 5.3. The dipped wicks wet themselves by capillary action. Moreover, the low heat capacity of the wicks favors rapid temperature rise immediately after the incidence of solar radiation over them. The increased temperature leads to enhanced evaporation resulting in 3.0 L/m$^2$d of condensate production even under cloudy weather condition. The condensate output of this modified system is about 18.0, 13.0, and 8.0% higher than the yield of conventional basin solar still, simple multi-wick solar still, and ordinary tubular solar still, respectively (Kumar & Anand, 1992).

### 5.3.2 *Corrugated Absorber Plate*

Tubular solar still incorporated with semi-circular corrugated absorber (shown in Fig. 5.4) produces condensate of about 4.30 L/m$^2$d which is nearly 26.5% higher than the condensate productivity of tubular still with flat absorber (Elshamy & El-Said, 2018). The enhancement in distillate yield is mainly due to the corrugated absorber's increased surface area which favors more solar radiation absorption in comparison to the flat absorber. This fact can be justified through the high basin water temperature observed in the system with semi-circular corrugated absorber in comparison to the system with flat absorber. Sathyamurthy et al. (2020) noticed 46.9% higher distillate yield in tubular solar still by incorporating thin mild steel fins to its absorber plate. Addition of fins inside the basin increases basin and water temperature by about 5.0% in comparison to the system without fins. Distillate yields of the system with and without fins are about 3.04 and 2.07 L/m$^2$d, respectively.

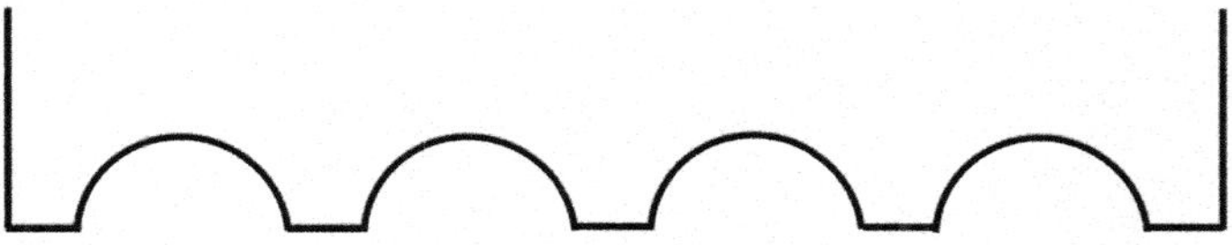

**Fig. 5.4** Schematic of semi-circular corrugated absorber (Elshamy & El-Said, 2018)

### 5.3.3 *Convex Absorber Plate*

Evaporating surface area of tubular still with convex absorber is higher in comparison to the surface area of flat absorber in tubular still. However, it is recommended to insulate the bottom surface of convex absorber to prevent heat loss and to line its top surface with wick as shown in Fig. 5.5, to improve feed water distribution, retention, and evaporation. Jute wick is found to be more advantageous than cotton wick. The convex height of the convex absorber may cause shadow over the evaporating surface. Hence, it is necessary to select optimum convex height for obtaining the best performance. For convex height up to 10 cm, no shadow is casted over the convex absorber throughout the year. On the contrary, around half of the absorber surface is shaded during the month of December for a convex height of 15 cm. In addition, for convex height of 20 cm, the absorber is subjected to shade for more than 7 months which can be seen from Fig. 5.6. Convex height of 15 cm gave maximum performance due to combined benefits of increased evaporation area of 21.3% in comparison to the flat surface and minimal shade in comparison to other convex heights. Distillate yield of tubular still with convex absorber and flat absorber is about 7.90 L/m$^2$d and 4.10 L/m$^2$d, respectively (Essa et al., 2021a).

**Fig. 5.5** Cross section of tubular solar still with wick-lined convex absorber (Essa et al., 2021a)

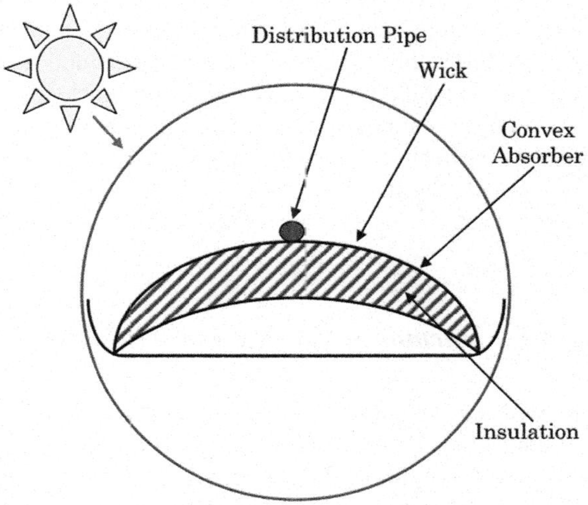

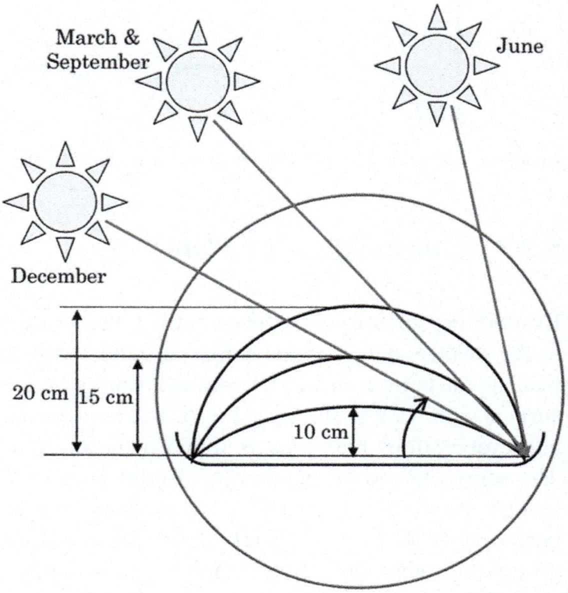


**Fig. 5.6** Shadow effect over the wick-lined convex absorber due to its convex height for various months (Essa et al., 2021a)

### *5.3.4 Hydrophilic Coating of Absorber's External Surface*

The schematic representing tubular solar thermal desalination system with external hydrophilic coated water trough is shown in Fig. 5.7. Hydrophilic coating over the external surface of tubular system's semi-circular water trough (absorber) increases the active evaporation surface area by two-folds. In this case, the overflowing excess water from the trough is also subjected to evaporation over the hydrophilic coated surface, thereby enhanced distillate production is made possible. Hydrophilic coating reduces the contact angle of water droplets and increases liquid-solid interaction, thereby wettability and formation of active nucleation sites also increases. Moreover, hydrophilic coatings also increase corrosion and scaling resistance, thereby the life of the desalination system can also be expected to increase. Distillate productivity of tubular solar desalination system with hydrophilic coated basin is about 25.0% higher than the system without hydrophilic coating (Tang et al., 2019).

### *5.3.5 Harmonic Vibration of Absorber*

El-said et al. (2020) installed a flat steel wire mesh (Fig. 5.8a) inside the basin water of tubular solar desalination system whose schematic is shown in Fig. 5.8b. The flat wire mesh is connected to an electric motor with eccentric mechanism as shown in Fig. 5.8c. The power consumption of the electric motor is about 50 W and it rotates at

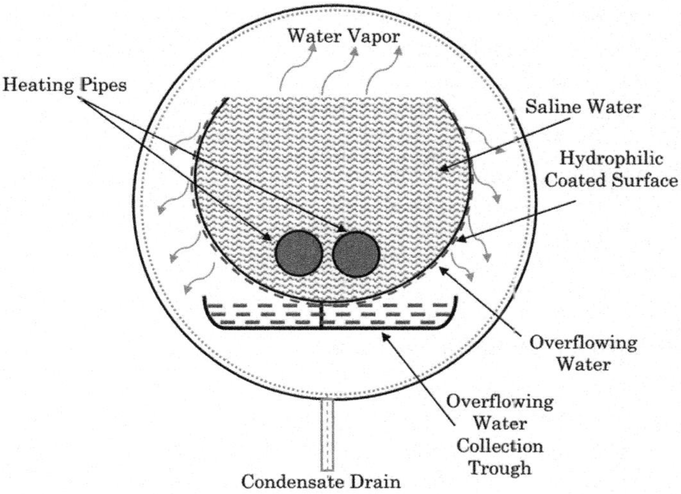


**Fig. 5.7** Schematic of tubular desalination system with hydrophilic coated water trough (Tang et al., 2019)

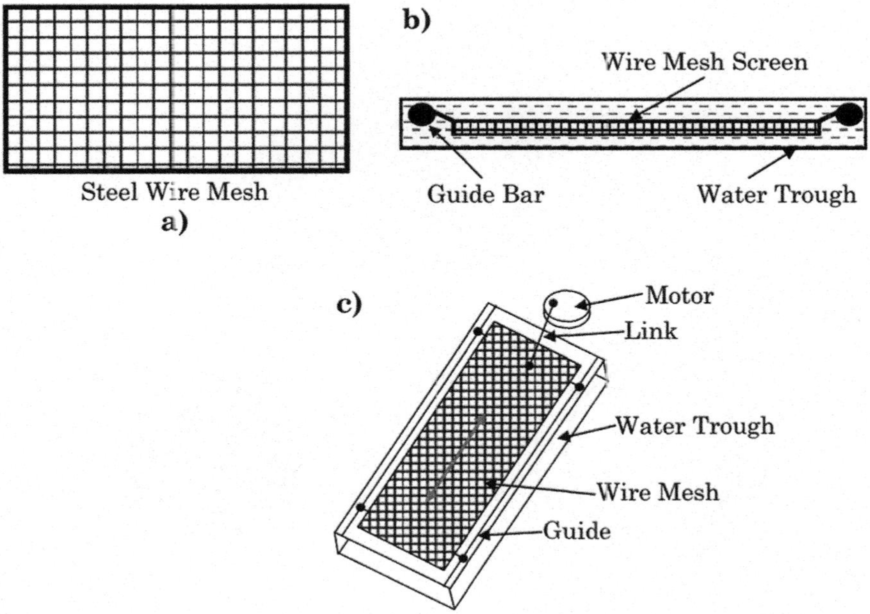


**Fig. 5.8** **a** Photograph of steel wire mesh, **b** Schematic of water thrcugh with wire mesh and guide bars, **c** Schematic showing the arrangement of wire mesh and motor with eccentricity mechanism in modified tubular solar desalination system (El-said et al., 2020)

a speed of about 50 rpm. The electric motor with eccentric mechanism moves the steel wire mesh by 10.0 cm to and fro horizontally during its each rotation. The vibration of the mesh thus achieved destroys the boundary layer and reduces the surface tension of the saline water, thereby convection heat transfer from the basin liner to the basin water is enhanced which leads to 34.0% higher distillate productivity in comparison to conventional tubular solar desalination system. The flat wire mesh acts as a porous media (porosity 61.0%) and it also absorbs, stores, and transfers heat energy to the basin water effectively.

### 5.3.6 Rotating Drum-Based Absorber

The photograph of tubular solar thermal desalination system with a rotating blackened drum to minimize basin water depth and to maintain continuous thin film of water over the evaporating surface for enhanced evaporation process is shown in Fig. 5.9a. The blackened rotating drum can have open ends or closed ends and can be bare or lined with wick as shown in Fig. 5.9b and c. The distillate productivities of

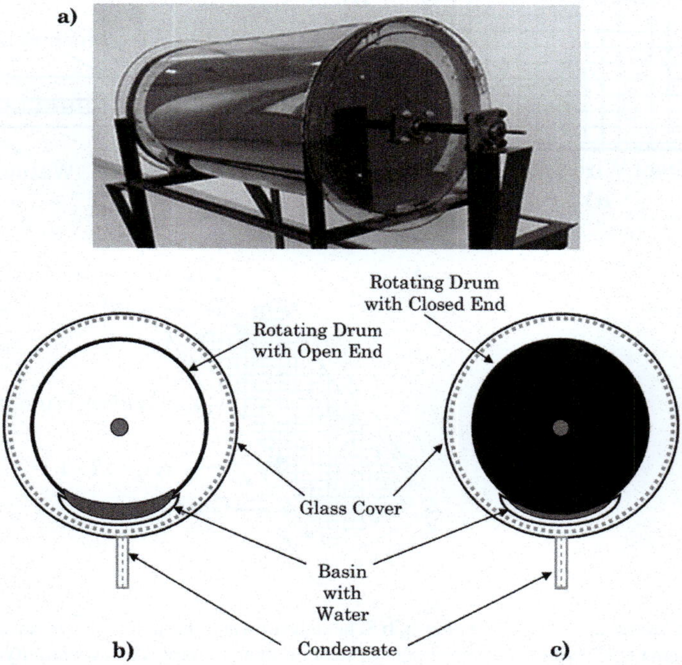


**Fig. 5.9** **a** Photograph of tubular solar desalination system with rotating drum, **b** Cross section of tubular solar desalination system with rotating open-ended drum **c** Cross section of tubular solar desalination system with rotating closed ended drum (Essa et al., 2021b)

tubular solar thermal desalination system with closed and open-ended rotating drum without wicks are about 121.0% and 136.0% higher than the distillate yield of conventional basin solar still, respectively. High distillate production with open-ended rotating drum is due to the additional exposure of inner surface of the drum to solar radiation during some hours of the day which results in increased temperature and enhanced evaporation. Lining the drum with wicks has increased the evaporation rate significantly. The optimum rotation speed of the drum without wick and with wick for maximum evaporation rate is about 0.10 rpm and 0.05 rpm, respectively. The distillate yield and thermal efficiency of tubular solar thermal desalination system with wick-lined open end drum at optimum rotation speed are about 6.60 L/m$^2$d and 58.0%, respectively (Essa et al., 2021b).

### 5.3.7 *Incorporation of Solar Concentrators*

The photograph and schematic of parabolic concentrator solar tracking system (PCST)-assisted tubular solar thermal desalination system is shown in Fig. 5.10. It consists of a transparent tube within which half-cylindrical black-coated aluminum trough is placed and assembled tightly. The assembled system is installed along the focus of PCST. The trough is then filled with saline water up to the required level. The black-coated aluminum trough with saline water is heated by the solar radiation reflected from the mirrors of PCST. Saline water begins to evaporate and the generated vapors begin to condense over the inner surface of the transparent tube. Provisions are made at the sidemost bottom of the transparent tube to drain the condensate. Tracking of the concentrator can be carried out manually by adjusting the angle of concentrator every 1 h (Elashmawy, 2017). Performance comparison of PCST-assisted tubular solar thermal desalination system with other tubular solar thermal desalination systems is shown in Table 5.3. Daily distillate yields of the PCST-assisted tubular solar thermal desalination system are about 676.0 and 493.0% higher than the distillate yield of tubular solar thermal desalination system with half-cylindrical trough and tubular solar thermal desalination system with rectangular trough and black cotton wick, respectively.

Tubular solar thermal desalination systems have also been used effectively in combination with compound parabolic concentrator (CPC) to increase its distillate yield. The schematic of CPC-integrated tubular solar thermal desalination system is shown in Fig. 5.11. Incorporation of CPC eliminates the need for tracking, and thereby makes the system simpler and cheaper in comparison to parabolic trough collector-integrated system. The feed water loading to the CPC-integrated tubular system's basin can be achieved by gravity by installing feed water tank at higher elevation from the desalination system. The distillate yield of tubular solar thermal desalination system with rectangular trough (2.0 × 0.03 m) placed in the focus of CPC with aperture area of 2.0 × 0.10 m is about 410.0 mL/d (Arunkumar et al., 2013).

**Fig. 5.10** **a** Photograph of parabolic concentrator solar tracking system (PCST)-assisted tubular solar thermal desalination system, **b** Schematic indicating the role of PCST in solar radiation concentration over tubular solar thermal desalination system (Elashmawy, 2020)

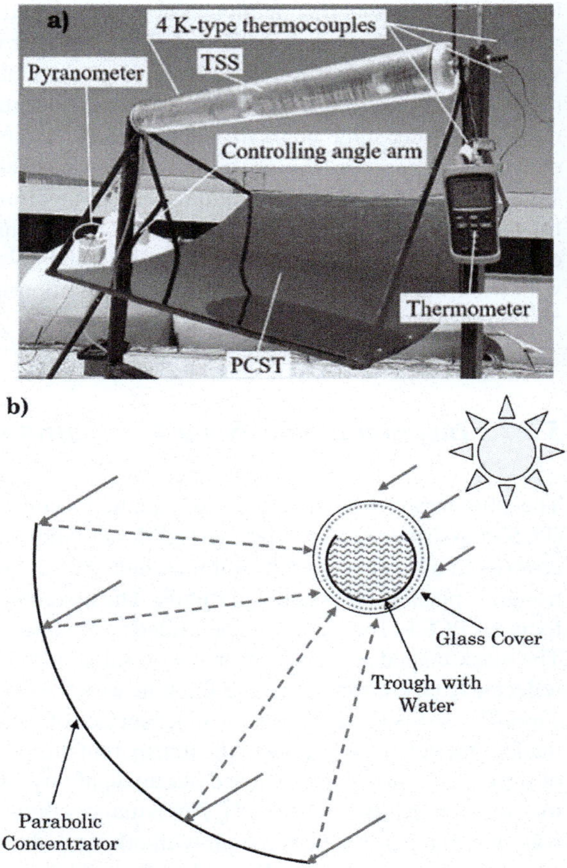


Kabeel et al. (2020a) enhanced the performance of cylindrical parabolic concentrator (1.2 $m^2$ aperture area)-assisted tubular solar thermal desalination system by incorporating a closed blackened half-cylindrical trough concentrically within the blackened half-cylindrical trough of the tubular solar thermal desalination system. The gap between the two concentric troughs is then filled with saline water as shown in Fig. 5.12. The level of water in the trough needs to be maintained constant by a feed water line aided with a control valve. The main aim of incorporating closed blackened half-cylindrical trough concentrically within the blackened half-cylindrical trough is to enhance the solar radiation collection area at its top surface and to increase the heat transfer between the absorber surface and the basin water. The distillate yields of the tubular solar thermal desalination system with water thicknesses of 1.0 cm, 2.0 cm, 3.0 cm, and 4.0 cm between the concentric troughs are about 7.37, 7.48, 7.13, and 6.59 L/$m^2$d, respectively. The distillate yield of the system with optimal water thickness of 2.0 cm is about 90.8% higher the yield of conventional tubular solar thermal

**Table 5.3** Comparison of parabolic concentrator solar tracking system (PCST)-assisted tubular solar thermal desalination system with other tubular solar thermal desalination system configurations (Elashmawy, 2017)

| Component | Features | Tubular solar thermal desalination system with rectangular trough and black wick | Tubular solar thermal desalination system with semi-circular trough | PCST-assisted tubular solar thermal desalination system with semi-circular trough |
|---|---|---|---|---|
| Transparent tube | Material | Polymethyl methacrylate (PMMA) | methacrylate (PMMA) | Polymethyl methacrylate (PMMA) |
| | Length | 0.54 m | 0.54 m | 0.54 m |
| | Inner and outer diameter | 0.10 × 0.11 m | 0.10 × 0.11 m | 0.10 × 0.11 m |
| Trough | Material | Aluminum | Aluminum | Aluminum |
| | Shape | Rectangular | Half-cylindrical | Half-cylindrical |
| | Dimension | 0.53 × 0.085 × 0.015 m | 0.53 × 0.08 m | 0.53 × 0.08 m |
| Wick | | Black cotton | – | – |
| Concentrator | Tracking | – | – | Manual |
| | Aperture dimension | – | – | 0.54 × 0.87 m |
| Actual yield (L/d) | | 0.28 | 0.22 | 1.66 |
| Thermal efficiency (%) | | 36.65 | 30.2 | 28.54 |
| Yield (L/m$^2$d) | | 4.71 | 3.60 | 3.53 (based on aperture area of PCST) |

desalination system with rectangular basin. This observation can be attributed to the increased absorber area and the presence of concentrator in the modified system. The thermal efficiencies of conventional tubular solar thermal desalination system and this modified system are about 32.0% and 61.4%, respectively.

### *5.3.8 Condensing Cover Geometry*

Rahbar et al. (2018) investigated the impact of cylindrical and triangular condensing cover geometry on distillate productivity of tubular solar thermal desalination system. The schematic of the tubular solar thermal desalination system with cylindrical and triangular condensing cover is shown in Fig. 5.13a, b. Triangular cover is cheap and easy to manufacture, while manufacturing cylindrical geometry is complex and

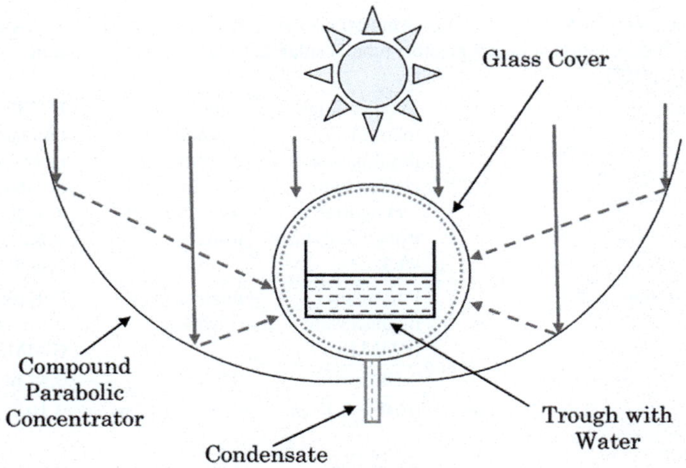


**Fig. 5.11** Schematic of tubular solar thermal desalination system placed in the focus of compound parabolic concentrator (CPC) (Arunkumar et al., 2013)

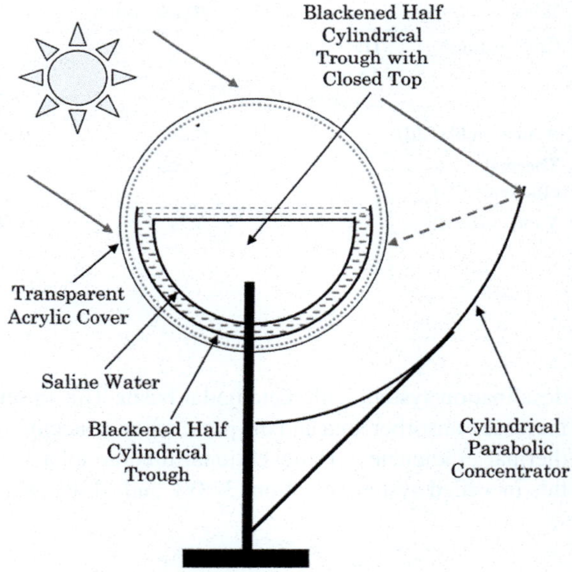


**Fig. 5.12** Schematic of cylindrical parabolic concentrator (CPC)-assisted tubular solar thermal desalination system with two concentric half-cylindrical troughs (Kabeel et al., 2020a)

costly. Geometry of condensing cover also affects the gap between the evaporating water surface and condensing cover which in turn affects the distillate productivity significantly. Distillate productivity of the system with cylindrical cover is observed to be about 20.0% higher than the system with triangular cover and this observation is mainly due to the higher strength of re-circulating zone inside the desalination

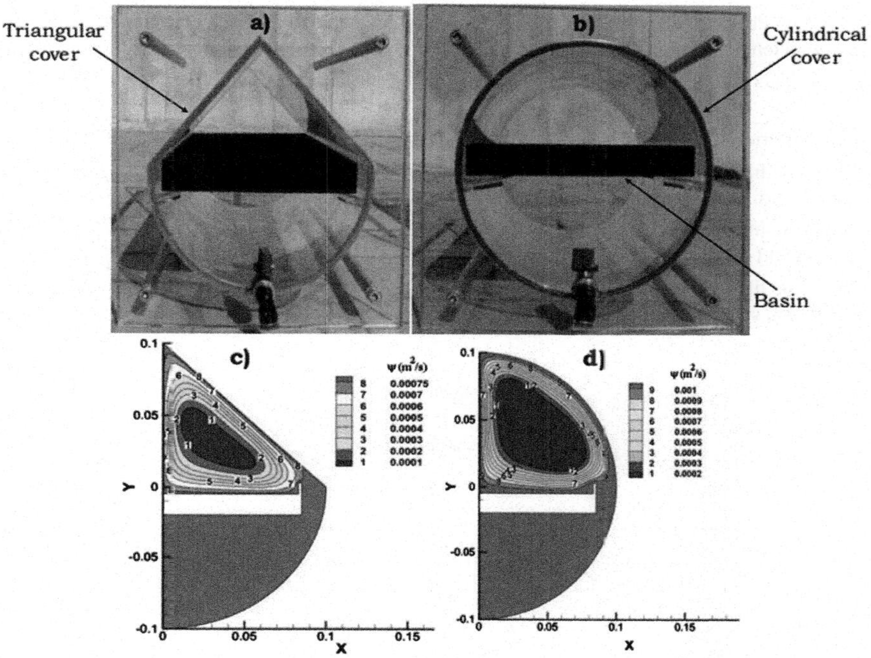


**Fig. 5.13** Tubular solar thermal desalination system with **a** triangular cover, **b** cylindrical cover, **c** streamlines inside desalination system with triangular cover, **d** streamlines inside desalination system with cylindrical cover (water temperature = 325 K; cover temperature = 318.5 K) (Rahbar et al., 2018)

system with cylindrical cover. The difference between the maximum and minimum values of streamlines inside the desalination system with cylindrical and triangular cover is about 0.008 $m^2$/s and 0.0065 $m^2$/s, respectively, which can be seen from Fig. 5.13c, d. The convective and evaporative heat transfer coefficients of the system with cylindrical cover are about 7.0% and 4.0% greater than those of the system with triangular cover, respectively.

### 5.3.9 Condensing Cover Material

Glass (3.0 mm thick), plexiglass (5.0 mm thick), vinyl chloride sheet (0.5 mm thick), polythene film sheet (0.15 mm thick), and polycarbonate sheet (2.0 mm thick) have been used as condensing cover for tubular solar thermal desalination systems (Rahbar et al., 2018; Ahsan et al., 2010, 2012). The properties of some of the condensing cover materials used in tubular solar thermal desalination system are tabulated in Table 5.4. Increasing condensing cover thickness from 3.0 to 5.0 mm reduces the transparency

**Table 5.4** Physical properties of various condensing cover materials of tubular solar thermal desalination system (Kabeel et al., 2010)

| Cover material | Thickness (mm) | Density ($kg/m^3$) | Mass (kg) | Durability (months) | Price ($INR/m^2$) |
|---|---|---|---|---|---|
| Vinyl chloride sheet | 0.05 | 1470 | 0.156 | 24 | 3656.00 |
| Polythene film sheet—Ordinary | 0.07 | 875 | 0.013 | 5 | 45.70 |
| Polythene film sheet—Soft | 0.10 | 905 | 0.019 | 24 | 65.30 |
| Polythene film sheet—Diastar (used for greenhouses) | 0.13 | 965 | 0.031 | 60 | 120.80 |

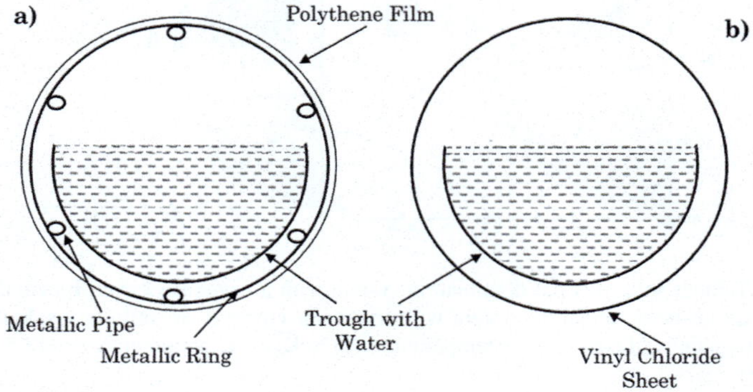


**Fig. 5.14** Cross section of tubular solar thermal desalination system with **a** polythene film sheet cover supported by metallic structure and **b** vinyl chloride sheet cover without any additional structural support (Ahsan et al., 2010)

of the cover and thermal efficiency of the system by about 67.6% (Elsmawy, 2019). Difficulties in tubular shape fabrication with glass and polyvinyl chloride sheet have made polythene film sheet a suitable cheap cover material for tubular solar thermal desalination system. Use of polythene film sheet instead of polyvinyl chloride sheet has reduced the cost and weight of the desalination system by about 92.0% and 61.0%, respectively. Polythene film sheet has also improved the durability of tubular solar thermal desalination system. However, proper support structure is needed to wrap the polythene film sheet to form the tubular cover as shown in Fig. 5.14. In this case, the support structure may obstruct the condensate flow and reduce the overall yield (Ahsan et al., 2010).

### 5.3.10 Condensing Cover Temperature

Temperature of tubular solar thermal desalination system's condensing cover can be reduced by spraying water or by circulating water or by blowing air over its surface. The schematic of spray cooled tubular solar thermal desalination system is shown in Fig. 5.15. Sprayed water carries away the heat energy of the cover, thereby the tubular cover is maintained at lower temperature which results in rapid condensation of vapors generated inside the distillation chamber. Cover cooled tubular system's distillate productivity is about 4.50 L/m$^2$d which is nearly 31.4% higher than conventional tubular solar thermal desalination system without water spraying. The optimum water depth in the basin and spraying water flow rate over the cover are about 0.5 cm and 2.0 L/h, respectively (Kabeel et al., 2019).

Elashmawy (2019) observed a drop in distillate productivity of parabolic trough collector (PTC)-assisted tubular solar thermal desalination system (0.977 m length × 0.08 m diameter) by about 10.0% and 43.8% when its condensing cover is cooled either by spraying 10 ml water every 30 min or by circulating water between the two covers at a rate of 40 ml/min, respectively. The concept of annulus cooling of PTC-assisted tubular solar thermal desalination system's cover is shown in Fig. 5.16. The main reason for the drop in distillate productivity is attributed to the compact geometry of PTC-assisted tubular solar thermal desalination system. Due to the

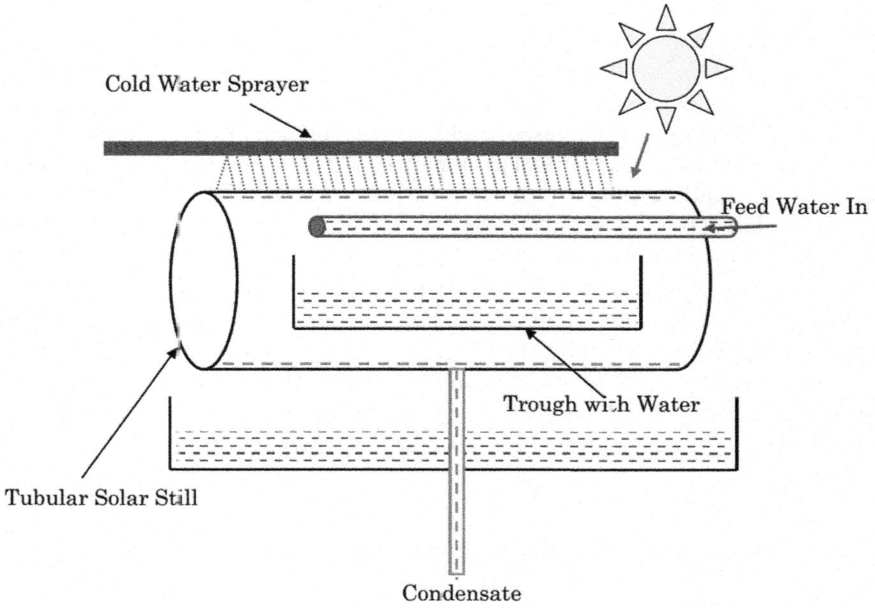


**Fig. 5.15** Schematic of spraying water-cooled tubular solar thermal desalination system (Kabeel et al., 2019)

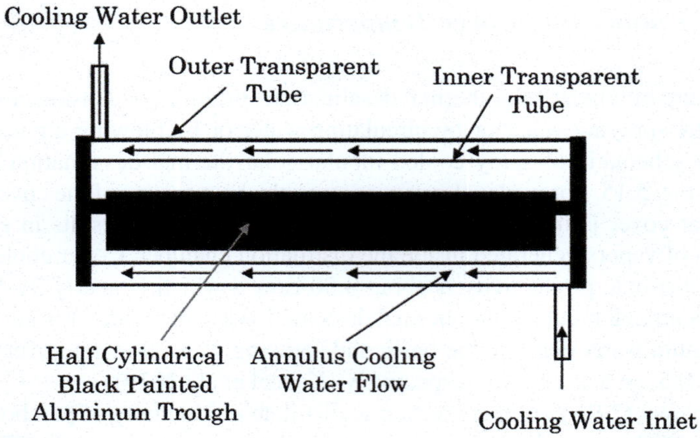


**Fig. 5.16** Side view of tubular solar thermal desalination system showing annular cooling water flow (Elashmawy, 2019)

compact geometry, both the basin water area and its heat capacity become lower than the cooling cover surface area and its heat capacity. Hence, cooling of the cover by circulating or spraying water reduces the temperature of the evaporating basin water leading to low distillate yield which is undesirable in solar thermal desalination systems.

Arunkumar et al. (2013) obtained about 143.0 and 63.9% higher distillate productivity for CPC-assisted water-cooled tubular solar thermal desalination system in comparison to CPC-assisted bare tubular and CPC-assisted air-cooled tubular solar thermal desalination system, respectively. The photograph of CPC-assisted annular water-cooled tubular solar thermal desalination system is shown in Fig. 5.17a, b. The concept of annular cooling of tubular solar desalination system has already been shown in Fig. 5.16. The distillate yields of CPC-assisted water-cooled and CPC-assisted bare tubular still are about 5.0 L/d and 2.05 L/d, respectively. Cooling water from an overhead tank is circulated through the annulus with the aid of gravity. The flow rate of cooling water can be adjusted using burettes. The cooling water leaving the annulus will be at an elevated temperature relative to the ambient. Hence, it can also been supplied as feed to the interconnected basin solar thermal desalination system to achieve additional distillate production if necessary (Arunkumar et al., 2016a, b).

### *5.3.11 Latent Heat Energy Storage Material*

The diurnal and nocturnal productivity of tubular solar thermal desalination system can be enhanced by modifying the basin of tubular solar thermal desalination system by installing copper tubes filled with paraffin wax (melting temperature around

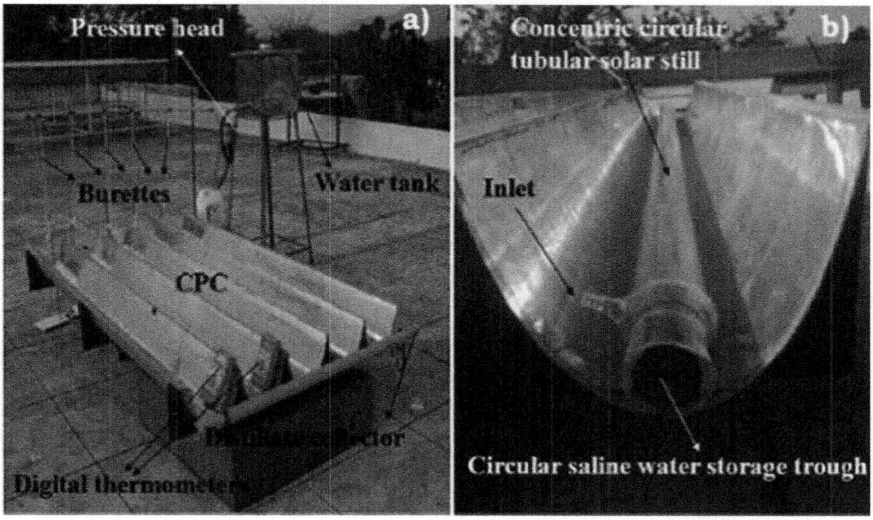


**Fig. 5.17** Photograph of **a** overall system, **b** water-cooled tubular solar thermal desalination system placed over the focus of compound parabolic concentrator (Arunkumar & Kabeel, 2017)

56.0 °C). Around 85.0% of the tube volume is filled with PCM and the remaining 15.0% volume is left free to facilitate the volumetric expansion of paraffin wax during phase change from solid-to-liquid state. The schematic of the tubular still's modified basin with phase change material (PCM) is shown in Fig. 5.18. Incorporation of copper tube increases surface area of the absorber and the PCM helps to extend the distillate production during non-sunshine hours by absorbing excess solar radiation in sunny hours. The modified system produces about 8.82–9.05 L/m$^2$d of distillate which is about 110.0–115.0% higher than the distillate production capacity of conventional tubular solar thermal desalination system without PCM (Kabeel et al., 2020b).

Increased thermal conductivity of PCM enhances heat energy absorption and heat energy release rate of PCM during sunny and non-sunny periods, respectively. Thermal conductivity of PCM can be improved by adding nano-particles. Thermal conductivities of bare PCM and 0.3 wt/wt% graphene nano-particle-loaded PCM are about 0.26 W/mK and 0.55 W/mK, respectively. The distillate yields of tubular solar still, tubular solar still with PCM, and tubular solar still with nano-particle-loaded PCM are about 2.45, 3.42, and 5.65 L/m$^2$d, respectively. The thermal efficiencies of tubular solar still, tubular solar still with PCM, and tubular solar still with nano-particle-loaded PCM are about 22.7%, 31.6%, and 52.3%, respectively (Kabeel et al., 2020c).

Arunkumar and Kabeel (2017) loaded paraffin wax between the 5.0 mm gap of two concentric half-cylindrical troughs and sealed their ends. The sealed trough is placed within the tubular glass cover and the entire system is then located at the focus

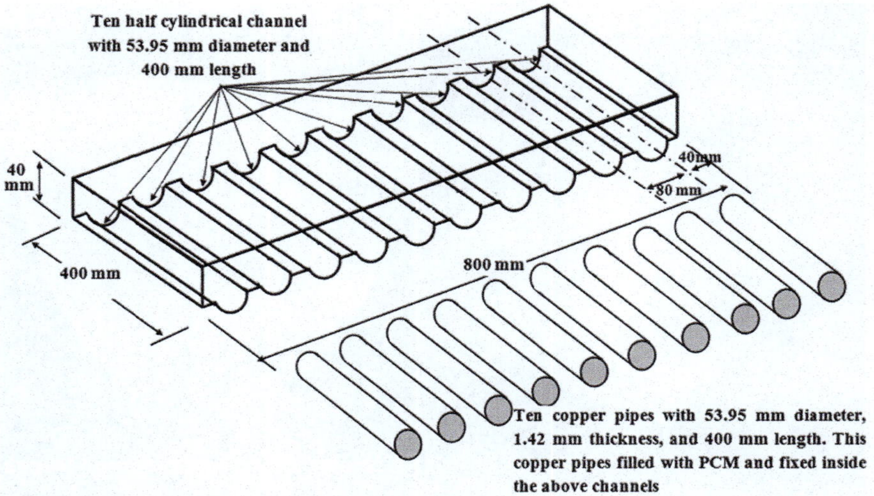


**Fig. 5.18** Schematic of modified basin and paraffin-wax-loaded tubes of tubular solar thermal desalination system (Kabeel et al., 2020b)

of compound parabolic concentrator (CPC). The distillate yield of the CPC-assisted tubular solar thermal desalination system with PCM is about 5.80 L/m$^2$d which is nearly 8.0% higher than the CPC-assisted tubular solar thermal desalination system without PCM. PTC-assisted tubular solar thermal desalination system loaded with 12 paraffin wax-filled aluminum tubes shown in Fig. 5.19 produced about 5.55 L/m$^2$d of distillate at a thermal efficiency of 44.1%. Moreover, the incorporation of tubes within the trough of tubular solar thermal desalination system has resulted in heterogeneous geometry leading to large surface contact area with the saline water (Elashmawy et al., 2021).

### *5.3.12 Hybrid Energy Storage Material*

Kabeel et al. (2020d) introduced the concept of hybrid thermal energy storage material in tubular solar thermal desalination systems. Sealed semi-circular copper trough (26.0 cm diameter and 86.0 cm length) containing 19.0 kg paraffin wax (as bottom layer) and 3.0-cm-thick graphite sheet (as top layer) is placed within the semi-circular basin of tubular solar thermal desalination system as shown in Fig. 5.20. Paraffin wax acts as latent heat energy storage medium while graphite acts as sensible heat energy storage medium. Hence, the system is termed as tubular solar thermal desalination system with hybrid energy storage medium. The gap between the two troughs is filled with feed water and the feed water level in the basin is maintained just above

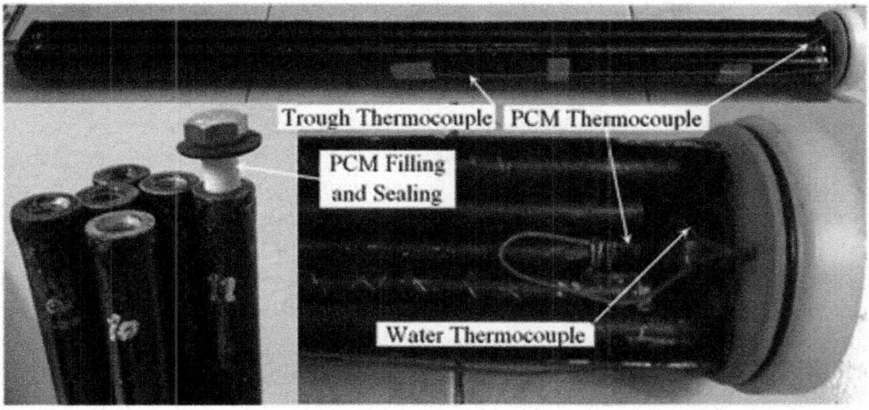


**Fig. 5.19** Photograph of PCM-loaded tubes within the water trough of parabolic trough collector-assisted tubular solar thermal desalination system (Elashmawy et al., 2021)

**Fig. 5.20** Schematic of parabolic trough collector-assisted tubular solar thermal desalination system with hybrid thermal energy storage (Kabeel et al., 2020d)

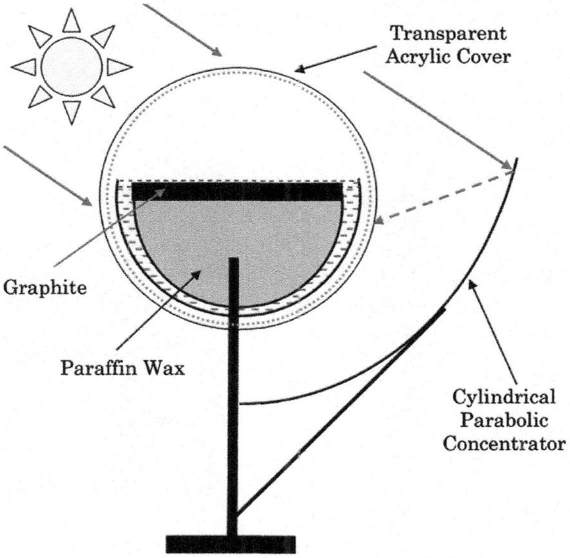


the sealed trough. The modified tubular solar thermal desalination system is then placed at the focus of parabolic trough collector. The distillate yield and thermal efficiency of the modified system are in the range of 9.56–9.81 L/m$^2$d and 44.2–45.3%, respectively.

### 5.3.13 *Sensible and Composite Sensible Heat Energy Storage Material*

Elashmawy (2020) incorporated low-cost gravel as sensible heat energy medium inside the trough of parabolic trough collector (PTC)-assisted tubular solar thermal desalination system whose photograph is shown in Fig. 5.21a. The distillate yield of gravel-loaded tubular solar thermal desalination system is about 4.51 L/m$^2$d at a thermal efficiency of 36.3%. The distillate yield is about 14.2% higher in comparison to the system without gravel and 890.4% higher in comparison to the system without concentrator and gravel. Elashmawy and Ahmed (2021) used 12 heterogeneous composite sensible energy storage tubes inside the trough of PTC-assisted tubular solar thermal desalination system. The composite sensible energy storage medium is formed by filling the aluminum tube with sand and inserting copper wire at center whose schematic is shown in Fig. 5.21b. The composite heat energy storage medium absorbs the heat energy from the trough by conduction and releases it to the water by convection. The tubular composite heat energy storage medium also increases the heat transfer surface area. Distillate yield of this system is about 4.90 L/m$^2$d at a

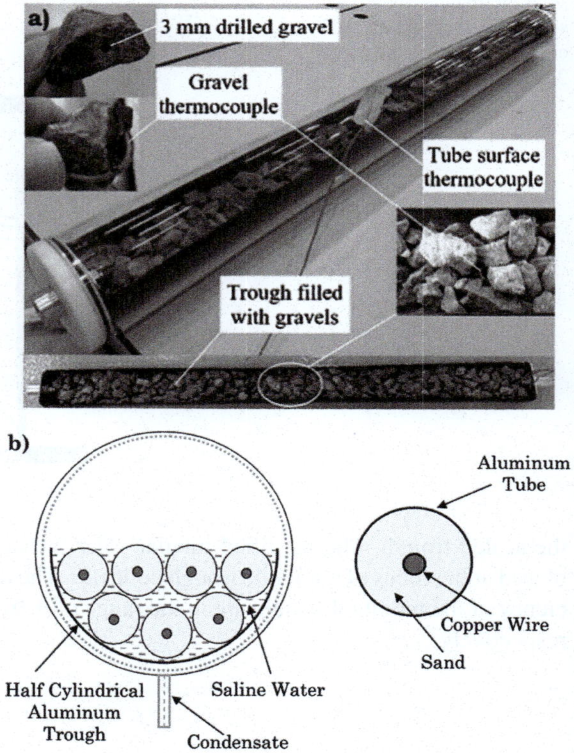


**Fig. 5.21** **a** Photograph of tubular solar thermal desalination system's basin with gravel as sensible heat energy storage medium (Elashmawy, 2020), **b** Schematic of tubular solar thermal desalination system with composite sensible heat energy storage medium (Elashmawy & Ahmed, 2021)

thermal efficiency of 38.31% which is about 24.05% higher than the distillate yield of similar system without any composite heat energy storage medium.

### 5.3.14 Operating Pressure

Reducing operation pressure is one of the ways to improve the distillate yield of solar thermal desalination systems. The operating pressure can be reduced by integrating the desalination system with vacuum pump. The vacuum pump operation mode adopted in conventional multi-effect desalination systems cannot be employed in small-scale tubular solar thermal desalination systems due to their sophisticated and complicated vacuum regulation methods. Constant optimum vacuum pressure operation is highly suited for small-scale tubular solar thermal desalination systems for improved performance (Xie et al., 2018). The schematic of tubular solar thermal desalination system coupled with solar collector and vacuum pump is shown in Fig. 5.22. The distillate production rates of the system with basin water temperatures of 50, 60, 70, and 80 °C at an operating pressure corresponding to the saturation pressure of basin water temperature are about 95.0, 69.0, 47.0, and 75.0% higher than the distillate yield observed at one atmosphere operating pressure. The distillate

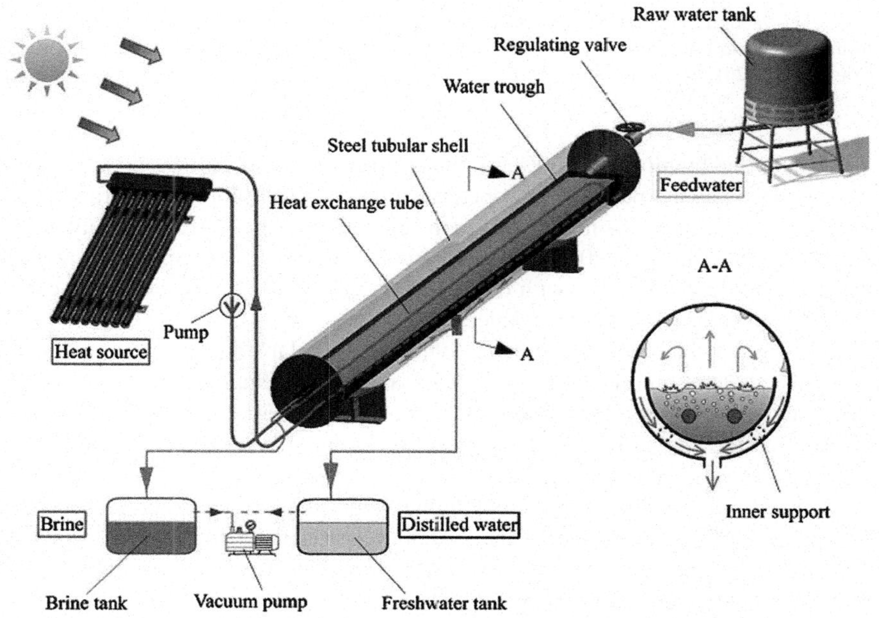


**Fig. 5.22** Schematic of tubular solar thermal desalination system integrated with solar collector and vacuum pump (Xie et al., 2018)

**Table 5.5** Variation of tubular cover temperature with basin water temperature and operating pressure (Xie et al., 2018)

| Fixed basin water temperature (°C) | Tubular cover temperature for various operating pressures (°C) (kPa) | | | | | | | | |
|---|---|---|---|---|---|---|---|---|---|
| | 20 | 30 | 40 | 50 | 60 | 70 | 80 | 90 | 100 |
| 50 | 47.1 | 46.1 | 45.3 | 44.7 | 43.9 | 43.6 | 42.9 | 43.1 | 42.9 |
| 60 | – | 55.9 | 55.4 | 54.8 | 54.3 | 53.7 | 53.2 | 52.9 | 52.6 |
| 70 | – | – | 65.5 | 64.3 | 63.7 | 63.7 | 63.6 | 63.2 | 62.4 |
| 80 | – | – | – | 75.8 | 75.0 | 74.4 | 73.9 | 73.5 | 73.2 |

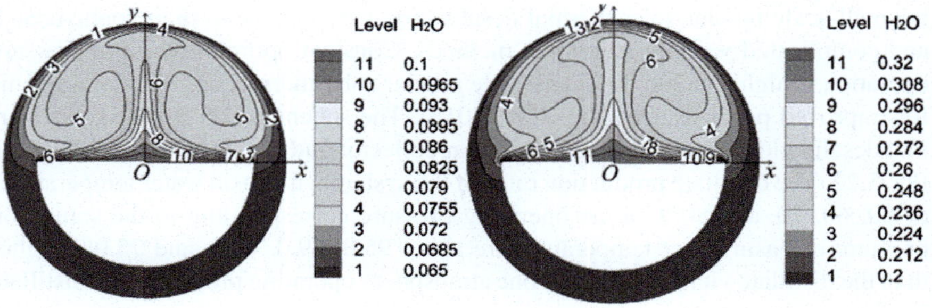


(a) (b)

**Fig. 5.23** Vapor mass fraction contours inside tubular solar desalination system with semi-circular trough and basin water temperature of 60.0 °C operating at pressures of **a** 101 kPa and **b** 40 kPa (Yan et al., 2020)

yield increased by more than 50% for operating pressures lower than 60 kPa and the increase rate was sharper when the operating pressure is closer to the saturation pressure. This phenomenon is mainly due to the increase in vapor diffusion coefficient at low pressures (Yan et al., 2020). The temperature difference between the basin water and condensing cover drops with the reduction in operating pressure which can be seen from the data displayed in Table 5.5.

The contours of vapor mass fraction within the tubular solar thermal desalination system with semi-circular trough holding basin water temperature at 60.0 °C and operating at pressures of 101 and 40 kPa are shown in Fig. 5.23a, b. It can be seen clearly that the vapors rose up from the basin water surface and then horizontally diffused from the center to the two sides. The horizontal diffusion was enhanced under vacuum conditions and thereby higher vapor mass fractions are observed on the lateral surface. The yield of the system was found to drop by 30% with the decrease in water depth in the trough from "radius $r$" to "0.80 times of radius $r$" and this observation can be attributed to the reduction in evaporation surface area and increased time required for the vapor to get transferred from the evaporating to the condensing surface (Yan et al., 2020).

## 5.4 Features of Tubular Solar Thermal Desalination Systems

(a) More compact and easier to fabricate than conventional basin solar still.
(b) Non-metallic materials can also be used for fabrication.
(c) Heat loss from the basin to the ambient is minimal due to no direct contact with the ambient.
(d) Cylindrical cover and semi-circular troughs favor easy utilization of solar radiation from any direction.
(e) No insulation requirement.
(f) Latent heat of condensation is left unutilized to the ambient.
(g) Fall back of condensate to the basin water is possible in systems with cylindrical condensing cover.

## References

Ahsan, A., & Fukuhara, T. (2010). Mass and heat transfer model of tubular solar still. *Solar Energy, 84*, 1147–1156.

Ahsan, A., Islam, K. M. S., Fukuhara, T., & Ghazali, A. H. (2010). Experimental study on evaporation, condensation and production of a new Tubular Solar Still. *Desalination, 260*, 172–179.

Ahsan, A., Imteaz, M., Rahman, A., Yusuf, B., & Fukuhara, T. (2012). Design, fabrication and performance analysis of an improved solar still. *Desalination, 292*, 105–112.

Anand, J. D., & Kumar, A. (1992). Modeling and performance of a tubular multiwick solar still. *Energy, 17*(11), 1067–1071.

Arunkumar, T., Jayaprakash, R., Ahsan, A., Denkenberger, D., & Okundamiya, M. S. (2013). Effect of water and air flow on concentric tubular solar water desalting system. *Applied Energy, 103*, 109–115.

Arunkumar, T., Velraj, R., Denkenberger, D. C., Sathyamurthy, R., Kumar, K. V., & Ahsan, A. (2016a). Productivity enhancements of compound parabolic concentrator tubular solar stills. *Renewable Energy, 88*, 391-7-400.

Arunkumar, T., Velraj, R., Denkenberger, D., Sathyamurthy, R., Vinothkumar, K., Porkumaran, K., & Ahsan, A. (2016b). Effect of heat removal on tubular solar desalting system. *Desalination, 379*, 24–33.

Arunkumar, T., & Kabeel, A. E. (2017). Effect of phase change material on concentric circular tubular solar still-Integration meets enhancement. *Desalination, 414*, 46–50.

El-said, E. M. S., Elshamy, S. M., & Kabeel, A. E. (2020). Performance enhancement of a tubular solar still by utilizing wire mesh packing under harmonic motion. *Desalination, 474*, 114165.

Elshamy, S. M., & El-Said, E. M. S. (2018). Comparative study based on thermal, exergetic and economic analyses of a tubular solar still with semi-circular corrugated absorber. *Journal of Cleaner Production.*

Essa, F. A., Alawee, W. H., Mohammed, S. A., Dhahad, H. A., Abdullah, A. S., & Omara, Z. M. (2021a). Experimental investigation of convex tubular solar still performance using wick and nanocomposites. *Case Studies in Thermal Engineering, 27*, 101368.

Essa, F. A., Abdullah, A. S., & Omara, Z. M. (2021b). Improving the performance of tubular solar still using rotating drum - Experimental and theoretical investigation. *Process Safety and Environmental Protection, 148*, 579–589.

Elashmawy, M. (2017). An experimental investigation of a parabolic concentrator solar tracking system integrated with a tubular solar still. *Desalination, 411*, 1–8.

Elashmawy, M. (2019). Effect of surface cooling and tube thickness on the performance of a high temperature standalone tubular solar still. *Applied Thermal Engineering, 156*, 276–286.

Elashmawy, M. (2020). Improving the performance of a parabolic concentrator solar tracking tubular solar still (PCST-TSS) using gravel as a sensible heat storage material. *Desalination, 473*, 114182.

Elashmawy, M., & Ahmed, M. M. Z. (2021). Enhancing tubular solar still productivity using composite aluminum/ copper/sand sensible energy storage tubes. *Solar Energy Materials & Solar Cells, 221*, 110882.

Elashmawy, M., Alhadri, M., & Ahmed, M. M. Z. (2021). Enhancing tubular solar still performance using novel PCM-tubes. *Desalination, 500*, 114880.

Kabeel, A. E., Sharshir, S. W., Abdelaziz, G. B., Halim, M. A., & Swidan, A. (2019). Improving performance of tubular solar still by controlling the water depth and cover cooling. *Journal of Cleaner Production, 233*(2019), 848–856.

Kabeel, A. E., Harby, K., Abdelgaied, M., & Eisa, A. (2020a). Performance of the modified tubular solar still integrated with cylindrical parabolic concentrators. *Solar Energy, 204*, 181–189.

Kabeel, A. E., Abdelgaied, M., Harby, K., & Eisa, A. (2020b). Augmentation of diurnal and nocturnal distillate of modified tubular solar still having copper tubes filled with PCM in the basin. *Journal of Energy Storage, 32*, 101992.

Kabeel, A. E., Sathyamurthy, R., Manokar, A. M., Sharshir, S. W., Essa, F. A., & Elshiekh, A. H. (2020c). Experimental study on tubular solar still using Graphene Oxide Nano particles in Phase Change Material (NPCM's) for fresh water production. *Journal of Energy Storage, 28*, 101204.

Kabeel, A. E., Harby, K., Abdelgaied, M., & Eisa, A. (2020d). Augmentation of a developed tubular solar still productivity using hybrid storage medium and CPC: An experimental approach. *Journal of Energy Storage, 28*, 101203.

Rahbar, N., Esfahani, J. A., & Fotouhi-Bafghi, E. (2015). Estimation of convective heat transfer coefficient and water-productivity in a tubular solar still - CFD simulation and theoretical analysis. *Solar Energy, 113*, 313–323.

Rahbar, N., Asadi, A., & Fotouhi-Bafghi, E. (2018). Performance evaluation of two solar stills of different geometries: Tubular versus triangular: Experimental study, numerical simulation, and second law analysis. *Desalination, 443*, 44–55.

Sathyamurthy, R., Mageshbabu, D., Madhu, B., Manokar, A. M., Prasad, A. R., & Sudhakar, M. (2020). Influence of fins on the absorber plate of tubular solar still- An experimental study. *Materials Today: Proceedings*. https://doi.org/10.1016/j.matpr.2020.11.355

Sharshir, S. W., Ellakany, Y. M., Algazzar, A. M., Elsheikh, A. H., Elkadeem, M. R., Edreis, E. M. A., Waly, A. S., Sathyamurthy, R., Panchal, H., & Elashry, M. S. (2019). A mini review of techniques used to improve the tubular solar still performance for solar water desalination. *Process Safety and Environmental Protection, 124*, 204–212.

Tang, J., Wang, C., Xie, W., Xia, Y., Yu, T., & Chen, Z. (2019). Study on the heat and mass transfer performance of a tubular still enhanced by hydrophilic surface modification. *Desalination, 469*, 114089.

Tiwari, G. N. (1988). Nocturnal water production by tubular solar stills using waste heat to preheat brine. *Desalination, 69*, 309–318.

Xie, G., Sun, L., Mo, Z., Liu, H., & Du, M. (2016). Conceptual design and experimental investigation involving a modular desalination system composed of arrayed tubular solar stills. *Applied Energy, 179*, 972–984.

Xie, G., Sun, L., Yan, T., Tang, J., Bao, J., & Du, M. (2018). Model development and experimental verification for tubular solar still operating under vacuum condition. *Energy*.

Yan, T., Xie, G., Liu, H., Wu, Z., & Sun, L. (2020). CFD investigation of vapor transportation in a tubular solar still operating under vacuum. *International Journal of Heat and Mass Transfer, 156*, 119917.

# Chapter 6
# Stepped Solar Thermal Desalination Systems

## 6.1 System Description and Working

Reduction of basin water depth is a key parameter to improve the distillate production rate of solar stills. Minimal basin water depth indicates minimal heat capacity, thereby water temperature rises rapidly leading to high evaporation rates. Incorporation of wicks, sponges, and other porous materials inside solar still's basin has been practiced to reduce the heat capacity of evaporating water. In case of stepped solar thermal desalination system, the presence of steps increases the evaporator area exposed to solar radiation. In addition, it also helps to maintain thin film of water over the evaporator surface. These advantages lead to high basin water temperature and better performance of stepped solar thermal desalination system in comparison to conventional basin solar still. Stepped solar thermal desalination system is also called stepped solar still. The schematic of stepped solar still is shown in Fig. 6.1. Higher temperature inside the stepped solar still in comparison to conventional basin still is due to the following reasons (Omara et al., 2013):

(a) Smaller air volume inside stepped solar still's chamber compared to basin still, hence heating rate of the trapped air will be high.
(b) Stepped basin provides more surface area, thereby more solar radiation is absorbed which leads to high absorber temperature.
(c) Thin film of water layer over the steps rather than thicker layer of water inside the basin still.

Stepped solar still consists of an insulated absorber plate with a number of steps, glass cover, distillate collection trough, and provisions for feed water entry and exit. Feed water path over the stepped solar still's absorber plate is shown in Fig. 6.2.



H. Sharon and K. Srinivas Reddy, *Solar Thermal Desalination Technologies for Potable Water*, https://doi.org/10.1007/978-3-031-49155-9_6

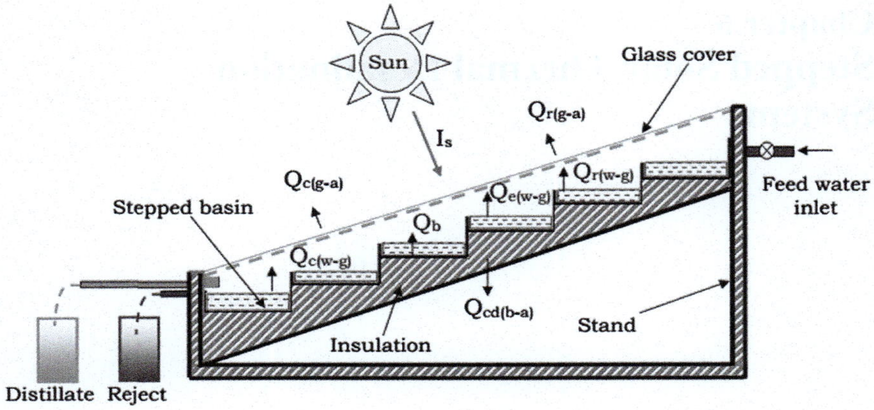


**Fig. 6.1** Schematic of stepped solar still along with its associated energy transport processes

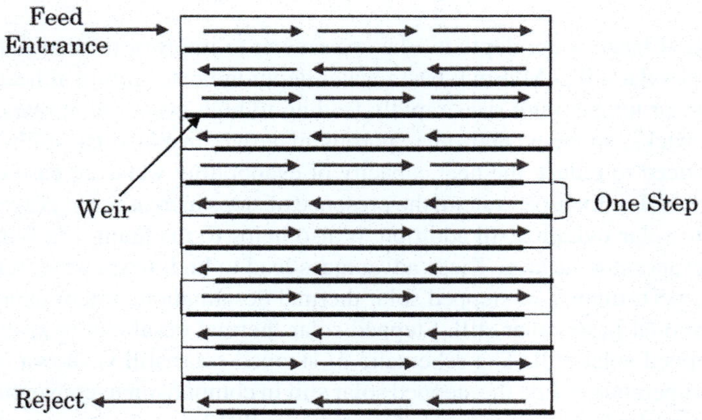


**Fig. 6.2** Feed water flow path over stepped solar still's absorber plate (Gawande & Bhuyar, 2010)

## 6.2 Thermodynamic Modeling of Stepped Solar Still

Assumptions considering for thermodynamic modeling of stepped solar still are listed below:

(a) The solar still is vapor tight and leak proof.
(b) Constant thickness of water layer over the stepped basin.
(c) No thermal gradient across water layer thickness.
(d) No thermal gradient across stepped basin liner thickness.
(e) Glass cover conductance resistance is negligible.
(f) No temperature gradient across glass cover thickness.

The energy transport processes occurring in stepped solar still are shown in Fig. 6.1.

Energy balance in stepped absorber plate,

$$\begin{pmatrix}\text{Rate of change of stepped}\\ \text{absorber temperature}\end{pmatrix} = \begin{pmatrix}\text{solar radiation absorbed}\\ \text{by the absorber}\end{pmatrix} - \begin{pmatrix}\text{convective heat transfer}\\ \text{from absorber to water}\end{pmatrix} - \begin{pmatrix}\text{heat loss from}\\ \text{absorber to ambient}\end{pmatrix}$$

$$(mc)_b \frac{dT_b}{dt} = \tau_g \tau_w \alpha_b A_b I_s - Q_{c(b-w)} - Q_{cd(b-a)} \tag{6.1}$$

Energy balance in water layer,

$$\begin{pmatrix}\text{Rate of change of basin}\\ \text{water temperature}\end{pmatrix} = \begin{pmatrix}\text{solar radiation absorbed}\\ \text{by the basin water}\end{pmatrix} + \begin{pmatrix}\text{convective heat transfer}\\ \text{from absorber to basin water}\end{pmatrix} - \begin{pmatrix}\text{convection, radiation, and evaporation}\\ \text{heat loss from basin water to glass cover}\end{pmatrix} - \begin{pmatrix}\text{energy required for preheating makeup}\\ \text{water to basin water temperature}\end{pmatrix}$$

$$(mc)_w \frac{dT_w}{dt} = \tau_g \alpha_w A_w I_s + Q_{c(b-w)} - (Q_c + Q_r + Q_e)_{(w-g)} - Q_{mw} \tag{6.2}$$

Energy balance in glass cover,

$$\begin{pmatrix}\text{Rate of change of glass}\\ \text{cover temperature}\end{pmatrix} = \begin{pmatrix}\text{solar radiation absorbed}\\ \text{by the glass cover}\end{pmatrix} + \begin{pmatrix}\text{convection, radiation and evaporation}\\ \text{heat loss from basin water to glass cover}\end{pmatrix} - \begin{pmatrix}\text{convection heat loss from}\\ \text{glass cover to ambient}\end{pmatrix} - \begin{pmatrix}\text{radiation heat loss from}\\ \text{glass cover to sky}\end{pmatrix}$$

$$(mc)_g \frac{dT_g}{dt} = \alpha_g A_g I_s + (Q_c + Q_r + Q_e)_{(w-g)} - Q_{c(g-a)} - Q_{r(g-sky)} \tag{6.3}$$

Convective heat transfer between stepped basin absorber and water layer ($Q_{c(b-w)}$ in W) is given as

$$Q_{c(b-w)} = 135.0 A_b (T_b - T_w) \tag{6.4}$$

where

$A_b$ —area of stepped basin ($m^2$),
$T_b$ —temperature of stepped basin (°C), and
$T_w$ —temperature of stepped basin water (°C).

Heat losses from stepped absorber to the ambient ($Q_{cd(b-a)}$ in W) are estimated by

$$Q_{cd(b-a)} = U_b A_b (T_b - T_a) \tag{6.5}$$

where

"$U_b$" represents bottom and side heat loss coefficient and is usually taken as 14.0 $W/m^2K$ for simplicity.
$T_a$ —ambient temperature (°C).

Convective heat transfer between the water in stepped basin and glass cover ($Q_{c(w-g)}$ in W) is estimated by

$$Q_{c(w-g)} = 0.884 \left[ T_w - T_g + \frac{(P_w - P_g)(T_w + 273.15)}{268900 - P_w} \right]^{1/3} A_w (T_w - T_g) \tag{6.6}$$

where

$P_w$, $P_g$ —partial vapor pressure of water vapor at water, glass cover temperature (Pa).

Evaporative heat transfer between water in stepped basin and glass cover ($Q_{e(w-g)}$ in W) is estimated by

$$Q_{e(w-g)} = 0.016237 h_{cw} A_w (P_w - P_g) \tag{6.7}$$

where

$h_{cw}$ —convective heat transfer coefficient between water and glass cover ($W/m^2K$).

Radiation heat transfer between water in stepped basin and glass cover ($Q_{r(w-g)}$ in W) is estimated by

$$Q_{r(w-g)} = \frac{\sigma A_w ([T_w + 273.15]^4 - [T_g + 273.15]^4)}{[(1/\varepsilon_w) + (1/\varepsilon_g) - 1]} \tag{6.8}$$

where

$\sigma$ —Stefan-Boltzmann constant,
$\varepsilon_w$ —emissivity of water, and
$\varepsilon_g$ —emissivity of glass cover.

Radiation heat transfer between glass cover and the sky ($Q_{r(g-sky)}$ in W) is given as

$$Q_{r(g-sky)} = \varepsilon_g A_g \sigma ([T_g + 273.15]^4 - [T_{sky} + 273.15]^4) \tag{6.9}$$

Temperature of the sky ($T_{sky}$) is given as

$$T_{sky} = T_a - 6.0 \tag{6.10}$$

Convection heat transfer between glass cover and the ambient ($Q_{c(g-a)}$ in W) is given as

$$Q_{c(g-a)} = h_{c(g-a)} A_g (T_g - T_a) \tag{6.11}$$

where

$A_g$ —surface area of glass cover ($m^2$).

Convective heat transfer coefficient between glass cover and the ambient ($h_{c(g-a)}$ in $W/m^2K$) is given as

$$h_{c(g-a)} = 2.8 + 3.0(V), \quad \text{for } V \leq 5.0 \text{ m/s} \tag{6.12}$$

$$h_{c(g-a)} = 6.15(V)^{0.8}, \quad \text{for } V > 5.0 \text{ m/s} \tag{6.13}$$

where

$V$ —wind speed (m/s).

Makeup water is added to the still to maintain constant water level in the stepped basin. Generally, makeup water equivalent to the water evaporated from the still is added.

Sensible heat energy required to heat this makeup water from the ambient temperature ($Q_{mw}$ in W) to the stepped basin water temperature is given as

$$Q_{mw} = m_e c_w (T_w - T_a) \tag{6.14}$$

If the glass cover of stepped solar still is cooled by flowing water over it, then the energy balance in glass cover (Eq. 6.3) needs to be modified as (El-Samadony & Kabeel, 2014)

$$\begin{aligned}\begin{pmatrix}\text{Rate of change of glass}\\ \text{cover temperature}\end{pmatrix} &= \begin{pmatrix}\text{solar radiation absorbed}\\ \text{by the glass cover}\end{pmatrix}\\ &+ \begin{pmatrix}\text{convection, radiation and evaporation}\\ \text{heat loss from basin water to glass cover}\end{pmatrix}\\ &- \begin{pmatrix}\text{convention heat transfer from glass}\\ \text{cover to flowing water film}\end{pmatrix}\end{aligned}$$

$$(mc)_g \frac{dT_g}{dt} = \tau_w \alpha_g A_g I_s + (Q_c + Q_r + Q_e)_{(w-g)} - Q_{cf} \tag{6.15}$$

Heat transfer from the glass cover to the water film flowing over it ($Q_{cf}$ in W) is given as (El-Samadony & Kabeel, 2014)

$$Q_{cd} = h_{cf} A_g (T_g - T_f) \tag{6.16}$$

where

$T_f$ —average temperature of the flowing water film (°C).

Convective heat transfer coefficient between the glass cover and flowing water film ($h_{cf}$ in W/m$^2$K) is estimated by

$$h_{cf} = \frac{K_f}{L} 0.664 Re_L^{0.5} Pr^{0.33}, \quad \text{if } Re_L \le 5 \times 10^5 \tag{6.17}$$

$$h_{cf} = \frac{K_f}{L} (0.037 Re_L^{4/5} - 871) Pr^{0.33}, \quad \text{if } Re_L > 5 \times 10^5 \tag{6.18}$$

where

$K_f$ —thermal conductivity of flowing water (W/mK),
$L$ —length of the glass cover (m),
$Re_L$ —Reynolds number based on length of the glass cover, and
$Pr$ —Prandtl number.

Reynolds number of the flowing water film ($Re_L$) is given as

$$Re_L = \frac{\rho_f L (Vol_f / x_f d)}{\mu_f} \tag{6.19}$$

where

$Vol_f$ —volumetric flow rate of flowing water film (m$^3$/s),
$d$ —width of the glass cover (m),
$x_f$ —cooling film thickness (m),
$\rho_f$ —density of flowing water (kg/m$^3$), and
$\mu_f$ —dynamic viscosity of flowing water (kg/ms).

Energy balance in flowing water film over the glass cover during film cooling is given as (El-Samadony & Kabeel, 2014)

$$\begin{pmatrix}\text{Rate of change of fluid} \\ \text{film temperature}\end{pmatrix} = \begin{pmatrix}\text{energy content} \\ \text{of inlet film}\end{pmatrix} - \begin{pmatrix}\text{energy content} \\ \text{of outlet film}\end{pmatrix} + \begin{pmatrix}\text{convection heat transfer between} \\ \text{glass cover and fluid film}\end{pmatrix} - \begin{pmatrix}\text{convection and radiation heat} \\ \text{loss between fluid film and sky}\end{pmatrix}$$

$$(mc)_f \frac{dT_f}{dt} = \dot{m}_{rf} c_{f1} T_{f1} - \dot{m}_{rf} c_{f2} T_{f2} + Q_{cf} - (Q_c + Q_r)_{(f-sky)} \tag{6.20}$$

Average temperature of the flowing fluid film ($T_f$) is estimated by

$$T_f = \frac{T_{f1} + T_{f2}}{2} \tag{6.21}$$

where

$T_{f1}$ —inlet temperature of the flowing film cooling water (°C) and
$T_{f2}$ —exit temperature of the flowing film cooling water leaving the glass cover (°C).

Convection heat transfer between the fluid film and the sky ($Q_{ca}$ in W) is given as

$$Q_{ca} = (5.7 + 3.8V)A_g(T_f - T_{sky}) \tag{6.22}$$

Radiation heat transfer between the fluid film and the sky ($Q_{rf}$ in W) is given as

$$Q_{rf} = \sigma A_g \varepsilon_f [(T_f + 273.15)^4 - (T_{sky} + 273.15)^4] \tag{6.23}$$

where

$\varepsilon_f$ —emissivity of the flowing cooling water film.

If the film cooling water after passing over the glass cover is used as makeup water for the still then Eq. (6.14) is modified as

$$Q_{mw} = \dot{m}_e c_w (T_w - T_{f2}) \quad \text{if } \dot{m}_{rf} \geq \dot{m}_e \tag{6.24}$$

$$Q_{mw} = \dot{m}_{rf} c_w (T_w - T_{f2}) + (\dot{m}_e - \dot{m}_{rf}) c_w (T_w - T_a) \quad \text{if } \dot{m}_{rf} < \dot{m}_e \tag{6.25}$$

where

$Q_{mw}$ —sensible heat energy required to heat makeup water inside the still (W),
$\dot{m}_e$ —evaporation rate (kg/s),
$\dot{m}_{rf}$ —cooling water mass flow rate (kg/s),
$T_w$ —temperature of the basin water (°C), and
$c_w$ —specific heat capacity of water (J/kgK).

In order to utilize the incident solar radiation effectively, and to increase stepped solar still's distillate productivity, phase change materials are incorporated beneath its absorber plate (Dashtban & Tabrizi, 2011).

Energy balance in stepped absorber plate of stepped solar still with phase change material is given as

$$\begin{pmatrix}\text{Rate of change of stepped}\\ \text{absorber temperature}\end{pmatrix} = \begin{pmatrix}\text{solar radiation absorbed}\\ \text{by the absorber}\end{pmatrix} - \begin{pmatrix}\text{convective heat transfer}\\ \text{from absorber to water}\end{pmatrix} - \begin{pmatrix}\text{conduction heat transfer from}\\ \text{absorber to phase change material}\end{pmatrix}$$

$$(mc)_b \frac{dT_b}{dt} = \tau_g \tau_w \alpha_b A_b I_s - Q_{c(b-w)} - \left(\frac{K_{pcm}}{x_{pcm}}\right)(T_b - T_{pcm})A_b \tag{6.26}$$

Energy balance in phase change material kept beneath the stepped absorber plate is given as

$$\begin{pmatrix}\text{Rate of change of phase}\\ \text{change material temperature}\end{pmatrix} = \begin{pmatrix}\text{conduction heat transfer from}\\ \text{absorber to phase change material}\end{pmatrix} - \begin{pmatrix}\text{convection heat loss from phase change}\\ \text{material to ambient via insulation}\end{pmatrix}$$

$$M_{pcm} = \frac{dT_{pcm}}{dt} = \left(\frac{K_{pcm}}{x_{pcm}}\right)(T_b - T_{pcm})A_b - \left(\frac{K_{ins}}{x_{ins}}\right)(T_{pcm} - T_a)A_b \tag{6.27}$$

where

$I_s$ —solar radiation intensity over the still (W/m$^2$),
$K_{pcm}$ —thermal conductivity of PCM (W/mK),
$x_{pcm}$ —thickness of PCM (m),
$K_{ins}$ —thermal conductivity of insulation (W/mK),
$x_{ins}$ —thickness of insulation (m),
$T_{pcm}$ —temperature of PCM (°C),
$\tau_g$ —transmissivity of glass cover,
$\tau_w$ —transmissivity of water, and
$\alpha_b$ —absorptivity of basin liner.

Heat capacity of the phase change material ($M_{pcm}$) is given as

$$M_{pcm} = m_{pcm} c_{s,pcm} \quad \text{if } (T_{pcm} < T_{melt,pcm}) \tag{6.28}$$

$$M_{pcm} = m_{pcm} L_{pcm} \quad \text{if } (T_{melt,pcm} \le T_{pcm} \le T_{melt,pcm} + \delta') \tag{6.29}$$

$$M_{pcm} = m_{pcm} c_{L,pcm} \quad \text{if } (T_{pcm} > T_{melt,pcm} + \delta') \tag{6.30}$$

where

$m_{pcm}$ —mass of PCM (kg),
$T_{melt,pcm}$ —melting temperature of PCM (°C),
$c_{s,pcm}$ —specific heat capacity of PCM at its solid state (J/kgK),
$c_{L,pcm}$ —specific heat capacity of PCM at its liquid state (J/kgK),

$L_{pcm}$ —latent heat of fusion of PCM (J/kg), and
$\delta$ —incremental rise in temperature from melting point of PCM (°C).

In some cases, reflectors are attached to the vertical wall of each step of the stepped solar still's absorber to effectively direct the incident solar radiation to the effective absorber area to cause enhanced evaporation (Omara et al., 2013).

Energy balance in absorber plate of stepped solar still with internal reflecting mirrors attached to the vertical surface of each of the steps is given as

$$\begin{pmatrix}\text{Rate of change of stepped}\\ \text{absorber temperature}\end{pmatrix} = \begin{pmatrix}\text{solar radiation absorbed}\\ \text{by the absorber}\end{pmatrix} - \begin{pmatrix}\text{solar radiation reflected by internal}\\ \text{mirrors and absorbed by the absorber}\end{pmatrix} - \begin{pmatrix}\text{convective heat transfer}\\ \text{from absorber to water}\end{pmatrix} - \begin{pmatrix}\text{heat loss from}\\ \text{absorber to ambient}\end{pmatrix}$$

$$(mc)_b \frac{dT_b}{dt} = \tau_g \tau_w A_b \alpha_b I_s + \tau_g \rho_m \tau_w \alpha_b W L_s I_s \times \frac{\tan\theta_s \cos\gamma}{\tan\alpha} N_p - Q_{c(b-w)} - Q_{cd(b-a)} \tag{6.31}$$

Energy balance in water layer of stepped solar still with internal reflecting mirrors attached to the vertical surface of each of the steps is given as

$$\begin{pmatrix}\text{Rate of change of basin}\\ \text{water temperature}\end{pmatrix} = \begin{pmatrix}\text{solar radiation absorbed}\\ \text{by the basin water}\end{pmatrix} + \begin{pmatrix}\text{solar radiation reflected by internal}\\ \text{mirrors and absorbed by the basin water}\end{pmatrix} + \begin{pmatrix}\text{convective heat transfer}\\ \text{from absorber to basin water}\end{pmatrix} - \begin{pmatrix}\text{convection, radiation and evaporation}\\ \text{heat loss from basin water to glass cover}\end{pmatrix} - \begin{pmatrix}\text{energy required for preheating makeup}\\ \text{water to basin water temperature}\end{pmatrix}$$

$$(mc)_w \frac{dT_w}{dt} = \tau_g \tau_w A_w I_s + \tau_g \rho_m \alpha_w W L_s I_s \frac{\tan\theta_s \cos\gamma}{\tan\alpha} N_p + Q_{c(b-w)} - (Q_c + Q_r + Q_e)_{(w-g)} - Q_{mw} \tag{6.32}$$

where

$W$ —width of each step (m),
$L_s$ —length of each step (m),
$\rho_m$ —reflectivity of mirror,
$N_p$ —number of steps in the basin,
$\alpha$ —solar altitude angle (degree),
$\gamma$ —solar azimuth angle (degree),
$\theta_s$ —tilt angle of glass cover (degree), and
$\alpha_w$ —absorptivity of water.

## 6.3 Parameters Influencing the Performance of Stepped Solar Still

### *6.3.1 Shape of Absorber Step*

Shape of the absorber plate's step has a significant impact on the distillate productivity of stepped solar still. Flat plate, convex type and concave type are some of the common shapes of existing stepped solar still absorber and are schematically represented in Fig. 6.3. Optimum width of the tray and water depth in stepped solar still with flat plate absorber plate is noticed to be about 120.0 mm and 5.0 mm, respectively (Kabeel et al., 2015). For fixed maximum tray depth, productivity of the system with convex and concave shaped trays is about 56.60% and 29.24% higher than the system with flat trays. This observation is mainly due to the minimum water depth at the middle and outer edges of convex and concave trays, respectively (Gawande and Bhuyar 2013).

### *6.3.2 Gap Between Water Surface and Glass Cover*

The gap between the water surface and glass cover also impacts the distillate yield of stepped solar still. At very low gap like 1.0 cm or below, the water surface is very much closer to the glass cover, thereby the water may get cooled down easily due to the associated heat loss from the glass cover and this can lead to low evaporation rates. At higher gaps beyond 2.0 cm, air resistance dominates, thereby the mass transfer within the system declines. At 2.0 cm gap, maximum distillate yield is observed. Hence, 2.0 cm can be considered as optimum gap for stepped solar still. Distillate yields at 1 cm, 2 cm, and 4 cm gap are about 6.10, 6.50 and 5.90 kg/m$^2$d, respectively (Keshtkar et al. 2020). The ratio between the gap and width of the still is called the **cavity aspect ratio**. Increase in cavity aspect ratio from 1.94 to 2.67 enhances distillate productivity and efficiency by 212% and three times, respectively (D'Cotha et al., 2021).

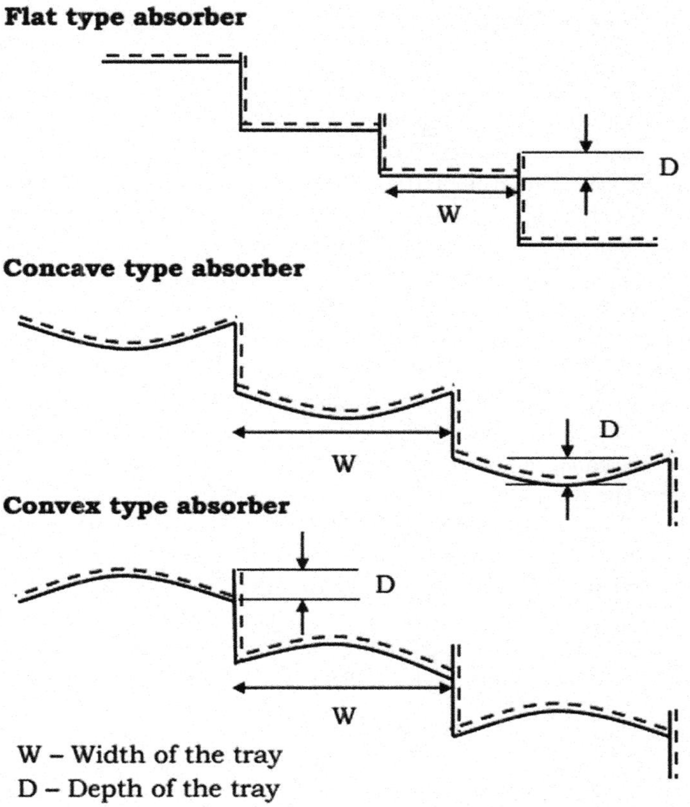


**Fig. 6.3** Schematic of various shapes of stepped solar still's absorber plate (Gawande and Bhuyar 2013)

### 6.3.3 *Number of Steps*

For a fixed absorber area, as the number of steps is increased, the distillate yield also increases. However, beyond six steps any increment in number of steps caused slight drop in distillate productivity of the system. Hence, six steps can be considered as optimum for stepped solar still. Distillate yield of stepped solar still with six steps and water surface to glass cover gap of 2.0 cm is about 17.4% higher than the yield of basin still of similar absorber area (Keshtkar et al., 2020).

### 6.3.4 Surface Modifications of Absorber Plate

Surface of solar still's absorber plate can be modified by coating, addition of fins, and by lining of wicks. Sathyamurthy et al. (2020) prepared nano-fluid by mixing fumed $SiO_2$ nano-particle of size 50 nm with turpentine oil. Thermal conductivities of the prepared nano-fluid with 10.0, 20.0, 30.0, and 40.0%wt concentrations of fumed $SiO_2$ are about 0.13, 0.14, 0.12, and 0.118 W/mK, respectively. The prepared nano-fluid is then added to black paint which is then coated over the absorber plate of stepped solar still. The distillate productivity of the stepped solar still coated with black paint consisting of 20.0 wt% nano-particle is about 4.30 L/m$^2$d which is nearly 34.3% higher than the distillate yield of the system with ordinary black paint coating.

Sangeetha et al. (2022) coated the absorber plate with activated carbon-ZnO nano-particle mixed black paint and obtained a distillate yield of about 16.91% higher than the system without coating. The increased yield due to nano-particle coating is attributed to the enhancement in heat transfer from the nano-particle-coated absorber to the water. The wettability and effective evaporation area of the absorber can be enhanced by lining the absorber surface with linen wick. The distillate productivity and thermal efficiency of this modified stepped solar still are around 3.81 L/m$^2$d and 52.8%, respectively. The obtained yield is around 15.4% higher than the yield of the similar system without linen wick lining (Sharshir et al., 2020). Blowing hot air below the absorber plate can enhance the distillate productivity of the stepped solar still by about 85% in comparison to conventional solar still (Abdullah, 2013).

### 6.3.5 Water Film Cooling of Glass Cover

The schematic of stepped solar still with water film cooling is shown in Fig. 6.4. Film cooling of glass cover increases the distillate productivity of stepped solar still by reducing its glass cover temperature and enhancing vapor condensation phenomena. However, increased film thickness can also have negative effect on still's distillate productivity. Increased film thickness for fixed volumetric flow rate indicates drop in speed of the flowing film, thereby cooling rate of the glass cover drops as a result glass cover temperature increases and the distillate productivity drops. Similarly, for fixed film thickness increasing volumetric flow rate of cooling film increases distillate production rate considerably. Wind speed was found to have no impact on productivity enhancement of water-film-cooled stepped solar still because the water film layer shields the glass cover from the blowing wind. Hence, the rate of heat energy removed by the flowing water film from the glass cover will be very much higher than the heat energy lost from the glass cover to the ambient via convection by flowing air. However, for stepped solar still without water film cooling increased wind speed has positive impact on still's distillate production enhancement.

Optimum water film thickness, water film volumetric flow rate, and glass cover length of stepped solar still are found to be in the range of $2.5 \times 10^{-4}$ to

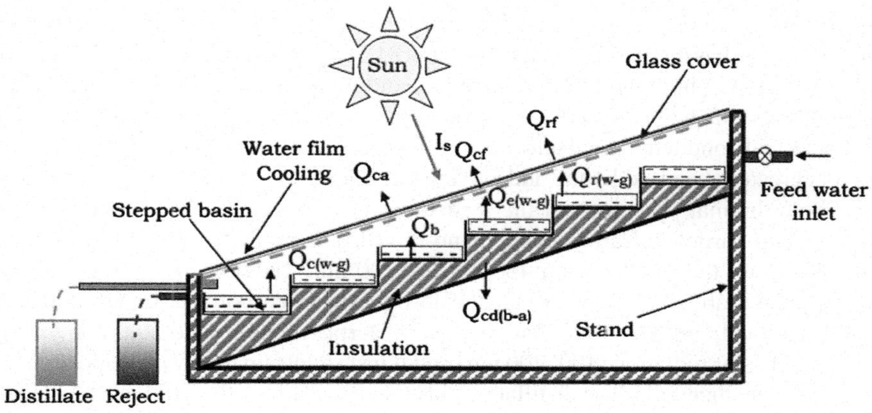


**Fig. 6.4** Schematic of stepped solar still with water film cover cooling

$5.5 \times 10^{-4}$ m, $4.0 \times 10^{-5}$ to $8.5 \times 10^{-5}$ $m^3/s$, and 2.0 to 2.8 m, respectively, for better performance. It is recommended to use water film at ambient temperature for glass cover cooling because reducing the cooling film temperature below ambient temperature may increase the operation cost and make the process significantly expensive. Distillate production enhancement of about 8.20% can be obtained by adopting water film cooling. The distillate productivity can be further increased by about 2.87% by utilizing the preheated glass cover cooling water as makeup water for the system (El-Samadony & Kabeel, 2014).

### 6.3.6 *Thermal Conductivity of Feed Water*

Jathar and Ganesan (2020) investigated the effect of feed water thermal conductivity on evaporation rate of stepped solar still. Thermal conductivity of feed water is improved by adding MgO, $Al_2O_3$, and $TiO_2$ nano-particles. Size of the nano-particles used for investigation is limited to 40 nm to prevent aggregation of nano-particles. The corresponding nano-particle is blended to the base fluid using magnetic stirrer at a stirring speed of 600 rpm for 30 min and further subjected to ultrasonic sonication for 2 h to produce the nano-fluid. The concentration range of nano-particles is about 0.1 to 0.20 vol% in the nano-fluids. Distillate yields of the stepped solar still with 0.2 vol% of MgO nano-fluid are about 10.6% and 18.7% higher than the stepped solar still with 0.2 vol% $Al_2O_3$ and $TiO_2$ nano-fluid, respectively. The highest yield obtained in still with MgO nano-fluid is due to the high thermal conductivity of MgO (48.5 W/mK) in comparison to $Al_2O_3$ (18.0 W/mK) and $TiO_2$ (11.2 W/mK). Yield of stepped solar still has increased by about 48.12% by adding 0.03 wt% of CuO nano-particle to the feed water. For the same concentration, graphene oxide (GO) nano-particle has increased the yield by about 81.59%. Hence, it can be concluded that incorporation of high thermal conductivity nano-particles in feed water can enhance heat distribution within the feed water and improve the evaporation rate significantly. The thermal conductivity enhancement of feed water is due to the Brownian motion and aggregation caused by the added nano-particles.

Photo-thermal properties of the feed water can be improved by adding carbon black nano-particles. Carbon black nano-fluid has nearly constant absorptivity of 99.9% for all the spectra from 400 nm to 1100 nm. The absorptivity of ordinary black paint is about 5.0% lower than the absorptivity of carbon black nano-fluid. On the other hand, ordinary water has very low absorptivity of around 20.0% in visible spectra of solar radiation (400–700 nm) and it just shoots up to 40% at a spectrum of 1100 nm and then falls. The distillate yield of stepped solar still with 1.5 wt% carbon black nano-particle of 40 nm size added to the feed water is about 4.15 L/m$^2$d which is nearly 26.0% higher than the yield of stepped solar still without any carbon black nano-particle added to its feed water (Sharshir et al. 2020). Increasing the mass flow rate of nano-particle-loaded feed water in the still from 30 to 68 L/h has dropped still's yield by about 88.54%. On contrary, reducing flow rate from 30 L/h to 8 L/h has increased yield by about 13.86%. These observations are due to the low and high residence time of nano-particle-loaded feed water over the absorber plate at high and low flow rates, respectively. High residence time indicates high-temperature rise and improved evaporation (Ajdari & Ameri, 2022).

### 6.3.7 *Reject Brine Recirculation*

The un-evaporated brine leaving the still will be relatively at high temperature compared to the ambient and just dumping it outside may indicate improper utilization of a useful resource. ElAgouz (2014) suggested recirculation of reject brine to increase evaporation and water recovery rate of stepped solar still. The schematic of the system proposed by ElAgouz (2014) to serve this purpose is shown in Fig. 6.5. Saline water in the bottom storage tank is allowed to flow over the cotton wick-lined absorber plate of stepped solar still with the aid of circulation pump (30 W power capacity). Part of the water flown over the absorber plate evaporates and is collected as distillate while the un-evaporated hot brine reaches back the bottom storage tank via the drain valve located at the bottom most edge of the stepped solar still.

The maximum distillate yield and efficiency was obtained at feed water circulation rate of 1.0 LPM. Recirculation of brine increases storage tank temperature, thereby it also increased the nocturnal distillate productivity of the still by about 5.0%. Distillate productivity of this modified system is about 43.0% higher than conventional basin still. The correlation relating total daily distillate production rate ($M_d$ in L/m$^2$d) and

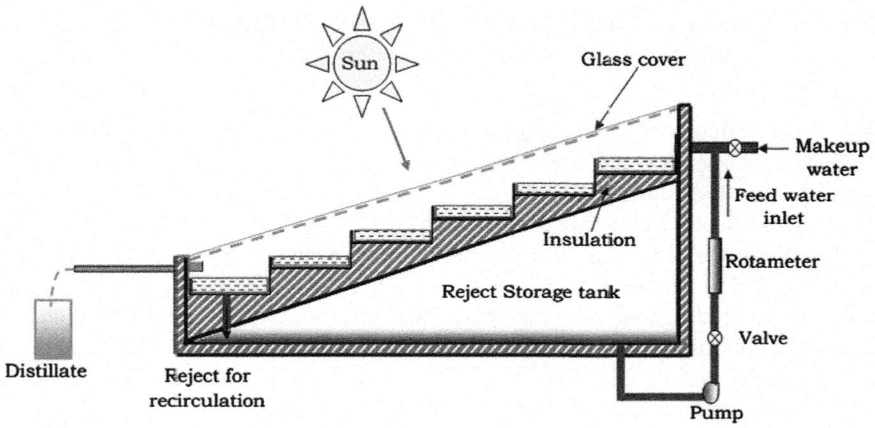


**Fig. 6.5** Schematic of stepped solar still with reject brine recirculation

daily thermal efficiency of the still (in %) with feed water mass in tank, feed water circulation rate, and solar radiation intensity is given as

$$M_d = 0.5684(m_{wt})^{-0.8916}(\dot{m}_{wc})^{0.0381}(I_s)^{0.5684} \tag{6.33}$$

$$\eta_{th} = 4.3723(m_{wt})^{-1.3171}(\dot{m}_{wc})^{0.0637}(I_s)^{4.3736} \tag{6.34}$$

$$50 \le m_{wt} \le 80kg, \; 1 \le \dot{m}_{wc} \le 4LPM, \; 535 \le I_s \le 855W/m^2$$

### 6.3.8 *Depth of Feed Water*

The time taken for evaporation of water depends on the depth of water in addition to the incident solar radiation intensity. Low depth indicates low energy requirement to break the hydrogen atoms in water to initiate evaporation. Diurnal distillate productivity will be higher at low-depth conditions while significant nocturnal productivity is observed at high-depth conditions. High depth also diminishes the positive impact of absorber surface modifications intended for enhanced distillate productivity (D'Cotha et al., 2021). The distillate productivities of the stepped solar still for a water depth of 2 cm, 3 cm, and 4 cm are about 1.32, 1.25, and 1.15 kg/m$^2$d, respectively (Alaudeen et al., 2014). The volumetric heat capacity of water increases with depth, thereby most portion of the incident solar radiation gets stored instead of initiating water evaporation resulting in low distillate production.

### *6.3.9 Integration with PVT (Photovoltaic-Thermal) Solar Collector*

Feed water circulation pump of stepped solar still requires electrical power for its operation. Naroei et al. (2018) suggested integration of PVT collector with stepped solar still in which the power required for operating recirculation pump is consumed from the produced electrical power output of the PVT unit. The photograph of this system is shown in Fig. 6.6. In addition, the feed water is circulated over the back surface of PVT collector, thereby it gets preheated and is then supplied to the stepped solar still for undergoing evaporation. The optimum feed water recirculation rate and optimum PVT area for maximum thermal efficiency of the distillation system are found to be 0.068 kg/min and 1.33 $m^2$, respectively. Distillate productivity of the stepped solar still integrated with PVT collector is about 5.71 L/$m^2$d which is about 20.0% higher than the yield of the stepped solar still without PVT collector and feed water recirculation.

Xiao et al. (2019) proposed a new type of PVT solar collector-integrated stepped solar still whose schematic is shown in Fig. 6.7. In this system, a bottom channel is created beneath the stepped absorber to minimize the heat loss from the absorber plate to the ambient via insulation. The feed water from the saline tank after passing below the PVT collector passes through this bottom channel and then flows over the stepped absorber plate as thin film for undergoing evaporation. The un-evaporated water leaving the still is disposed to the saline water tank which is then recirculated. Increasing the channel depth decreases the distillate productivity of the system due to reduced absorber temperature caused by increased heat capacity of water within

**Fig. 6.6** Photograph of stepped solar still utilizing preheated feed water from PVT solar collector (Naroei et al., 2018)

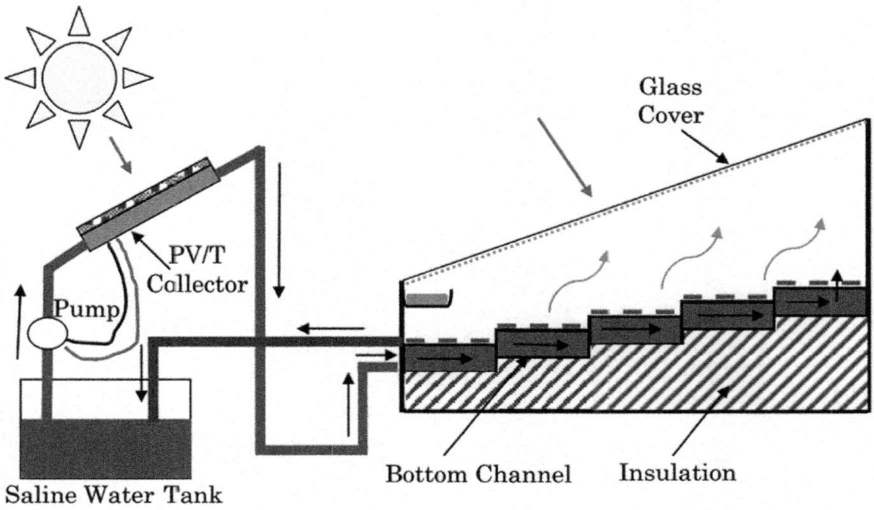


**Fig. 6.7** Schematic of PVT collector-integrated stepped solar still with bottom channel (Xiao et al., 2019)

the channel. At a bottom channel depth of 0.01 m, the heat transfer from the absorber plate to water, temperature of flowing water, and daily yield have increased by 44%, 16.4%, and 51.7%, respectively. However, at a bottom channel depth of 0.03 m, the daily yield has dropped by about 17% in comparison to the earlier case with a bottom channel depth of 0.01 m. The maximum distillate productivity of this system is about 1.55 kg/m$^2$-h and 1.10 kg/m$^2$-h with and without bottom channel, respectively. Hence, provision of bottom channel is proved to be fruitful for stepped solar still.

### *6.3.10 Thermal Energy Storage Materials*

Thermal energy storage materials can be of phase change type and non-phase change type and they can be incorporated either above or below the absorber plate of stepped solar still. When the absorber plate temperature becomes higher than the temperature of the incorporated thermal energy storage material, heat energy is transferred to the energy storage material. This process is called **charging**. When the absorber plate temperature drops below the temperature of the energy storage material, then the stored energy from the thermal energy storage material gets transferred to the absorber plate. This process is called **discharging**. In case of phase change materials (PCM), heat energy is stored as sensible heat energy up to its melting point (solid state) and the heat energy further added gets stored as latent heat energy until the PCM gets fully melted (phase change state). Heat energy added beyond melting of the PCM will be stored as sensible heat energy (liquid state).

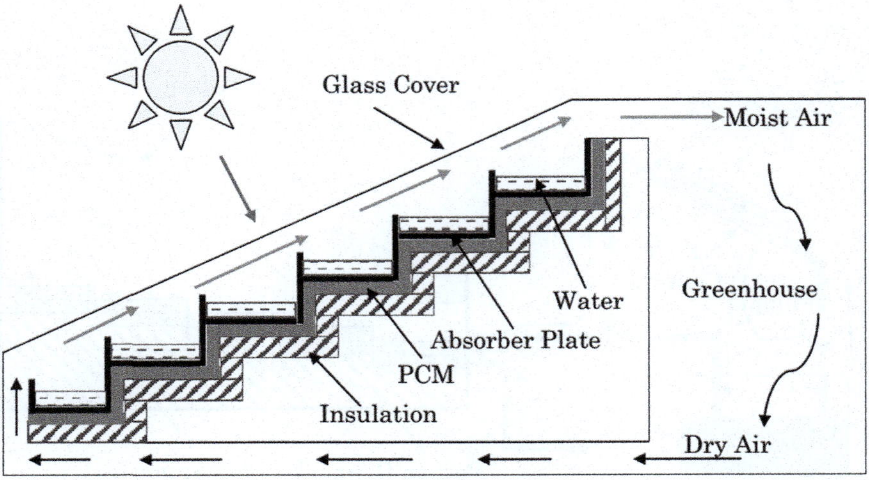


**Fig. 6.8** Schematic of stepped solar still with phase change material (PCM) used for humidifying greenhouse (Radhwan, 2004b)

Radhwan (2004a) used stepped solar still with built-in 10.0-cm-thick paraffin wax layer below its absorber plate for heating and humidifying a greenhouse whose schematic is shown in Fig. 6.8. The distillate yield and thermal efficiency of this still are about 4.60 L/m$^2$d and 57.0%, respectively. The distillate productivity and thermal efficiency of the still with PCM are about 6.52% and 4.0%, higher than the still without PCM. Tabrizi et al. (2010) conducted experiments on stepped solar still with 3.0-cm-thick PCM layer below its absorber plate. Average solar radiation intensities for sunny day and partially cloudy day are about 610 and 540 W/m$^2$, respectively. On sunny day, distillate yield of the still is about 4.85 L/m$^2$d and is lower than the system without PCM by 5.64%. On cloudy day, distillate yield of the system with PCM is about 61.9% higher than the system without PCM.

Dashtban and Tabrizi (2011) observed a distillate yield of about 6.75 kg/m$^2$d in stepped solar still with 2.0-cm-thick PCM beneath its absorber plate which is about 31% higher than the yield from the system without PCM. However, the productivity of the system dropped to 4.12 kg/m$^2$d with increase in water depth from 2.0 to 10.0 cm. The evaporative heat transfer coefficients observed in stepped solar still with 675 g, 1350 g, and 2700 g PCM encapsulated in copper tubes (each tube can hold about 45 g) are about 19.98, 33.58, and 34.16 W/m$^2$K, respectively (Grewal & Kumar, 2022). The distillate yield of stepped solar still with 0.03 wt% composite CuO/GO nano-particle (30/70 volume %)-loaded feed water has increased by about 32.8% with the addition of PCM beneath its absorber plate (Ajdari & Ameri, 2022). Dispersing graphene oxide nano-particle in PCM (0.6 wt%) has increased distillate productivity of stepped solar still with only PCM by about 25%. Moreover, the added nano-particle has also decreased the melting point of PCM (Safaei et al., 2019).

### 6.3.11 Reflectors

Omara et al. (2013) attached mirrors/reflectors to the vertical surfaces of steps inside the stepped solar still to enhance the system's solar radiation exposure. The addition of internal reflecting mirrors over the vertical surface of each step enhances the solar radiation intensity incident over the horizontal tray with water, thereby the productivity of this stepped solar still is about 18.0% higher than the stepped solar still without internal mirrors. Thermal efficiencies of conventional solar still, stepped solar still, and stepped solar still with internal mirrors are about 34.0%, 53.0%, and 56.0%, respectively. The schematic of stepped solar still with internal reflectors, top and bottom external reflectors is shown in Fig. 6.9. Distillate productivity of stepped solar still with external top and bottom reflectors and internal reflectors is about 125.0% higher than the yield of conventional solar still (Omara et al., 2014).

Inclination angle of external reflectors must be adjustable such that it can be fixed at optimum tilt angle for different seasons to collect and reflect more solar radiation. The optimum external reflector inclinations in stepped solar still for different seasons are tabulated in Table 6.1. During winter season, the altitude angle of the sun will be low, therefore the top reflector must be slightly tilted forward to collect and reflect solar radiation over the aperture of the still. Similarly, during summer the sun rays will be overhead (i.e., altitude angle will be high), hence the top reflector must be tilted backward to collect and reflect solar radiation over the aperture of the still. It has been advised to tilt the bottom reflector upwards from horizontal to have more distillate production throughout the year.

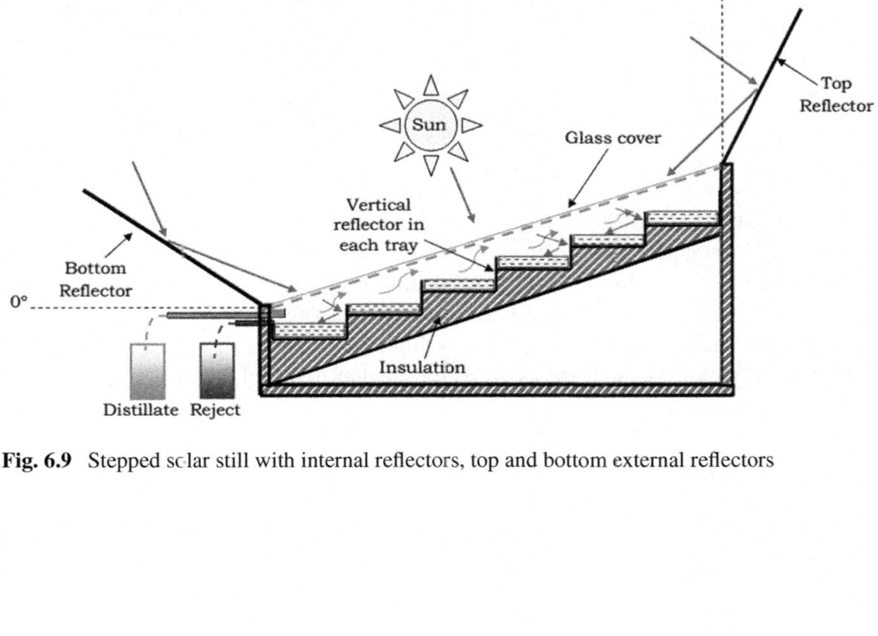


**Fig. 6.9** Stepped solar still with internal reflectors, top and bottom external reflectors

**Table 6.1** Optimum external reflector inclinations for various seasons (Omara et al., 2014)

| Season | Reflector angle | |
|---|---|---|
| | Top (from vertical) | Bottom (from horizontal) |
| Spring | 7° | 30° |
| Summer | 15° | 50° |
| Autumn | 8° | 30° |
| Winter | 15° | 20° |

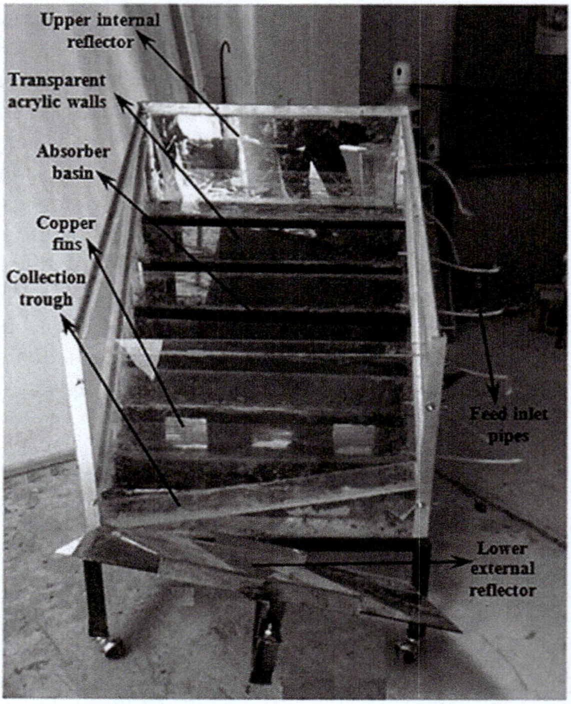


**Fig. 6.10** Photograph of transparent acrylic stepped solar still with copper fins, bottom external reflector, and upper internal reflector (Shmroukh and Ookawara 2020)

Shmroukh and Ookawara (2020) used copper fins over the absorber plate of transparent acrylic stepped solar still incorporated with one internal reflector and one bottom reflector. The photograph of the system is shown in Fig. 6.10. The distillate yield of this system is about 8.28 L/m$^2$d which is about 129% higher than yield of conventional basin solar still. The thermal efficiency of the system is about 63% and its nocturnal yield is around 0.64 L/m$^2$d for a water depth of 3.0 cm.

### 6.3.12 Additional Condenser

The schematic of the stepped solar still with inbuilt passive condenser is shown in Fig. 6.11. In passive condenser, the flow of vapors from the distillation chamber towards the condensing surface is achieved only by buoyancy effect. Muftah et al. (2017) obtained a distillate yield of about 8.9 L/m$^2$d from the stepped solar still with fins, internal and external reflectors, and passive inbuilt condenser. The yield of the system with passive condenser is about 29.0% higher than the stepped solar still without any modifications. The schematic of the stepped solar still with active external condenser is shown in Fig. 6.12. Essa et al. (2020) developed a new type of stepped solar still with suspended trays attached over the vertical face of each step and used $Al_2O_3$ nano-particle (size: 10–14 nm) mixed paraffin wax as energy

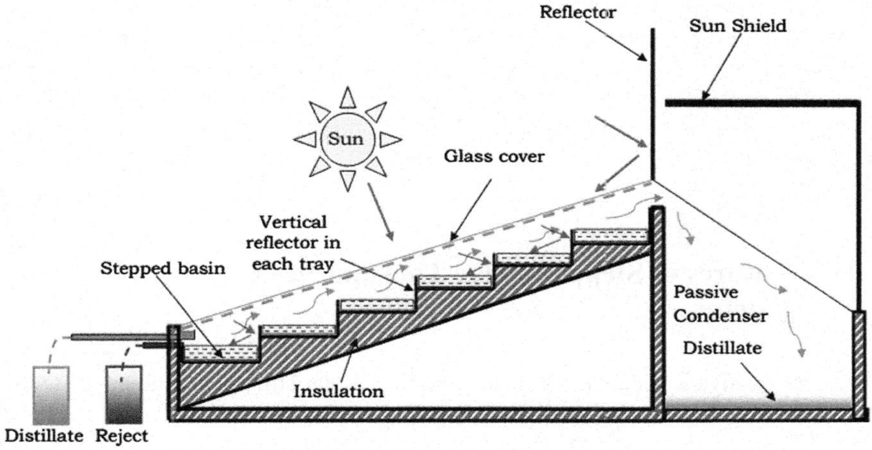


**Fig. 6.11** Schematic of stepped solar still with inbuilt passive condenser

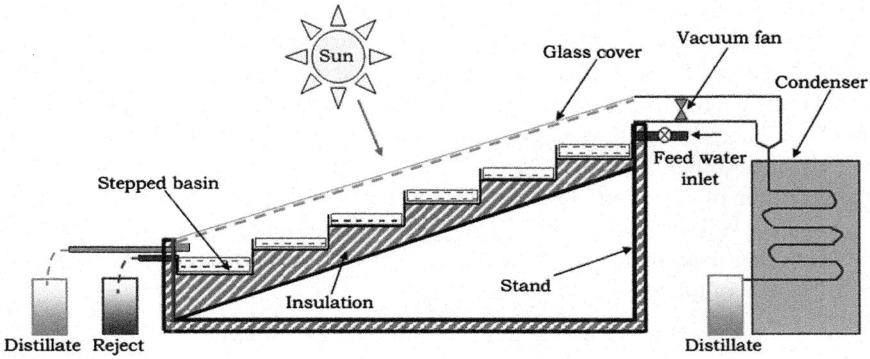


**Fig. 6.12** Schematic of stepped solar still with active external condenser

storage material below its absorber plate. Moreover, an external water-based active condenser was connected to the distillation chamber to facilitate forced condensation by sucking vapors from the distillation chamber using a small fan having a maximum speed of 10000 rpm. The role of suspended trays is to increase the evaporation surface area and to tap more solar radiation. Energy storage material is incorporated to extend operation of the system even after sunset hours. The role of active external condenser section is to prevent vapor accumulation inside the distillation chamber, thereby both evaporation and condensation rate are expected to enhance. Thermal efficiencies of the system with suspended trays and suspended trays with phase change material are about 46.0% and 48.7%, respectively. The distillate productivities of the system with suspended trays, energy storage, and external condenser for a fan speed of 0 rpm (no external condensation), 400 rpm, and 2000 rpm are about 4.18 L/m$^2$d, 5.72 L/m$^2$d, and 6.20 L/m$^2$d, respectively. For fan speed above 2000 rpm, distillate productivity drops and is mainly due to the fact that most of the generated vapors are carried out of the system without condensing either over the glass cover or over the condensing tube, thereby it is finally lost. Moreover, the thermal efficiency of the system also drops from 52.4% to 50.0% with the increase in fan speed from 2000 rpm to 2800 rpm. The utilization of external forced condenser in addition to stearic acid PCM-loaded absorber has resulted in 104% higher yield than conventional stepped solar still (Toosi et al., 2021).

## 6.4 Features of Stepped Solar Thermal Desalination Systems

(a) The positive aspects of both basin and tilted solar still are combined in stepped solar still.
(b) Higher absorber area per ground area occupancy compared to basin and tilted solar still.
(c) Construction is little bit complex than basin, tubular, and tilted solar still.
(d) Incorporation of internal and external mirrors is a suitable option for obtaining significant yield improvement in a cost-effective way.
(e) Adoption of PCMs in stepped solar still is observed to be beneficial only during cloudy conditions while in summer it affects yield of the system negatively.
(f) Continuous makeup water circulation is essential to prevent accidental dry out of the absorber steps.
(g) Addition of nano-particles to the feed water seems to enhance the yield tremendously. However, there will be negative impacts on the human health while handling nano-particles. In addition, no proper methods are available for safe disposal of the nano-particle-loaded reject.

# References

Abdullah, A. S. (2013). Improving the performance of stepped solar still. *Desalination, 319*, 60–65.
Ajdari, H., & Ameri, A. (2022). Performance assessment of an inclined stepped solar still integrated with PCM and CuO/GO nanocomposite as a nanofluid. *Journal of Building Engineering, 49*, 104090.
Alaudeen, A., Johnson, K., Ganasundar, P., Abuthahir, A. S., & Srithar, K. (2014). Study on stepped type basin in a solar still. *Journal of King Saud University-Engineering Sciences, 26*, 176–183.
Dashtban, M., & Tabrizi, F. F. (2011). Thermal analysis of a weir-type cascade solar still integrated with PCM storage. *Desalination, 279*, 415–422.
D'Cotha, J. S., Sajeesh, P., Suresh, P. R., & Jithu, J. (2021). Inherent configuration characteristics altering the distillate enhancement of passive stepped solar still: A review. *Journal of King Saud University-Engineering Sciences*. https://doi.org/10.1016/j.jksues.2021.10.001.
El-Agouz, S. A. (2014). Experimental investigation of stepped solar still with continuous water circulation. *Energy Conversion and Management, 86*, 186–193.
El-Samadony, Y. A. F., & Kabeel, A. E. (2014). Theoretical estimation of the optimum glass cover water film cooling parameters combinations of a stepped solar still. *Energy* 1–7. https://doi.org/10.1016/j.energy.2014.01.080.
El-Samadony, Y. A. F., El-Maghlany, W. M., & Kabeel, A. E. (2016). Influence of glass cover inclination angle on radiation heat transfer rate within stepped solar still. *Desalination, 384*, 68–77.
Essa, F. A., Omara, Z. M., Abdullah, A. S., Shanmugan, S., Panchal, H., Kabeel, A. E., Sathyamurthy, R., Alawee, W. H., Manokar, A. M., & Elsheikh, A. H. (2020). Wall-suspended trays inside stepped distiller with $Al_2O_3$/paraffin wax mixture and vapor suction: Experimental implementation. *Journal of Energy Storage, 32*, 102008.
Gawande, J., & Bhuyar, L. B. (2013). Effect of shape of the absorber surface on the performance of stepped type solar still. *Energy and Power Engineering, 5*, 489–497.
Grewal, R., & Kumar, M. (2022). Performance evaluation of a concatenated stepped solar still system loaded with different masses of energy storage material. *Energy, 259*, 125005.
Jathar, L. D., & Ganesan, S. (2020). Assessing the performance of concave type stepped solar still with nanoparticles and condensing cover cooling arrangement: An experimental approach. Groundwater for Sustainable Development. https://doi.org/10.1016/j.gsd.2020.100539 .
Kabeel, A. E., Omara, Z. M., & Younes, M. M. (2015). Techniques used to improve the performance of the stepped solar still-A review. *Renewable and Sustainable Energy Reviews, 46*, 178–188.
Keshtkar, M., Eslami, M., & Jafarpur, K. (2020). Effct of design parameters on performance of passive basin solar stills considering instantaneous ambient conditions: A transient CFD modeling. *Solar Energy, 201*, 884–907.
Muftah, A, F., Sopian, K., & Alghoul, M. A. (2017). Performance of basin type stepped solar still enhanced with superior design concepts. Desalination. https://doi.org/10.1016/j.desal.2017.07.017.
Naroei, M., Sarhaddi, F., & Sobhnamayan, F. (2018). Efficiency of a photovoltaic thermal stepped solar still: Experimental and numerical analysis. *Desalination, 441*, 87–95.
Omara, Z. M., Kabeel, A. E., & Younes, M. M. (2013). Enhancing the stepped solar still performance using internal reflectors. *Desalination, 314*, 67–72.
Omara, Z. M., Kabeel, A. E., & Younes, M. M. (2014). Enhancing the stepped solar still performance using internal and external reflectors. *Energy Conversion and Management, 78*, 876–881.
Radhwan, A. M. (2004a). Transient performance of a stepped solar still with built-in latent heat thermal energy storage. *Desalination, 171*, 61–76.
Radhwan, A. M. (2004b). Transient analysis of a stepped solar still for heating and humidifying greenhouses. *Desalination, 161*, 89–97.
Safaei, M. R., Goshayeshi, H. R., & Chaer, I. (2019). Solar still efficiency enhancement by using graphene oxide/Paraffin Nano-PCM. *Energies, 12*, 1–13.

Sangeetha, A., Shanhugan, S., & Gorjian, S. (2022). Experimental evaluation and thermodynamic Gibbs free energy analysis of a double-slope U-shaped stepped basin solar still using activated carbon with ZnO nanoparticles. *Journal of Cleaner Production, 380*, 135118.

Sarhaddi, F., Tabrizi, F. F., Zoori, H. A., & Mousavi, S. A. H. S. (2017). Comparative study of two weir type cascade solar stills with and without PCM storage using energy and exergy analysis. *Energy Conversion and Management, 133*, 97–109.

Sathyamurthy, R., Kabeel, A. E., Balasubramanian, M., Devarajan, M., Sharshir, S. W., & Manokar, A. M. (2020). Experimental study on enhancing the yield from stepped solar still coated using fumed silica nanoparticle in black paint. *Materials Letters, 272*, 127873.

Shmroukh, A. N., & Ookawara, S. (2020). Evaluation of transparent acrylic stepped solar still equipped with internal and external reflectors and copper fins. *Thermal Science and Engineering Progress, 18*, 100518.

Sharshir, S. W., Eltawil, M. A., Algazzar, A. M., Sathyamurthy, R., & Kandeal, A. W. (2020). Performance enhancement of stepped double slope solar still by using nanoparticles and linen wicks: Energy, exergy and economic analysis. *Applied Thermal Engineering, 174*, 115278.

Tabrizi, F. F., Dashtban, M., & Moghaddam, H. (2010). Experimental investigation of a weir-type cascade solar still with built-in latent heat thermal energy storage system. *Desalination, 260*, 248–253.

Toosi, S. S. A., Goshayeshi, H. R., & Heris, S. Z. (2021). Experimental investigation of stepped solar still with phase change material and external condenser. *Journal of Energy Storage, 40*, 102681.

Xiao, L., Shi, R., Wu, Shuang-Ying., & Chen, Zhi-Li. (2019). Performance study on a photovoltaic thermal (PV/T) stepped solar still with a bottom channel. *Desalination, 471*, 114129.

# Chapter 7
# Multi-effect/Stage Solar Thermal Desalination Systems

## 7.1 Basics of Multi-effect/Stage Solar Thermal Desalination Systems

Heat energy that is being lost from the condensing surface of the solar thermal desalination system to the ambient can be recovered by utilizing multi-effects/stages to enhance distillate productivity (Adhikari et al., 1995; Goosen et al., 2000). In this concept, thin film of feed water is allowed to flow over the lower condensing cover of the still. This thin film evaporates by utilizing the latent heat of the condensing vapor from the bottom effect thereby, additional distillate is obtained. This modified solar still where heat energy lost via the condensing surface is recovered and used for further distillation is termed as **multi-effect distillation system**. The overall efficiency of the multi-effect still is about 20.0–30.0% higher than the single effect still (Tiwari and Lawrence, 1992). In the case of multi-effect systems, the feed water is distributed as thin film over each evaporating surface (Diaf et al., 2016) and, in the case of multi-stage systems, feed water is allowed to accumulate as bulk in each of the basins. Generally, multi-stage systems are bulkier and heavier than multi-effect systems. The schematic of passive double effect solar still and passive double stage solar still is shown in Figs. 7.1 and 7.2, respectively.

## 7.2 Classification of Multi-effect/Stage Solar Still

Multi-effect/stage solar still can be broadly classified into passive and active types. In passive type, no external solar collectors or heating systems are utilized for feed water heating/pre-heating. Feed water is heated only within the distillation chamber with the aid of solar radiation that is incident over the desalination system's aperture. The schematic of passive multi-effect/stage desalination system is shown in Figs. 7.1 and 7.2, respectively. In passive-type multi-stage/effect desalination system, the stages/effects need to be transparent to receive the solar radiation thereby, it



H. Sharon and K. Srinivas Reddy, *Solar Thermal Desalination Technologies for Potable Water*, https://doi.org/10.1007/978-3-031-49155-9_7

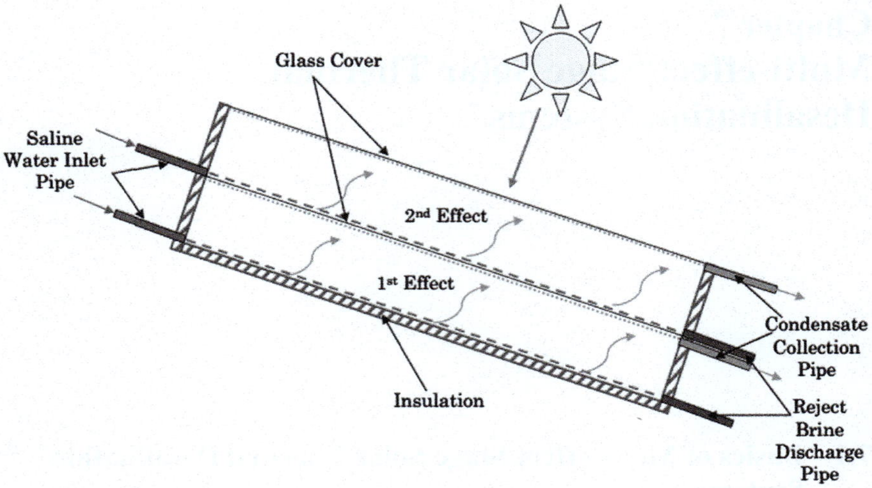


**Fig. 7.1** Schematic of passive double effect solar still

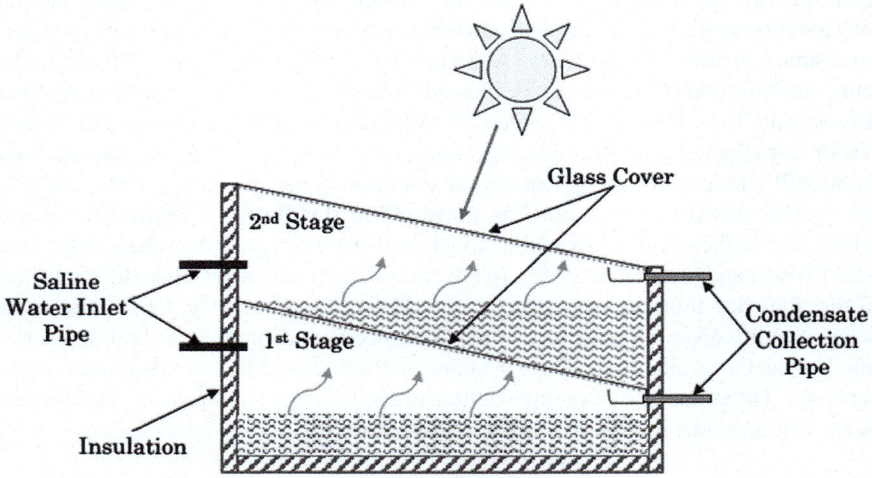


**Fig. 7.2** Schematic of passive double stage solar still

must be fabricated with glass which is more fragile. In addition, salt deposition over the evaporating surface of the topmost stage/effect affects the solar radiation intensity reaching the basin liner of the bottom stage thereby, distillate yield of passive multi-stage/effect desalination system drops with operation time (Adhikari et al., 1995). These drawbacks are eliminated in active-type multi-stage/effect desalination systems. In active type, solar collectors or waste heat are utilized to heat the feed water supplied to the desalination system either directly or indirectly. Hence, these

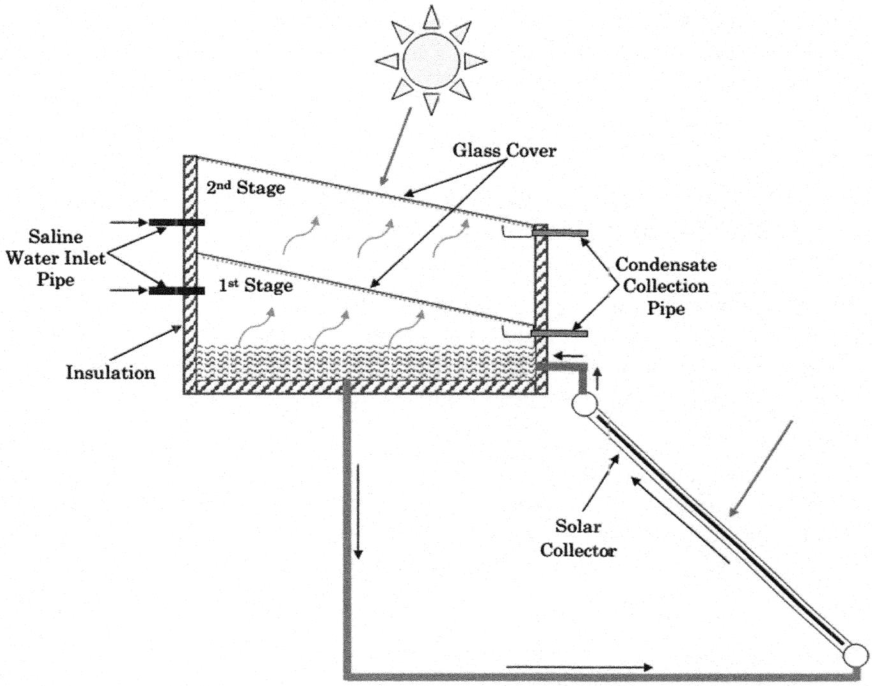


**Fig. 7.3** Schematic of double effect active solar still

systems need not to be transparent and can be made even with metals. Therefore, active systems are more preferred over passive-type multi-stage/effect desalination systems for practical applications. The schematic of active multi-effect solar thermal desalination system is shown in Fig. 7.3.

Multi-stage/effect desalination systems can also be classified based on feed water flow type as parallel flow type and series flow type. In parallel flow type, feed water is distributed separately to each effect. In series flow type, feed water first enters the topmost stage/effect and then moves to other lower effects/stages subsequently by the aid of gravity. The brine leaving the bottommost stage/effect of series flow-type desalination system will be highly concentrated than the reject brine of parallel flow-type desalination system. Multi-stage/effect desalination system has also been classified based on the operating pressure as non-evacuated type (ambient pressure operation) and evacuated type (operating pressure lower than ambient). Classification has also been carried based on the shape of the system as horizontal/tilted multi-effect/basin solar still, vertical multi-effect solar still, and concentric/tubular multi-effect solar still. Generally, the term multi-effect/stage desalination system refers to the horizontal/tilted types, while other systems are named based on their particular shape. In this chapter, focus has been made only on active type horizontal/tilted multi-effect/stage solar stills, while other types are discussed separately in the upcoming chapters.

## 7.3 Thermodynamic Modeling of Active Multi-effect/Stage Solar Thermal Desalination System

The schematic indicating the energy balance in active multi-stage solar thermal desalination system with corrugated stages is shown in Fig. 7.4.

The major assumptions considered during thermodynamic modeling of active multi-effect/stage solar thermal desalination system are as follows:

(1) The average temperature of circulating fluid is same as the temperature of the feed water in the first stage
(2) The temperature of the feed water stacked in each stage is similar to the temperature of the corresponding stage
(3) The feed water mass in each stage is nearly the same

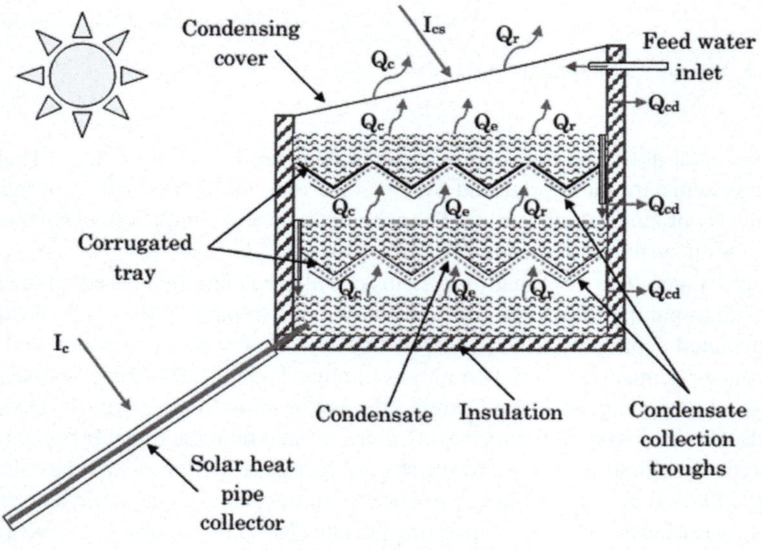


**Fig. 7.4** Energy transport process in active multi-stage solar thermal desalination system with corrugated trays (Xiong et al., 2013)

Energy balance in the first stage is given as

$$\begin{pmatrix}\text{rate of temperature}\\ \text{change of 1st stage}\end{pmatrix} = \begin{pmatrix}\text{heat energy input}\\ \text{from solar collectors}\end{pmatrix} - \begin{pmatrix}\text{convention heat transfer}\\ \text{from 1st stage to 2nd stage}\end{pmatrix} - \begin{pmatrix}\text{radiation heat transfer}\\ \text{from 1st stage to 2nd stage}\end{pmatrix} - \begin{pmatrix}\text{evaporation heat transfer}\\ \text{from 1st stage to 2nd stage}\end{pmatrix} - \begin{pmatrix}\text{conduction heat loss}\\ \text{from 1st stage to ambient}\end{pmatrix}$$

$$(mc)_{s1}\frac{dT_{s1}}{dt} = \eta_c I_c A_c - Q_{c(s1-s2)} - Q_{r(s1-s2)} - Q_{e(s1-s2)} - Q_{cd(s1-a)} \tag{7.1}$$

Energy balance in intermediate stages other than the last stage is given as

$$\begin{pmatrix}\text{rate of temperature}\\ \text{change of } i\text{th stage}\end{pmatrix} = \begin{pmatrix}\text{convention heat transfer from}\\ (i-1)\text{th stage to } i\text{th stage}\end{pmatrix} + \begin{pmatrix}\text{radiation heat transfer from}\\ (i-1)\text{th stage to } i\text{th stage}\end{pmatrix} + \begin{pmatrix}\text{evaporation heat transfer from}\\ (i-1)\text{th stage to } i\text{th stage}\end{pmatrix} - \begin{pmatrix}\text{convection heat transfer from}\\ i\text{th stage to } (i+1)\text{th stage}\end{pmatrix} - \begin{pmatrix}\text{radiation heat transfer from}\\ i\text{th stage to } (i+1)\text{th stage}\end{pmatrix} - \begin{pmatrix}\text{evaporation heat transfer from}\\ i\text{th stage to } (i+1)\text{th stage}\end{pmatrix} - \begin{pmatrix}\text{conduction heat loss from}\\ i\text{th stage to ambient}\end{pmatrix}$$

$$\begin{aligned}(mc)_{si}\frac{dT_{si}}{dt} =& Q_{c(s(i-1)-s(i))} + Q_{r(s(i-1)-s(i))} + Q_{e(s(i-1)-s(i))} - Q_{c(s(i)-s(i+1))} \\ &- Q_{r(s(i)-s(i+1))} - Q_{e(s(i)-s(i+1))} - Q_{cd(s(i)-a)}\end{aligned} \tag{7.2}$$

where value of "$i$" starts from 2 to $(N-1)$; "$N$" represents last stage.

Energy balance in the $N$th stage is given as

$$\begin{pmatrix}\text{rate of temperature}\\ \text{change of } N\text{th stage}\end{pmatrix} = \begin{pmatrix}\text{solar radiation absorbed}\\ \text{by the } N\text{th stage}\end{pmatrix} - \begin{pmatrix}\text{convention heat transfer from}\\ (N-1)\text{th stage to } N\text{th stage}\end{pmatrix} + \begin{pmatrix}\text{radiation heat transfer from}\\ (N-1)\text{th stage to } N\text{th stage}\end{pmatrix} + \begin{pmatrix}\text{evaporation heat transfer from}\\ (N-1)\text{th stage to } N\text{th stage}\end{pmatrix} - \begin{pmatrix}\text{convection heat transfer from}\\ N\text{th stage to glass cover}\end{pmatrix} - \begin{pmatrix}\text{radiation heat transfer from}\\ N\text{th stage to glass cover}\end{pmatrix} - \begin{pmatrix}\text{evaporation heat transfer}\\ \text{from } N\text{th stage to glass cover}\end{pmatrix} - \begin{pmatrix}\text{conduction heat loss}\\ \text{from } N\text{th stage to ambient}\end{pmatrix}$$

$$(mc)_{sN}\frac{dT_{sN}}{dt} = Q_{h(N)} + Q_{c(s(N-1)-s(N))} + Q_{r(s(N-1)-s(N))} + Q_{e(s(N-1)-s(N))} - Q_{c(s(N)-g)} - Q_{r(s(N)-g)} - Q_{e(s(N)-g)} - Q_{cd(s(N)-a)} \tag{7.3}$$

Energy balance in the glass cover of the system is given as

$$\begin{pmatrix}\text{rate of temperature}\\ \text{change of glass cover}\end{pmatrix} = \begin{pmatrix}\text{solar radiation absorbed}\\ \text{by the glass cover}\end{pmatrix} + \begin{pmatrix}\text{convention heat transfer from}\\ N\text{th stage to glass cover}\end{pmatrix} + \begin{pmatrix}\text{radiation heat transfer from}\\ N\text{th stage to glass cover}\end{pmatrix} + \begin{pmatrix}\text{evaporation heat transfer}\\ \text{from } N\text{th stage to glass cover}\end{pmatrix} - \begin{pmatrix}\text{convection heat loss from}\\ \text{glass cover to ambient}\end{pmatrix} - \begin{pmatrix}\text{radiation heat loss from}\\ \text{glass cover to ambient}\end{pmatrix}$$

$$(mc)_g\frac{dT_g}{dt} = Q_{h(g)} + Q_{c(s(N)-g)} + Q_{r(s(N)-g)} + Q_{e(s(N)-g)} - Q_{c(g-a)} - Q_{r(g-a)} \tag{7.4}$$

The convective and evaporative heat transfer occurs simultaneously in solar thermal desalination systems. The mass transfer rate of the humid air ($\dot{m}_{ha}$ in kg/m$^2$·s) due to convective heat transfer from the evaporating to the condensing surface is estimated as

$$\dot{m}_{ha} = \frac{h_{c(e-c)}A_e(T_e - T_c)}{c_{p,ha}A_e(T_e - T_c)} = \frac{h_{c(e-c)}}{c_{p,ha}} \tag{7.5}$$

where

$A_e$—surface area of evaporating surface (m$^2$);
$h_{c(e-c)}$—Convective heat transfer coefficient between the evaporating and condensing surface (W/m$^2$K);
$c_{p,ha}$—specific heat capacity of humid air (J/kgK);
$T_e$—evaporating surface temperature (°C);
$T_c$—condensing surface temperature (°C).

The convective heat transfer coefficient between the evaporating and condensing surface cannot be estimated accurately for active multi-stage desalination systems using the correlation formulated by Dunkle which is listed in Chap. 3 (Eqs. 3.11 and 3.14). Because Dunkle's correlation is valid only for a mean air temperature of 50 °C (within the distillation chamber) and equivalent temperature difference of within 17 °C between the evaporating and condensing surfaces. Generally, active multi-stage/effect solar stills operate beyond 50 °C and in addition their Grashoff number also exceeds $2.51 \times 10^5$, hence modified correlations are needed for evaporation rate estimation in active multi-stage desalination systems (Adhikari et al., 1990).

Mass of humid air ($\dot{m}_{ha}$) is the summation of mass of dry air ($\dot{m}_{da}$) and mass of water vapor ($\dot{m}_{mv}$)

$$\dot{m}_{ha} = \dot{m}_{da} + \dot{m}_{mv} \tag{7.6}$$

Mass transfer rate of water vapor on the evaporating surface (($\dot{m}_{mv})_e$ in kg/m$^2$·s) is given as

$$(\dot{m}_{wv})_e = \frac{M_v P_{we}}{M_v P_{we} + M_a(P_T - P_{we})} \frac{h_{c(e-c)}}{c_{p,ha}} \tag{7.7}$$

Mass transfer rate of water vapor on the condensing surface (($\dot{m}_{mv})_c$ in kg/m$^2$·s) is given as

$$(\dot{m}_{wv})_c = \frac{M_v P_{wc}}{M_v P_{wc} + M_a(P_T - P_{wc})} \frac{h_{c(e-c)}}{c_{p,ha}} \tag{7.8}$$

where

$M_v$—molar mass of water vapor (18 kg/kmol);
$M_a$—molar mass of air (28.97 kg/kmol);
$P_{wc}$, $P_{we}$—partial vapor pressure of water vapor at condensing, evaporating surface temperature (Pa);
$P_T$—Atmospheric pressure (101325 Pa).

Net mass transfer of water vapor (($\dot{m}_e$ in kg/m$^2$·s) from the evaporating to condensing surface in each stage is given as

$$\dot{m}_e = (\dot{m}_{mw})_e - (\dot{m}_{mw})_c \tag{7.9}$$

Evaporative heat transfer from the evaporating (one stage) to the condensing surface (next stage) ($Q_{e(e-c)}$ in W) is estimated as

$$Q_{e(e-c)} = \dot{m}_e A_e h_{fg} \tag{7.10}$$

where

$h_{fg}$—Latent heat of evaporation (J/kg).

Convective heat transfer from the evaporating (one stage) to the condensing surface (next stage) ($Q_{c(e-c)}$ in W) is estimated as

$$Q_{c(e-c)} = 0.20\frac{K_{ha}}{x}\left[\frac{g\beta x^3\left[(T_e - T_c) + \left[\frac{\frac{(P_{we} - P_{wc})}{\frac{M_a P_T}{M_a - M_v} - P_{we}}}{}(T_e + 273.15)\right]\right]}{v_{ha}^2}\right]^{0.26} \tag{7.11}$$

where

$K_{ha}$—Thermal conductivity of humid air (W/mK);
$x$—Gap between evaporating and condensing surface (m);
$\beta$—Volumetric thermal expansion coefficient (1/K);
$v_{ha}$—Kinematic viscosity of humid air (m$^2$/s).

Assuming both the evaporating and condensing surfaces are nearly parallel to each other, the radiation heat transfer from the evaporating (one stage) to the condensing surface (next stage) ($Q_{r(e-c)}$ in W) is estimated as

$$Q_{r(e-c)} = \frac{\sigma A_e (T_e^4 - T_c^4)}{(1/\varepsilon_e) + (1/\varepsilon_c) - 1} \tag{7.12}$$

where

$\varepsilon_e$—Emissivity of the evaporating surface;
$\varepsilon_c$—Emissivity of the condensing surface;
$\sigma$—Stefan-Boltzmann constant ($5.67 \times 10^{-8}$ W/m$^2$K$^4$).

Heat loss from each stage of the desalination system's wall to the ambient via insulation ($Q_{cd(s(i)-a)}$ in W) is estimated as

$$Q_{cd(s(i)-a)} = \frac{T_{s(i)} - T_a}{\left[\frac{x_{cd}}{K_{cd} A_{wall(s(i))}}\right] + \left[\frac{1}{h_{ca} A_{wall(s(i))}}\right]} \tag{7.13}$$

where

$T_{s(i)}$—Temperature of ith stage (°C);
$T_a$—ambient temperature (°C);
$x_{cd}$—Thickness of insulation (m);

$K_{cd}$—thermal conductivity of insulation (W/mK);
$h_{ca}$—convective heat transfer coefficient between the distillation stage and ambient ($W/m^2K$);
$A_{wall(s(i))}$—Surface area of the wall of ith stage ($m^2$).

For the bottommost stage, heat loss to the ambient occurs through its base in addition to its wall and is estimated as

$$Q_{cd(s1-a)} = \frac{T_{s1} - T_a}{\left[\frac{x_{cd}}{K_{cd}(A_{wall(s1)} + A_{base(s(1))})}\right] + \left[\frac{1}{h_{ca}(A_{wall(s1)} + A_{base(s(1))})}\right]} \tag{7.14}$$

where

$A_{wall(s(1))}$—Surface area of the wall of the first stage ($m^2$);
$A_{base(s(1))}$—Surface area of the base of the first stage ($m^2$);
$Q_{cd(s1-a)}$—Heat loss from the first stage of the desalination system to the ambient via insulation (W);
$T_{s(1)}$—Temperature of the first stage (°C).

Heat energy supplied by the vacuum tube collector to the bottom-most stage ($Q_{hc}$ in W) of the desalination system is estimated as

$$Q_{hc} = \eta_c I_c A_{sc} \tag{7.15}$$

The instantaneous efficiency of vacuum tube solar collector ($\eta_c$) of aperture area ($A_{sc}$ in $m^2$) exposed to solar radiation intensity ($I_c$ in $W/m^2$) is given as (Chen et al., 2017)

$$\eta_c = 0.635 - 1.96\left(\frac{T_{s1} - T_a}{I_c}\right) \tag{7.16}$$

$$\eta_c = 0.803 - 2.01\left(\frac{T_{s1} - T_a}{I_c}\right) - 0.034\left(\frac{(T_{s1} - T_a)^2}{I_c}\right) \tag{7.17}$$

Average water temperature inside the solar collector is assumed to be identical to the water temperature of stage 1 to which it is connected.

The solar radiation intensity absorbed by the black coated last (i.e.) the uppermost stage ($Q_{h(N)}$ in W) is estimated by

$$Q_{h(N)} = A_{ls}\alpha_{ls}\tau_g I_{ls} \tag{7.18}$$

where

$A_{ls}$—absorber surface area of last stage (m$^2$);
$I_{ls}$—solar radiation incident over the absorber of last stage (W/m$^2$);
$\alpha_{ls}$—absorbtivity of last stage;
$\tau_g$—Transmissivity of glass cover.

The solar radiation intensity absorbed by the glass cover of the uppermost stage ($Q_{h(g)}$ in W) is estimated by

$$Q_{h(g)} = A_g \alpha_g I_g \tag{7.19}$$

where

$A_g$—surface area of glass cover (m$^2$);
$I_g$—solar radiation incident over the glass cover (W/m$^2$);
$\alpha_g$—absorbtivity of glass cover.

The cumulative daily distillate yield ($M_{total}$ in kg/d) obtained from the desalination system is estimated by

$$M_{total} = \int_{t=0h}^{t=24h} \dot{m}_e A_e dt \tag{7.20}$$

In most of the multi-stage solar thermal desalination systems, the glass cover is replaced by metal condensing cover such that the uppermost stage do not receive any direct solar radiation over it. In addition, the condensing cover might also be shaded. These are practiced to prevent the reversal of heat flow from the topmost stage to the bottom stage which may affect the system's distillate productivity in a negative way. Hence, in this case, Eqs. 7.3, and 7.4, can be modified as

Energy balance in the $N$th stage is given as

$$\begin{pmatrix}\text{rate of temperature}\\ \text{change of } N\text{th stage}\end{pmatrix} = \begin{pmatrix}\text{convention heat transfer from}\\ (N-1)\text{th stage to } N\text{th stage}\end{pmatrix}$$
$$+ \begin{pmatrix}\text{radiation heat transfer from}\\ (N-1)\text{th stage to } N\text{th stage}\end{pmatrix} + \begin{pmatrix}\text{evaporation heat transfer from}\\ (N-1)\text{th stage to } N\text{th stage}\end{pmatrix}$$
$$- \begin{pmatrix}\text{convection heat transfer from}\\ N\text{th stage to condensing cover}\end{pmatrix} + \begin{pmatrix}\text{radiation heat transfer from}\\ N\text{th stage to condensing cover}\end{pmatrix}$$
$$- \begin{pmatrix}\text{evaporation heat transfer from}\\ N\text{th stage to condensing cover}\end{pmatrix} - \begin{pmatrix}\text{conduction heat loss from}\\ N\text{th stage to ambient}\end{pmatrix}$$

$$(mc)_{sN}\frac{dT_{sN}}{dt} = Q_{c(s(N-1)-s(N))} + Q_{r(s(N-1)-s(N))} + Q_{e(s(N-1)-s(N))} \\ - Q_{c(s(N)-cc)} - Q_{r(s(N)-cc)} - Q_{e(s(N)-cc)} - Q_{cd(s(N)-a)} \quad (7.21)$$

Energy balance in the last condensing cover of the desalination system is given as

$$\begin{pmatrix}\text{rate of temperature}\\ \text{change of condensing stage}\end{pmatrix} = \begin{pmatrix}\text{convention heat transfer from}\\ N\text{th stage to condensing cover}\end{pmatrix} \\ + \begin{pmatrix}\text{radiation heat transfer from}\\ N\text{th stage to condensing cover}\end{pmatrix} + \begin{pmatrix}\text{evaporation heat transfer from}\\ N\text{th stage to condensing cover}\end{pmatrix} \\ - \begin{pmatrix}\text{convection heat loss from}\\ \text{condensing cover to ambient}\end{pmatrix} - \begin{pmatrix}\text{radiation heat loss from}\\ \text{condensing cover to ambient}\end{pmatrix}$$

$$(mc)_{cc}\frac{dT_{cc}}{dt} = Q_{c(s(N)-cc)} + Q_{r(s(N)-cc)} + Q_{e(s(N)-cc)} - Q_{e(s(N)-cc)} \\ - Q_{c(cc-a)} - Q_{r(cc-a)} \quad (7.22)$$

The convection heat losses ($Q_{c(cc-a)}$ in W) and radiation heat losses ($Q_{r(cc-a)}$ in W) from the outermost condensing cover of the desalination system to the ambient is estimated by

$$Q_{c(cc-a)} = A_{cc}(5.7 + 3.8V_{wind})(T_{cc} - T_a) \quad (7.23)$$

$$Q_{r(cc-a)} = \sigma\varepsilon_{cc}A_{cc}(T_{cc}^4 - T_a^4) \quad (7.24)$$

where

$A_{cc}$—surface area of outermost condensing cover (m$^2$);
$T_{cc}$—Temperature of outermost condensing cover (K);
$T_a$—Ambient temperature (K);
$V_{wind}$—Wind speed (m/s);
$\varepsilon_{cc}$—Emissivity of the last condensing surface.

Sometimes, the outermost condensing cover can also be lined with wetted wick to enhance evaporation and condensation rates in the stages. Hence, in addition to convective and radiative heat loss to the ambient, evaporative heat loss to the ambient also takes place.

Evaporative heat transport occurring between the condensing cover of the last stage lined with wetted wick to the ambient ($Q_{e(cc-a)}$ in W) can be quantified by

$$Q_{e(cc-a)} = (0.016)(2.8 + 3.0(V_{wind}))A_{cc}(P_{cc} - (\varphi P_a)) \quad (7.25)$$

where

$P_{cc}$—partial vapor pressure of water vapor at last condensing cover temperature (Pa);
$P_a$—partial vapor pressure of water vapor at ambient temperature (Pa).

If the presence of NaCl in feed water is considered, its role in reducing the partial vapor pressure of feed water at the evaporating surface is accounted by taking into account the mole fraction of water vapor. The modified partial vapor pressure at the evaporating surface by considering the salt concentration is quantified by

$$\begin{pmatrix}\text{actual partial pressure of water}\\ \text{vapor at evaporating surface}\end{pmatrix} = \begin{pmatrix}\text{mole fraction of}\\ \text{water vapor}\end{pmatrix} \times \begin{pmatrix}\text{saturation vapor pressure of water vapor at evaporating}\\ \text{surface corresponding to stage temperature}\end{pmatrix} \tag{7.26}$$

Performance ratio (PR) is an important parameter used to assess the actual performance of multi-stage/effect solar thermal desalination systems. Performance ratio is generally a measure of how effectively the supplied heat energy is re-used for fresh water production in multi-effect distillation systems. Mathematically, it can be defined as the ratio of the total energy required to evaporate feed water to produce fresh water to the net energy supplied by the solar collectors.

$$PR = \frac{\sum_{7:30}^{19:30}\left(\sum_{i=1}^{i=n} m_i h_{fg}\right)}{\sum_{7:30}^{19:30} Q_{s(ab)}} \tag{7.27}$$

Consider a multi-stage/effect desalination system with "$N$" stages/effects supplied with heat energy "$Q$" by the solar thermal collectors to the first effect. The approximate performance ratio of this system if the entire supplied energy is utilized for evaporation without any loss to the ambient or via other forms of heat transfer like convection and radiation heat transfer is given as (Li et al., 2018)

$$PR = \frac{Q_1 + Q_2 + \cdots + Q_N}{Q} \tag{7.28}$$

For first stage/effect heat energy input is "$Q$"

$$Q_1 = Q \tag{7.29}$$

For the second stage/effect heat energy input will be the latent heat of condensation from the first stage/effect which is "$Q$" (under ideal case)

$$Q_2 = Q \tag{7.30}$$

Similarly, for the $N$th stage/effect the heat energy supplied will be

$$Q_N = Q \tag{7.31}$$

Therefore, $PR$ of the ideal system with "$N$" effects/stages will be

$$PR = \frac{NQ}{Q} = N \tag{7.32}$$

However, in general, the $PR$ will not be "$N$" because of the loss of the supplied heat energy to the ambient thereby, all the supplied heat energy cannot be used for evaporation. In addition, there will be convection and radiation heat transfer fractions in the total heat transferred from the evaporating to the condensing surface. If additional heat energy is supplied to any of the effects it must be reflected in the above calculations to estimate $PR$ accurately.

## 7.4 Parameters Influencing the Performance of Active Multi-effect/Stage Solar Still

### *7.4.1 Number of Effects/Stages*

The distillate yield of active multi-stage/effect solar thermal desalination system increases with increase in the number of stages/effects up to 4 or 5 due to effective re-utilization of the latent heat of condensation. The distillate yield of two-effect system is about 70% higher than the yield of single-effect system (Kalbasi et al., 2018). Beyond 4 or 5 stages/effects, resistance offered by the metal partitions, water layer/wetted wick, and air gap becomes more significant thereby, the transfer of latent heat from the previous stage to the additional stage/effect diminishes. This causes reduced temperature rise in additional stages/effects resulting in negligible temperature difference between the adjacent stages/effects which in turn leads to no significant distillate yield enhancement. The optimum number of stages/effects is also dependent on the climatic conditions of the location (Adhikari and Kumar, 1993). The schematic of active multi-stage and multi-effect solar thermal desalination system is shown in Figs. 7.5 and 7.6, respectively.

In the case of multi-stage systems, feed water in the bottom-most stage will be heated by the solar collector directly while the feed water in other stages utilizes only the released latent heat thereby, the temperature difference and yield drops as we move from the bottom-most stage to the topmost stage. In the case of multi-effect systems, the temperature difference between the effects and yield in each effect increases as we move from the bottom-most effect to the topmost effect if the feed water heated in solar collector is distributed simultaneously to all the effects (Reddy et al., 2012). Cumulative distillate yield of the four-stage and four-effect solar thermal desalination

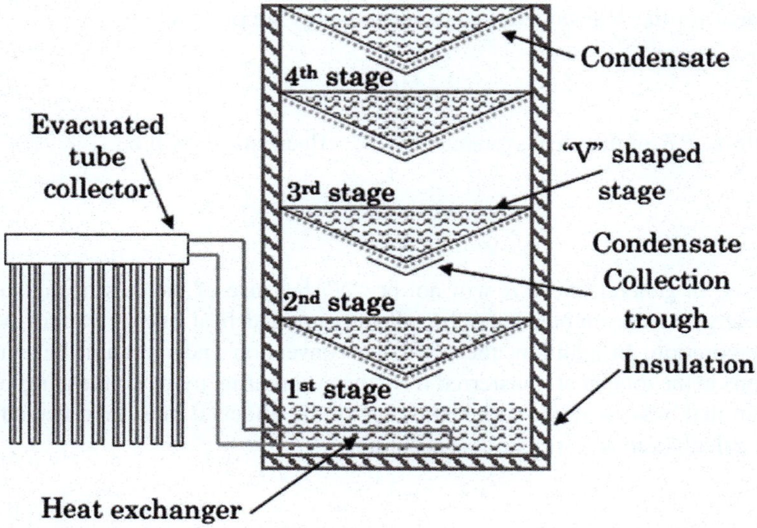


**Fig. 7.5** Schematic of active multi-stage solar thermal desalination system with "*V*-"shaped stages (Shatat & Mahkamov, 2010)

system is about 11.0 and 28.0 kg/m$^2$d, respectively. Distillate yield of multi-effect system is higher because its feed water flow rate can be controlled, while, in the case of multi-stage system, feed water is held as a bulk thereby, chance of enhanced evaporation is limited (Shatat & Mahkamov, 2010; Reddy et al., 2012).

Condensing surface temperature of various stages in active four-stage solar still estimated based on experiments is given as (Shatat and Mahkamov, 2010).

Condensing surface temperature of the first stage is

$$T_{c1} = T_{s2} - 2.0(°C) \tag{7.33}$$

Condensing surface temperature of the second stage is

$$T_{c2} = T_{s3} - 2.7(°C) \tag{7.34}$$

Condensing surface temperature of the third stage is

$$T_{c3} = T_{s4} - 1.11(°C) \tag{7.35}$$

Condensing surface temperature of the fourth stage is

$$T_{c4} = T_{s4} - (0.00007T_{s4}^3 - 0.015T_{s4}^2 + 0.9763T_{s4} - 10.324)(°C) \tag{7.36}$$

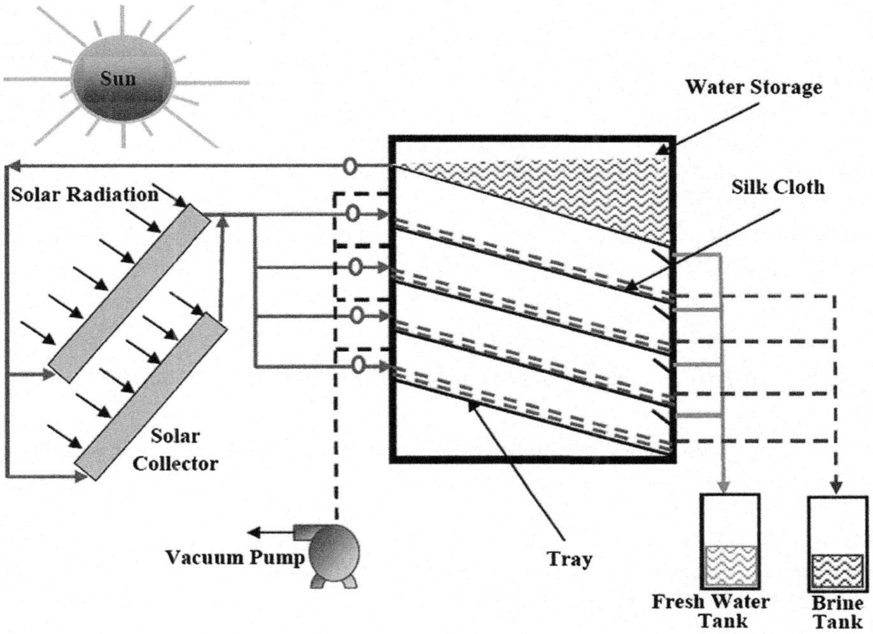


**Fig. 7.6** Schematic of active multi-effect solar thermal desalination system with parallel trays (Sharon and Reddy, 2015)

## 7.4.2 *Mode of Operation*

Nocturnal distillate yield occupies around 40.0% of the cumulative distillate yield of active multi-stage solar stills with high basin water depth. Hence, operation and investigation of active multi-stage solar thermal desalination system round the clock (24 h) is highly recommended. The desalination system can be operated in non-continuous mode and in continuous mode.

In **non-continuous mode**, each of the stages of desalination system is filled with feed water once and is operated without any further addition of makeup feed water to compensate for the evaporated water.

Daily distillate yield of multi-stage solar still ($M_d$ in kg/m$^2$d) with "$N$" stages under non-continuous mode of operation is given as (Estahbanati et al., 2015)

$$M_d = -0.7169N^2 + 7.2516N + 5.3496 \tag{7.37}$$

In **continuous mode** of operation, the evaporated water is compensated by adding make-up feed water at the end of 24 h.

Daily distillate yield of multi-stage solar still ($M_d$ in kg/m$^2$d) with "$N$" stages under continuous mode operation is given as (Estahbanati et al., 2015)

$$M_d = -0.2539N^2 + 6.1976N + 6.4149 \quad (7.38)$$

Continuous mode of operation is highly recommended for industrial applications as the operation of the desalination system will be continuous and it also produces more distillate than non-continuous mode operated multi-stage desalination system. Distillate yield of first, second, third, and fourth stages of the multi-stage desalination system operating under continuous mode is about 4.0%, 5.0%, 9.0%, and 16.0%, respectively, higher than the yield of the corresponding stages of the system operating under non-continuous mode (Estahbanati et al., 2015).

### 7.4.3 Gap Between Evaporating and Condensing Surfaces

Distance between the evaporating water surface/wick and the adjacent vapor condensing surface is called the gap between the stages/effects. Schematic representing the estimation of the gap between the stages/effects of different shapes is shown in Fig. 7.7. As gap increases, air trapped between the stages/effects also increases thereby, resistance for movement of vapors from the evaporating to the condensing surface increases and the yield drops considerably. Increasing gap from 0.26 m to 0.36 m has reduced the distillate yield of evacuated three-stage desalination system by approximately 93.3% (Ahmed et al., 2009). Reduction in gap generally, increases distillate yield. However, reduction in gap beyond 0.10 m is not recommended for practical purposes in horizontal/tilted multi-stage/effect desalination systems due to the difficulties observed during maintenance activities (Reddy et al., 2012). In addition, gap reduction also enhances the chances of distillate contamination with the incoming saline feed water in horizontal/tilted multi-stage/effect desalination systems (Reddy & Sharon, 2016).

### 7.4.4 Stage/Effect Construction Materials

Materials used for stage/effect construction have a huge impact on the desalination system's performance. The material used for stage/effect construction must have good thermal conductivity to facilitate heat transfer to other stages. In addition, it must also have suitable surface roughness to drain the formed condensate in an efficient way. Moreover, the material must be less expensive, less weight, and possess high mechanical resistance to handle basin water weight and temperature without undergoing deformation. The surface roughness of the stage/effect cannot be too high as it causes formation of bigger condensate drops which may enhance fall back of the condensate over the evaporating surface. On the other hand, too low surface roughness prevents formation and growth of condensate drops.

The impact of tray material and roughness on the performance of multi-stage solar thermal desalination system for a given heat input is tabulated in Table 7.1. Aluminum

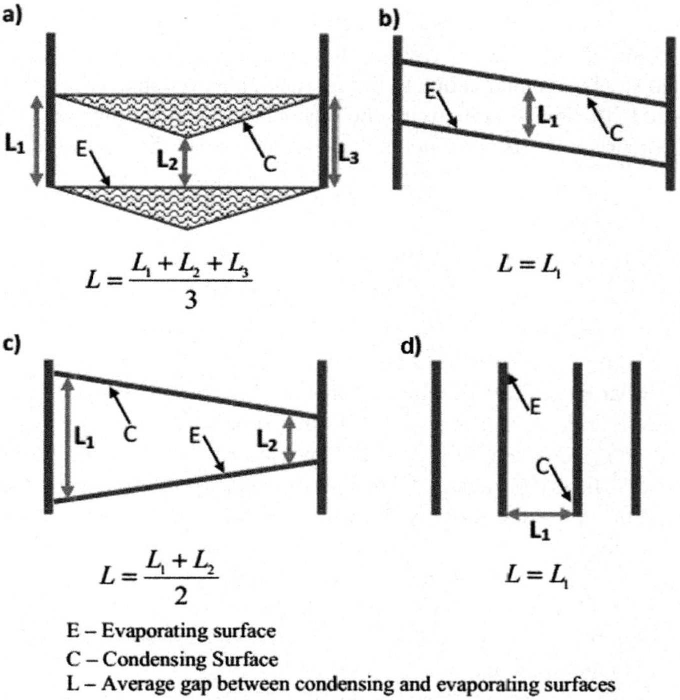


**Fig. 7.7** Estimation of gap between the evaporating and condensing surface of stages/effects of various desalination systems

**Table 7.1** Performance of desalination system with stages made of different materials (Vieira da Silva et al., 2015)

| Stage material | Thickness (mm) | Mean surface roughness (μm) | Water evaporated per kilo Joule of energy supplied (mL/kJ) | Difference between evaporated and collected condensate (%) |
|---|---|---|---|---|
| Aluminum | 1.20 | 0.189 | 0.402 | 0.57 |
| Polished steel | 0.50 | 0.073 | 0.304 | 9.15 |
| Stainless steel | 1.20 | 0.092 | 0.336 | 2.66 |
| Chromed steel | 1.20 | 0.010 | 0.300 | 1.76 |

stage produced the highest condensate and chromed steel stage produced the lowest condensate per kilo Joule of energy supplied. Moreover, loss in condensate is higher for polished steel stage and is due to the fall back phenomena. This observation can be attributed to the high conductivity and reasonable surface roughness of aluminum stage in comparison to the chromed steel stage (Vieira da Silva et al., 2015).

### *7.4.5 Shape of Stage/Effect*

The optimum tilt angle/slope recommended for the stage/tray surfaces employed in multi-effect/stage desalination system is about 12° to prevent condensate fall back (Vieira da Silva et al., 2015). The schematic of shapes of various stages used in multi-stage/effect desalination systems is shown in Fig. 7.8. Stages generally used in desalination systems are "V-" shaped, "inverted V-" shaped, and corrugated type. In "V-shaped", "inverted V-shaped", and corrugated-type stages, feed water is stored as bulk. In tray-type stages/effects, feed water is allowed to flow as a thin film and is distributed uniformly with the aid of wicks lined over the trays. Distillate yield of systems with "V-shaped" stages is higher than the yield of stages with "inverted V-shaped" stages. In "inverted V-shaped" stages, most of the feed water is directly in contact with the system's wall thereby, the heat energy of the feed water is lost to the ambient via conduction.

In "V-shaped" stages, bulk feed water is not in contact with the walls directly thereby, undesired heat loss to the ambient is prevented/limited thus evaporation rate is enhanced. Moreover, in "inverted V-shaped" stages, condensate has to be collected at two locations thereby, potential condensate loss will be experienced during condensate collection. Moreover, the cost of construction of "V-shaped" stage is less expensive than the construction cost of "inverted V-shaped" stage. In addition, the desalination system with "V-shaped" stages requires less solar collector area for producing the required quantity of distillate in comparison to the desalination system with "inverted V-shaped" stages (Abdessemed et al., 2019). Corrugated stages have benefits like reduced condensation resistance of the water vapor, reduced condensate loss during collection, and increased condensing surface area (Xiong et al., 2013). Few drawbacks of corrugated stages are high fabrication cost and need for a number of condensate collection troughs. Effects generally used in desalination systems are normally flat trays (Fig. 7.8c) which can be stacked one above the above or arranged in a zigzag manner (Fig. 7.8d).

### *7.4.6 Surface Area of Evaporator*

In general, cumulative distillate yield of solar still increases with increase in surface area of evaporation. However, it is necessary to identify the optimum evaporation area based on the maximum yield per $m^2$ of evaporation area to operate the sys-

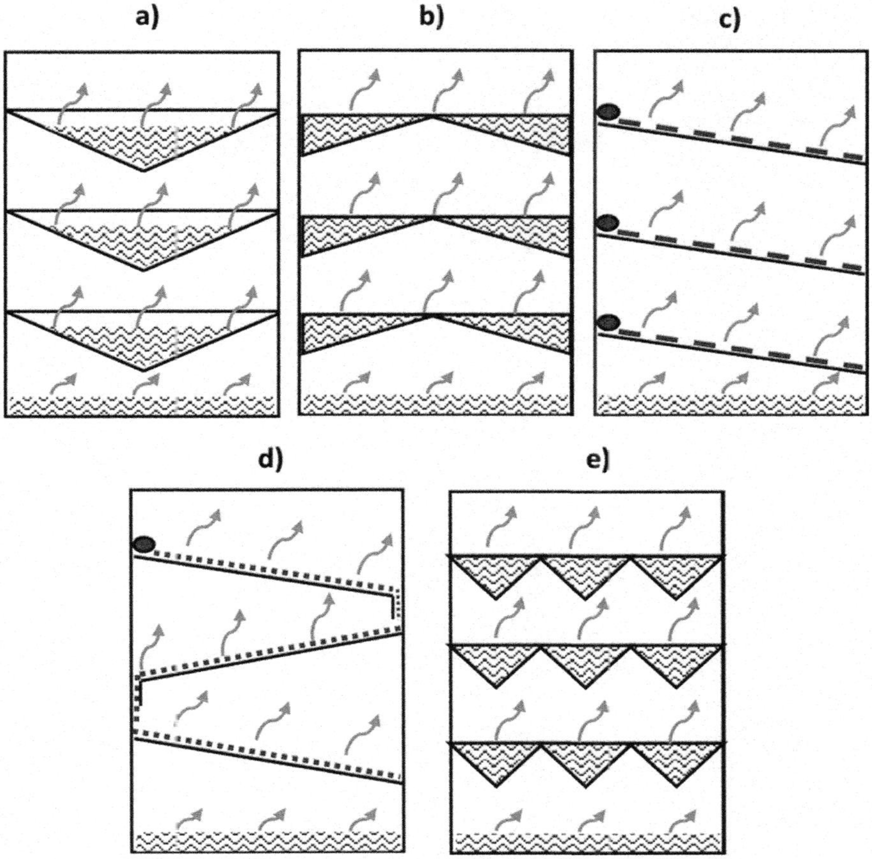


**Fig. 7.8** Shapes of effect/stage **a** V-shaped (Bait and Si-Ameur, 2016) **b** Inverted V-shaped **c** parallel flat tray type (Diaf et al., 2016) **d** zigzag flat tray type (Reddy and Sharon, 2017) **e** corrugated (Chen et al., 2017)

tem economically. Variation of yield per $m^2$ of basin area and efficiency of single-, double-, and triple-stage solar stills with basin area is shown in Fig. 7.9a and b, respectively. Distillate yield per $m^2$ of basin area drops with increase in basin area and it can be attributed to the increased heat losses to the ambient with increase in basin area and this comment can also be justified from the efficiency trend shown in Fig. 7.9b. Hence, stage/effect area of 1.0 $m^2$ is optimum for enhanced performance of multi-stage/effect desalination systems.

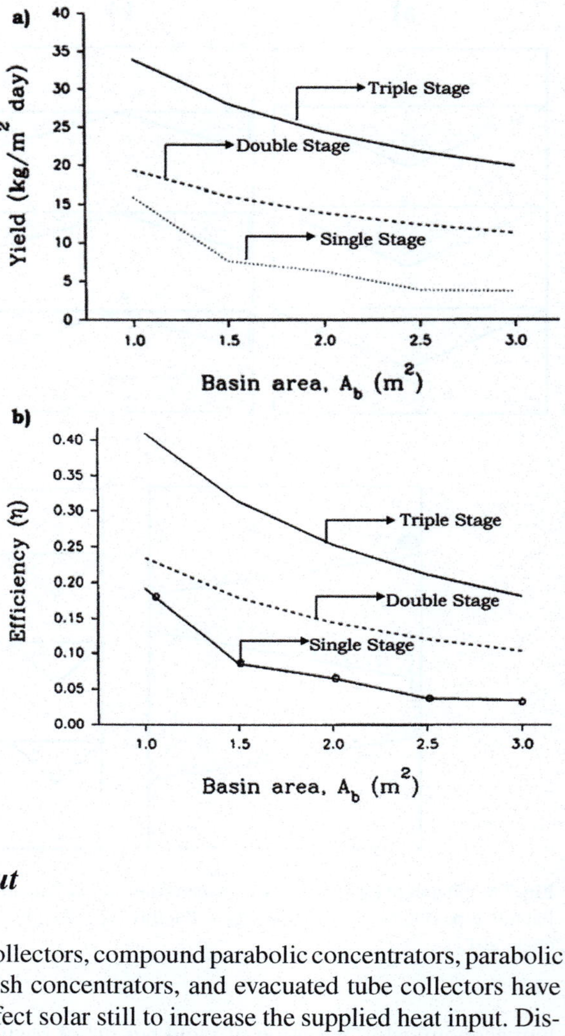


**Fig. 7.9** Impact of basin area on **a** distillate yield and **b** efficiency of single-, double-, and triple-stage active solar stills (Kumar and Tiwari, 1999)

### *7.4.7 Heat Energy Input*

Solar collectors like flat plate collectors, compound parabolic concentrators, parabolic trough collectors, parabolic dish concentrators, and evacuated tube collectors have been utilized in multi-stage/effect solar still to increase the supplied heat input. Distillate production of double effect still has enhanced by about 240% by increasing the heat energy input from 200 to 500 $W/m^2$ (Kalbasi et al., 2018). Photograph of multi-stage solar still integrated with evacuated heat pipe collector and parabolic trough collector is shown in Fig. 7.10a and b, respectively. Flat plate collectors are widely preferred for multi-stage/effect solar still because of their simplicity, low cost, low maintenance requirement, and robustness (Fernandez and Chargoy, 1990). Increasing the number of flat plate collectors integrated with single-, double-, and triple-stage solar stills increases the cumulative distillate yield/$m^2$ of basin area. The optimum number of solar collectors for single- (10 nos.) and double-stage stills (6 nos.) is higher than the optimum number of solar collectors for triple-stage still (4 nos.). Efficiency of double- and triple-stage solar stills generally drops with the addi-

**Fig. 7.10** Photograph of **a** Triple-stage solar still integrated with evacuated heat pipe collector (Xiong et al., 2013) **b** Four-stage solar still integrated with parabolic trough collector (Abdessemed et al., 2019)

tion of solar collectors. However, for single-stage still, efficiency increased with the addition of collectors up to 3 nos. and further addition of collectors beyond three has shown no significant improvement in the efficiency of the still (Kumar and Tiwari, 1999). Mode of solar collector connection has also a significant impact on the performance of multi-stage/effect solar stills. Outlet temperature of solar collectors in parallel connection will be higher than in series connection; hence, the required number of collectors can be reduced by adopting parallel connection (Reddy et al., 2012). The schematic depicting parallel and series connection of solar flat plate collector is shown in Fig. 7.11.

Optimum number of collectors required for enhanced distillate production in multi-stage solar thermal desalination system is about 5 and 3 in series and parallel configuration, respectively. Beyond this optimum number, heat energy loss to the

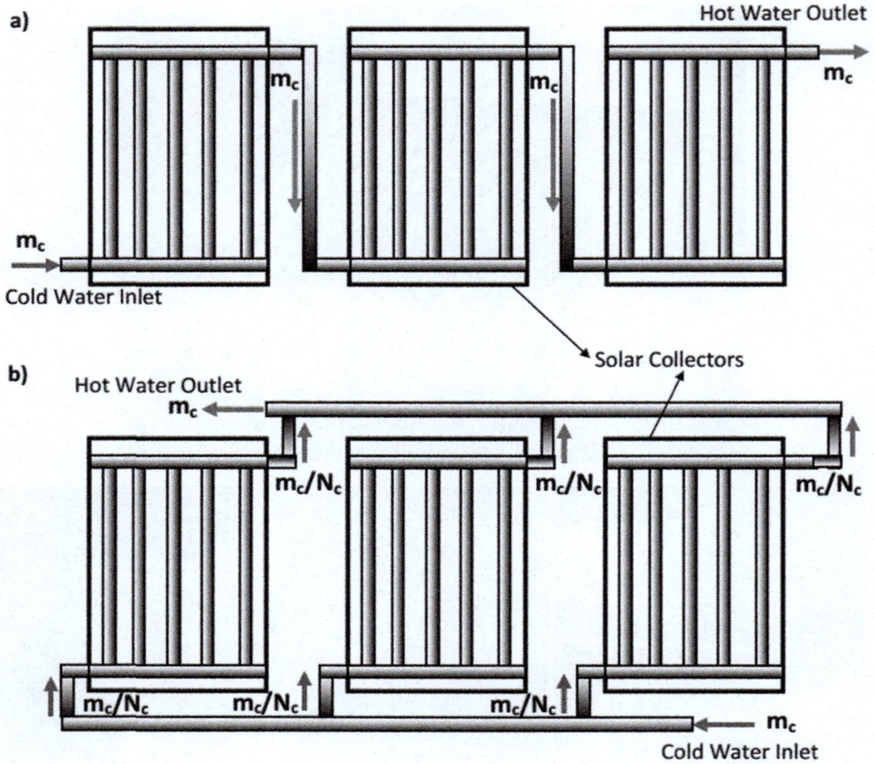


**Fig. 7.11** Flat plate solar collector connection in **a** Series and **b** Parallel mode

ambient increases with increase in collector area which in turn causes a negligible rise in distillate productivity (Reddy and Sharon, 2017). Distillate productivity enhancement by about 93.0% and 48.0% is observed during winter and summer season, respectively, by increasing the collector-to-basin area ratio from 3.45 to 6.90. Thermal efficiency of multi-stage still drops from 70.0% to 55.0% and 60.0% to 50.0% during summer and winter, respectively with increase in collector to basin area ratio from 3.45 to 6.90 which reflects high heat losses from the desalination system. Another inference is that, if the evaporating surface area remains constant the rate of evaporation cannot be doubled by just doubling the solar collector area alone (Feilizadeh et al., 2015). Adhikari and Kumar (1993) recommended an optimum ratio of solar collector area to basin area to be around 4–6 due to negligible enhancement in yield beyond this value. It is also recommended to disconnect the solar collector from the multi-stage/effect solar desalination system during non-sunshine hours to prevent heat loss from the basin water through the collector. Solar collectors operating in forced circulation mode are better than thermo siphon-operated collectors. However, for less populated, remote regions lacking electricity and skilled labors thermo

**Table 7.2** Performance of various non-evacuated active multi-stage solar stills

| S.no | Type of non-evacuated active multi-stage solar distillation unit | Type of solar collector | Distillate yield (kg/d) |
|---|---|---|---|
| 1. | Double-effect solar still (Prasad and Tiwari, 1996) | Compound parabolic collector—1.0 $m^2$; Concentration ratio—16 | 14.68 |
| 2. | Triple-basin solar still (AlBaharna et al., 1993) | Flat plate collector—1.5 $m^2$ | 24.48 |
| 3. | Triple-effect solar still with enhanced condensation surface (Xiong et al., 2013) | Heat pipe vacuum tube collector—3.76 $m^2$ | 43.00 |
| 4. | Four-stage solar distillation unit (Shatat and Mahkamov, 2010) | Evacuated tube solar collector—1.7 $m^2$ | 11.00 |
| 5. | Six-stage solar distillation unit (Schwarzer et al., 2001) | Flat plate collector—2.0 $m^2$ | 50.00 |
| 6. | Seven-stage solar distillation unit (Schwarzer et al., 2009) | Vacuum tube collector—2.2 $m^2$ | 37.00 |

siphon mode operation is highly recommended (Rajaseenivasan et al., 2013). Among concentrators, compound parabolic concentrator is highly recommended due to its requirement of minimal seasonal tracking which can be carried out even manually (Prasad and Tiwari, 1996). Summary of few non-evacuated active multi-stage solar still's performance is tabulated in Table 7.2.

### *7.4.8 Depth of Water*

Depth of water in each stage is an important parameter impacting the performance of multi-stage desalination system. Efficiency of active double-stage still increases with basin water depth, while the efficiency drops with increase in water depth for passive single stage still (Tiwari and Lawrence, 1992). Optimum water depth of the bottom-most stage is fixed based on the criteria of maximum yield and complete submersion of heat exchanger coil in the basin water even at the end of distillation process. At very low water depth, the heat exchanger coil will not be completely submerged in basin water thereby, the supplied heat energy is not effectively utilized for carrying out the evaporation process. At very high water depth, the heat capacity of basin water is increased thereby, for the fixed heat energy supplied, the temperature rise is not sufficient to cause enhanced evaporation (Shatat and Mahkamov, 2010). Increasing water depth from 1 to 3 cm declines the yield of the double effect system by about 26% (Kalbasi et al., 2018).

### 7.4.9 Feed Water Flow Rate

In the case of multi-effect desalination systems, the yield is found to increase with the reduction in feed water flow rate from 150.0 kg/m$^2$d to 55.0 kg/ m$^2$d. This observation is due to the reduction in heat capacity and fluid film thickness at low flow rates. Further, drop below 55.0 kg/m$^2$d has resulted in low yield and it can be attributed to the reduced availability of feed water for the evaporation process (Reddy et al., 2012). Optimum feed water flow rate is also season-dependent. For summer season and winter season, the optimum daily flow rate is about 135.0 kg/d and 75.0 kg/d, respectively, for multi-stage series flow desalination system. Feed flow rate beyond this optimum value reduces the operating temperature of the system, while feed flow rate below optimum value results in reduced temperature difference between the stages, salt/slit deposition, and release of non-condensable gases (Reddy and Sharon, 2017). The optimum length of the jute wick lined effect is about 2.5 m for a fixed feed water flow rate of 1.0 kg/m$^2$h. Further, increase in length of the effect for this fixed flow rate causes dry out of the lined jute wick. Hence, optimum flow rate is also dependent on the length of the effect (Tiwari et al., 1989).

### 7.4.10 Feed Water Salinity

Salinity of feed water has a significant impact on the performance of the multi-stage/effect desalination system. Increased salinity of the feed water indicates increased boiling point elevation and surface tension thereby, the vapor pressure at the evaporating surface decreases. The decreased vapor pressure results in reduced partial pressure difference between the evaporating and condensing surface leading to decreased yield. Distillate yield of four-effect desalination system is found to drop by about 40.0% with the increase in salt concentration from 0.0 wt% to 5.0 wt% (Reddy et al., 2012). Distillate yield of non-evacuated air-cooled and wetted wick-cooled five-stage series flow desalination system is found to drop by about 13.0% and 18.0%, respectively, for every 5.0 wt% increase in salt concentration. Similarly, for evacuated air-cooled and wetted wick-cooled systems the drop in distillate yield is about 8.80% and 9.61%, respectively, for every 5.0 wt% increase in salt concentration. Annual average distillate yield and performance ratio of the wetted wick-cooled evacuated five-stage series flow desalination system desalting saline water with 5.0 wt% salinity is around 44.07 kg/d and 11.78, respectively (Reddy and Sharon, 2017).

### 7.4.11 Operating Pressure

Evaporation process within the desalination system can be enhanced by reducing the operating pressure. As operating pressure drops, the saturation temperature also drops thereby, the preheated feed water in each of the effects/stages which is relatively at a temperature higher than the saturation temperature vaporizes with ease thereby, more distillate is obtained. However, there is always an optimum operating pressure for each system to achieve maximum yield. Below the optimum operating pressure, the feed water evaporates abruptly in all the stages and these vapors cannot condense immediately due to high temperature of the top stages/effects in comparison to the lower stages/effects thereby, leads to poor yield. For four-effect desalination system, this scenario has been observed for an operating pressure below 0.02 bar. The distillate yield of the system treating saline water increased from 19.35 kg/m$^2$d to 40.26 kg/m$^2$d with the reduction in operating pressure from atmospheric pressure to 0.02 bar (Reddy et al., 2012).

The operating pressure of desalination system can be lowered with the help of vacuum pump. Batch mode evacuation is preferred over continuous mode evacuation process. Sometimes, evacuation will be carried out for few minutes before the start of distillation process and, after that, no further evacuation will be carried out. Operating pressure can be decreased from the bottom to the top stages using suitable valve/orifice controls to have better yield (Jubran et al., 2000). Cylindrical-type stages are highly recommended for desalination systems under low-pressure operation rather than rectangular stages. At an operating pressure of 0.30 bar, the rectangular stage of 2.3 mm thickness fails due to high stresses, while the deflection and stresses in cylindrical stage with wall stiffeners are within the safer limit making it a suitable configuration for low-pressure operation. Photograph of cylindrical triple-stage solar thermal desalination system is shown in Fig. 7.12, and its distillate yield at an operating pressure of 0.3 bar is around 17.0 kg/m$^2$d. Operating pressure of about 0.20 bar is highly recommended for desalination system operating around 60–90 °C due to minimal power consumption and high specific yield (Reddy & Sharon, 2016). About 16.0% enhancement in yield is observed for five-stage series flow desalination system for every 0.20 bar drop in operating pressure (Reddy and Sharon, 2017). The relationship between the operating pressure (P in bar) and distillate yield (Md in kg/m$^2$d) of cylindrical triple stage desalination system is given as (Ahmed et al., 2009)

$$M_d = -6.9189 \ln(P) + 10.018 \tag{7.39}$$

Performance summary of few evacuated active multi-stage/effect solar thermal desalination systems is tabulated in Table 7.3.

**Fig. 7.12** Photograph of evacuated cylindrical triple-stage desalination system (Ahmed et al., 2009)

### *7.4.12 Surface Temperature of Last Condensing Stage/Effect*

In multi-stage desalination systems, the temperature difference between the bottom stage and the last condensing surface acts as the driving force for the occurrence of sustained evaporation. The temperature of the last condensing surface can be reduced by flowing water over it (Bapeshwararao et al., 1983) or by lining its outer surface with wetted wick to facilitate evaporative cooling (Reddy and Sharon, 2017). The lowest temperature close to the wet bulb temperature can be achieved by evaporative cooling. The maximum temperature difference between the bottom stage and last condensing surface of wetted wick-cooled five-stage series flow desalination system is about 33.5 °C and 13.6 °C, during summer and winter season respectively. Distillate yield of wetted wick-cooled desalination system is about 28.6 kg/d which is around 36.6% higher than the yield obtained by the air-cooled system. Evaporative cooling is a cheap and effective option for enhancing the yield of desalination systems in hot and low-humid sites (Reddy and Sharon, 2017).

**Table 7.3** Summary of few evacuated active multi-stage solar thermal desalination systems performance

| S.No | Type of evacuated multi-stage solar distillation unit | Type of solar collector | Evacuation mode/Vacuum pressure (bar) | Distillate yield (kg/d) |
|---|---|---|---|---|
| 1. | Triple-effect evacuated solar still (Nishikawa et al., 1998) | Thermo-siphon solar collector—7.8 $m^2$ | Evacuation at regular intervals | 100.00 |
| 2. | Two-stage evacuated flash desalination unit (Maroo and Yogi Goswami, 2009) | Flat plate collector—1.0 $m^2$ | Evacuation using natural forces | 8.66 |
| 3. | Triple-basin evacuated solar still (Abakar and Ismail, 2005) | Flat plate collector—NA | 0.50 | 14.20 |
| | | | 0.30 | 17.00 |
| 4. | Evacuated four-stage tray type solar still Ahmed et al. (2009) | Flat plate collector—2.7 $m^2$ | 0.02 | 53.20 |
| 5. | Evacuated five-stage series flow solar still (Reddy and Sharon, 2017) | Flat plate collector—4.0 $m^2$ | 0.20 | 21.00 to 33.00 |

NA—not available

## 7.5 Features of Multi-effect/Stage Solar Thermal Desalination Systems

(a) Performance of multi-effect/stage desalination systems is far better than single-effect/stage desalination systems due to the effective re-utilization of latent heat of condensation.
(b) Distillate yield is about 5–6 times higher than conventional basin solar stills.
(c) Evacuated multi-effect desalination systems are more reliable and compact.
(d) Suitable for meeting the drinking and cooking water demands of a single family in remote and rural regions.
(e) Relatively high initial and operation cost than single-effect/stage desalination systems.

## References


Abakar, Y. A., & Ismail, A. F. (2005). Theoretical and experimental investigation of a novel multi-stage evacuated solar still. *Journal of Solar Energy Engineering, 127*, 381–385.

Abdessemed, A., Bougriou, C., Guerraiche, D., & Abachi, R. (2019). Effects of tray shape of a multi-stage solar still coupled to a parabolic concentrating solar collector in Algeria. *Renewable Energy, 132*, 1134–1140.

Adhikari, R. S., Kumar, A., & Kumar, A. (1990). Estimation of mass transfer rates in solar stills. *International Journal of Energy Research, 14*, 737–744.

Adhikari, R. S., & Kumar, A. (1993). Transient simulation studies on a multi-stage stacked tray solar still. *Desalination, 91*, 1–20.

Adhikari, R. S., Kumar, A., & Sootha, G. D. (1995). Simulation studies on a multi-stage stacked tray solar still. *Solar Energy, 54*(5), 317–325.

Ahmed, M. I., Hrairi, M., & Ismail, A. F. (2009). On the characteristics of multistage evacuated solar distillation. *Renewable Energy, 34*, 1471–1478.

Al Baharna, N. S., Al Mahdi, N., & Zaky, F. (1993). Performance analysis of a triple basin still integrated with a natural circulation solar heater. *Energy Convers Manage, 34*(7), 545–56.

Bait, O., & Si-Ameur, M. (2016). Numerical investigation of a multi-stage solar still under Batna climatic conditions: Effect of radiation term on mass and heat energy balances. *Energy, 98*, 308–323.

Bapeshwararao, V. S. V., Singh, U., & Tiwari, G. N. (1983). Transient analysis of double basin solar still. *Energy Conversion and Management, 23*(2), 83–90.

Chen, Z., Peng, J., Chen, G., Hou, L., Yu, T., Yao, Y., & Zheng, H. (2017). Analysis of heat and mass transferring mechanism of multi-stage stacked-tray solar seawater desalination still and experimental research on its performance. *Solar Energy, 142*, 278–287.

Diaf, A., Cherfa, A., Karadaniz, L., & Tigrine, Z. (2016). A technical - economical study of solar desalination. *Desalination, 377*, 123–127.

Estahbanati, M. R. K., Feilizadeh, M., Jafarpur, K., Feilizadeh, M., & Rahimpour, M. R. (2015). Experimental investigation of a multi-effect active solar still: The effect of the number of stages. *Applied Energy, 137*, 46–55.

Feilizadeh, M., Estahbanati, M. R. K., Jafarpur, K., Roostaazad, R., Feilizadeh, M., & Taghvaei, H. (2015). Year-round outdoor experiments on a multi stage active solar still with different numbers of solar collectors. *Applied Energy, 152*, 39–46.

Fernandez, J. L., & Chargoy, N. (1990). Multi-stage, indirectly heated solar still. *Solar Energy, 44*(4), 215–223.

Goosen, M. F. A., Sablani, S. S., Shayya, W. H., Paton, C., & Al-Hinai, H. (2000). Thermodynamic and economic considerations in solar desalination. *Desalination, 129*, 63–89.

Jubran, B. A., Ahmed, M. I., Ismail, A. F., & Abakar, Y. A. (2000). Numerical modelling of a multi-stage solar still. *Energy Conversion and Management, 41*, 1107–1121.

Kalbasi, R., Alemrajabi, A. A., & Afrand, M. (2018). Thermal modeling and analysis of single and double effect solar stills: An experimental validation. *Applied Thermal Engineering, 129*, 1455–1465.

Kumar, S., & Tiwari, G. N. (1999). Triple basin active solar still. *International Journal of Energy Research, 23*, 529–542.

Li, Shuang-Fei., Liu, Zhen-Hua., Shao, Zhi-Xiong., Xiao, Hong-shen, & Xia, N. (2018). Performance study on a passive solar seawater desalination system using multi-effect heat recovery. *Applied Energy, 213*, 343–352.

Liu, Zhem-hua, Hu, Ren-Lin., & Chen, Xiu-juan. (2014). A novel integrated solar desalination system with multi-stage evaporation/heat recovery processes. *Renewable Energy, 64*, 26–33.

Maroo, S. C., & Yogi Goswami, D. (2009). Theoretical analysis of a single-stage and two stage solar driven flash desalination based on passive vacuum generation. *Desalination, 249*, 635–46.

Nishikawa, H., Tsuchiya, T., Narasaki, Y., Kamiya, I., & Sato, H. (1998). Triple effect evacuated solar still system for getting fresh water from seawater. *Applied Thermal Engineering, 18*, 1067–75.

Prasad, B., & Tiwari, G. N. (1996). Analysis of double effect active solar distillation. *Energy Conversion and Management, 37*(11), 1647–1656.

Rajaseenivasan, T., Murugavel, K. K., Elango, T., & Hansen, R. S. (2013). A review of different methods to enhance the productivity of the multi-effect solar still. *Renewable and Sustainable Energy Reviews, 17*, 248–259.

Reddy, K. S., Kumar, K. R., O'Donovan, T. S., & Mallick, T. K. (2012). Performance analysis of an evacuated multi-stage solar water desalination system. *Desalination, 288*, 80–92.

Reddy, K. S., & Sharon, H. (2016). Active multi-effect vertical solar still: Mathematical modeling, performance investigation and enviro-economic analyses. *Desalination, 395*, 99–120.

Reddy, K. S., & Sharon, H. (2017). Energy-environment-economic investigations on evacuated active multiple stage series flow solar distillation unit for potable water production. *Energy Conversion and Management, 151*, 259–285.

Schwarzer, K., Eugenia Vieira, M., Faber, C., & Muller, C. (2001). Solar thermal desalination system with heat recovery. *Desalination, 137*, 23–29.

Schwarzer, K., Vieira da Silva, E., Hoffschmidt, B., & Schwarzer, T. (2009). A new solar desalination system with heat recovery for decentralized drinking water production. *Desalination, 248*, 204–211.

Sharon, H., & Reddy, K. S. (2015). A review on solar energy driven desalination technologies. *Renewable and Sustainable Energy Reviews, 41*, 1080–1118.

Shatat, M. I. M., & Mahkamov, K. (2010). Determination of rational de sign parameters of a multi-stage solar water desalination still using transient mathematical modelling. *Renewable Energy, 35*, 52–61.

Tiwari, G. N., Lawrence, S. A., & Gupta, S. P. (1989). Analytical study of multi-effect solar still. *Energy Conversion and Management, 29*(4), 259–263.

Tiwari, G. N., & Lawrence, S. A. (1992). Thermal Evaluation of High Temperature Distillation under an Active Mode of Operation. *Desalination, 85*, 135–145.

Tiwari, G. N., Singh, A. K., Saxena, P., & Rai, S. N. (1992). The performance of multi-effect active distillation system. *International Journal of Solar Energy, 13*(4), 277–287.

Vieira da Silva, M. E., Schwarzer, K., Pinheiro, F. N., Rocha, P. A. C., & Freitas de Andrade, C. (2015). Experimental study of tray materials in a thermal desalination tower with controlled heat source. *Desalination, 374*, 38–46.

Xiong, J., Xie, G., & Zheng, H. (2013). Experimental and numerical study on a new multi-effect solar still with enhanced condensation surface. *Energy Conversion and Management, 73*, 176–185.

# Chapter 8
# Multi-effect Tubular Solar Thermal Desalination Systems

## 8.1 Description and Working

Tubular solar stills are compact, easy to fabricate, and have reasonably good distillate productivity. The performance and yield of the tubular solar still can be enhanced by recovering and reusing its latent heat of condensation by incorporating multiple effects. The schematic of two-effect tubular solar desalination system is shown in Fig. 8.1. The main components of the system are two semi-circular troughs and two tubular condensing covers. Semi-circular trough is inserted through a tubular condensing cover and then the two open ends of the tubular condensing cover are tightly closed. This arrangement is called an effect. The first effect thus fabricated is then placed inside another larger semi-circular trough which is then inserted through another large tubular condensing cover and the two ends are closed tightly.

Provisions are made on the system for supplying feed water to the troughs and collecting condensate from the tubular condensing cover which can be seen in Fig. 8.1. Feed water supply to each effect of the multi-effect tubular solar desalination system is achieved by cascaded feed water flow arrangement. The feed water in the first effect is heated by a heat exchanger coil circulating high-temperature fluid from heater/solar collectors to initiate evaporation process. The vapors generated from the evaporating water surface of the first effect condense over the inner surface of the first effect's tubular condensing cover by releasing its latent heat. The released latent heat is utilized by the feed water of the second effect's trough to cause evaporation. The same process continues in all other available effects. The latent heat released over the last effect's condensing cover is dissipated to the ambient by convection and radiation heat transfer.



H. Sharon and K. Srinivas Reddy, *Solar Thermal Desalination Technologies for Potable Water*, https://doi.org/10.1007/978-3-031-49155-9_8

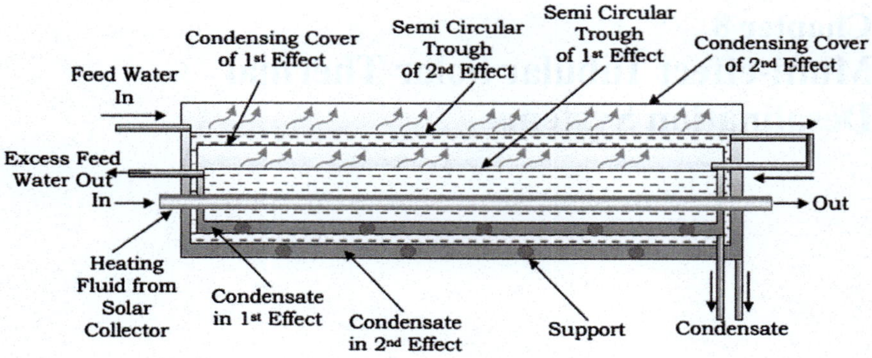


**Fig. 8.1** Schematic of multi-effect tubular solar desalination system (Zheng et al., 2013a)

## 8.2 Thermodynamic Modeling of Multi-effect Tubular Solar Thermal Desalination System

Assumptions considered for thermodynamic modeling of multi-effect tubular solar thermal desalination system are listed below (Yan et al., 2019):

(a) Condensation rate in each effect is equivalent to the evaporation rate in each effect.
(b) Heat lost through the end plates of each effect is nearly negligible (only 3.0% of the total heat input).
(c) No thermal gradients across the thickness of water, trough, and cover.

The energy transport process occurring in a triple-effect tubular solar thermal desalination system is shown in Fig. 8.2.

Energy balance in the first effect is given as

$$\begin{pmatrix}\text{rate of change of}\\ \text{first temperature}\end{pmatrix} = \begin{pmatrix}\text{heat input to}\\ \text{first effect}\end{pmatrix} - \begin{pmatrix}\text{radiation heat transfer}\\ \text{from first to second effect}\end{pmatrix} - \begin{pmatrix}\text{evaporation heat transfer}\\ \text{from first to second effect}\end{pmatrix} - \begin{pmatrix}\text{convection heat transfer}\\ \text{from first to second effect}\end{pmatrix}$$

$$(\rho c V)_1 \frac{dT_1}{dt} = Q_{in} - Q_{r(1-2)} - Q_{e(1-2)} - Q_{c(1-2)} \tag{8.1}$$

Energy balance in the second effect is given as

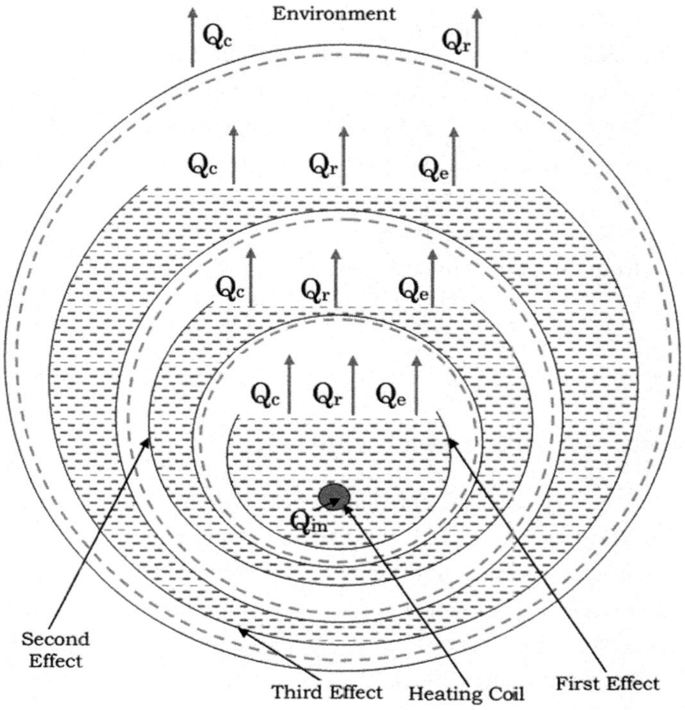


**Fig. 8.2** Energy transport process in a triple-effect tubular solar desalination system (Chen et al., 2013)

$$\begin{pmatrix}\text{rate of change of second}\\ \text{effect temperature}\end{pmatrix} = \begin{pmatrix}\text{radiation heat transfer}\\ \text{from first to second effect}\end{pmatrix} + \begin{pmatrix}\text{evaporation heat transfer}\\ \text{from first to second effect}\end{pmatrix} + \begin{pmatrix}\text{convection heat transfer}\\ \text{from first to second effect}\end{pmatrix} - \begin{pmatrix}\text{radiation heat transfer}\\ \text{from second to third effect}\end{pmatrix} - \begin{pmatrix}\text{evaporation heat transfer}\\ \text{from second to third effect}\end{pmatrix} - \begin{pmatrix}\text{convection heat transfer}\\ \text{from second to third effect}\end{pmatrix}$$

$$(\rho cV)_2\frac{dT_2}{dt} = Q_{r(1-2)} + Q_{e(1-2)} + Q_{c(1-2)} - Q_{r(2-3)} - Q_{e(2-3)} - Q_{c(2-3)} \quad (8.2)$$

Energy balance in the third effect is given as

$$\begin{pmatrix}\text{rate of change of third}\\ \text{effect temperature}\end{pmatrix} = \begin{pmatrix}\text{radiation heat transfer}\\ \text{from second to third effect}\end{pmatrix}$$
$$+\begin{pmatrix}\text{evaporation heat transfer}\\ \text{from second to third effect}\end{pmatrix} + \begin{pmatrix}\text{convection heat transfer}\\ \text{from second to third effect}\end{pmatrix}$$
$$-\begin{pmatrix}\text{radiation heat transfer from}\\ \text{third effect to outer cover}\end{pmatrix} - \begin{pmatrix}\text{evaporation heat transfer from}\\ \text{third effect to outer cover}\end{pmatrix}$$
$$-\begin{pmatrix}\text{convection heat transfer}\\ \text{from third effect to outer cover}\end{pmatrix}$$

$$(\rho c V)_3 \frac{dT_3}{dt} = Q_{r(2-3)} + Q_{e(2-3)} + Q_{c(2-3)} - Q_{r(3-c)} - Q_{e(3-c)} - Q_{c(3-c)} \quad (8.3)$$

Energy balance in the outer condensing cover is given as

$$\begin{pmatrix}\text{rate of change of outer}\\ \text{cover temperature}\end{pmatrix} = \begin{pmatrix}\text{radiation heat transfer from}\\ \text{third effect to outer cover}\end{pmatrix}$$
$$+\begin{pmatrix}\text{evaporation heat transfer}\\ \text{from third effect to outer cover}\end{pmatrix} + \begin{pmatrix}\text{convection heat transfer from}\\ \text{third effect to outer cover}\end{pmatrix}$$
$$-\begin{pmatrix}\text{radiation heat transfer from}\\ \text{outer cover to the ambient}\end{pmatrix} - \begin{pmatrix}\text{convection heat transfer from}\\ \text{outer cover to the ambient}\end{pmatrix}$$

$$(\rho c V)_c \frac{dT_c}{dt} = Q_{r(3-c)} + Q_{e(3-c)} + Q_{c(3-c)} - Q_{r(c-a)} - Q_{c(c-a)} \quad (8.4)$$

The distillate production rate ($\dot{m}_e$ in kg/s) in each effect is computed by (Zheng et al., 2013a, b)

$$\dot{m}_e = h_w A_w (\rho_e - \rho_c) \quad (8.5)$$

The density of water vapor at the evaporating surface ($\rho_e$ in kg/m$^3$) and at the condensing surface ($\rho_c$ in kg/m$^3$) is estimated by

$$\rho_e = \frac{P_e M_w}{\bar{R} T_e} \quad (8.6)$$

$$\rho_c = \frac{P_c M_w}{\bar{R} T_c} \quad (8.7)$$

where

"$P_e$" and "$P_c$" are saturation water vapor pressure at the evaporating and condensing surface temperature, respectively, in Pa.
"$M_w$" molar mass of water vapor (18.0 kg/kmol).

"$\bar{R}$" Universal gas constant (8314 J/KmolK).
"$T_e$" and "$T_c$" are evaporating and condensing surface temperature (K).

The mass transfer coefficient ($h_w$ in m/s) in each effect of the tubular solar thermal desalination system is computed by

$$h_w = \frac{Sh_m \times D_{vm}}{x} \tag{8.8}$$

"$x$" is the characteristic space between the evaporating and condensing surface of each effect of the distillation chamber in "$m$".

The Sherwood number of water vapor-gas mixture ($Sh_m$) inside the desalination chamber is estimated by (Chen et al., 2013)

$$Sh_m = 0.397(Gr_m \times Sc_m)^{0.25} \tag{8.9}$$

The modified Grashof number of water vapor-gas mixture ($Gr'_m$) inside the desalination chamber is computed as (Zheng et al., 2013b)

$$Gr'_m = \frac{x^3 \rho_m^2 g}{\mu_m^2 (T_c + 273.15)} \left[ (T_e - T_c) + \frac{(P_e - P_c)(T_e + 273.15)}{[(M_a P_T)/(M_a M_w)] - P_e} \right] \tag{8.10}$$

where

"$\rho_m$" Density of vapor-gas mixture (kg/m$^3$).
"$M_a$" molar mass of gas (kg/kmol).
"$P_T$" total or atmospheric pressure (101325 Pa).
"$g$" acceleration due to gravity (9.81 m/s$^2$).

The dynamic viscosity of gas-water vapor mixture ($\mu_m$ in kg/ms) will be in the magnitude of $10^{-7}$ Pa.s. It can also be computed by the following relation (Zheng et al., 2013b)

$$\mu_m = \frac{Y_a \mu_a}{Y_a + Y_w(\sqrt{(M_a/M_w)})} + \frac{Y_w \mu_w}{Y_w + Y_a(\sqrt{(M_w/M_a)})} \tag{8.11}$$

where

"$Y_a$" and "$Y_w$" represent the mole fractions of gas medium and water vapor, respectively.
"$M_a$" and "$M_w$" represent the molar mass of gas and water vapor (kg/kmol), respectively.
"$\mu_a$" and "$\mu_w$" represent the dynamic viscosity of air and water vapor (kg/ms), respectively.

The mass diffusivity of water vapor in various gas media at different temperatures ($D_{vm}$ in cm$^2$/s) can be computed by (Zheng et al., 2013b)

**Table 8.1** Molecular diffusion volume ($\sum v$) and mass diffusivity of water vapor in different gas media (Zheng et al., 2013b)

| Molecular diffusion volume of various gases ($\sum v$) | | | | |
|---|---|---|---|---|
| Water | Oxygen | Carbon di-oxide | Helium | Air |
| 13.1 | 16.3 | 26.9 | 2.67 | 19.7 |
| Temperature (°C) | Mass diffusivity coefficient of water vapor ($\times 10^{-7}$ m$^2$/s) | | | |
| | Oxygen | Carbon di-oxide | Helium | Air |
| 60 | 300.2 | 237.3 | 937.6 | 284.7 |
| 65 | 306.6 | 244.9 | 962.1 | 292.7 |
| 70 | 315.3 | 251.1 | 982.1 | 299.2 |
| 75 | 325.8 | 259.0 | 1011.8 | 306.2 |
| 80 | 330.9 | 265.7 | 1038.2 | 313.4 |
| 85 | 340.8 | 273.2 | 1059.6 | 319.2 |

$$D_{vm} = \frac{0.00143 T_{av}^{1.75}}{P_t \sqrt{2/[(1/M_a) + (1/M_w)]} \left[ (\sum v)_a^{1/3} + (\sum v)_w^{1/3} \right]^2} \tag{8.12}$$

where

"$P_T$" represents the total or atmospheric pressure (bar).
"$T_{av}$" represents the average temperature of evaporating and condensing surface (K).

The molecular diffusion volume ($\sum v$) and mass diffusivity of water vapor in different gas media is listed in Table 8.1.

The mass diffusivity of water vapor in air ($D_{va}$ in m$^2$/s) for different temperatures ($T$ in Kelvin) can be computed by (Zheng et al., 2013b)

$$D_{va} = 0.256 \times 10^{-4} \left( \frac{P_o}{P} \right) \left( \frac{T}{T_o} \right)^{3/2} \tag{8.13}$$

"$P_o$" and "$T_o$" are the standard pressure (101.325 kPa) and the temperature (298 K), respectively.

The Schmidt number of water vapor-gas mixture ($Sc_m$) inside the desalination chamber is computed as

$$Sc_m = \frac{\mu_m}{D_{vm} \rho_m} \tag{8.14}$$

The evaporative heat transfer between the evaporating and condensing surface ($Q_{e(e-c)}$ in W) of each effect in the distillation chamber is estimated by

$$Q_{e(e-c)} = \dot{m}_e h_{fg} \tag{8.15}$$

The convective heat transfer between the evaporating and condensing surface ($Q_{c(e-c)}$ in W) of each effect in the distillation chamber is estimated by

$$Q_{c(e-c)} = h_{c(e-c)} A_w (T_e - T_c) \tag{8.16}$$

The convective heat transfer coefficient between the evaporating and condensing surface ($h_{c(e-c)}$ in W/m$^2$K) is given as (Yan et al., 2019; Gang et al., 2019)

$$h_{c(e-c)} = \frac{h_w \lambda_m}{D_{vm}} \left( \frac{Pr_m}{Sc_m} \right)^{0.25} \tag{8.17}$$

where

"$h_{fg}$" represents the latent heat of evaporation (J/kg).
"$\lambda_m$" represents the thermal conductivity of vapor-air mixture (W/mK).
"$A_w$" represents the surface area of water in the stage (m$^2$).
" $Pr_m$" represents the Prandtl number of vapor-air mixture.

The radiation heat transfer between the evaporating and condensing surface ($Q_{r(e-c)}$ in W) of each effect in the distillation chamber is estimated by (Yan et al., 2019)

$$Q_{r(e-c)} = \frac{\sigma A_w (T_e^4 - T_c^4)}{\frac{1}{\varepsilon_w} + \frac{(1-\varepsilon_c)}{\varepsilon_c} \left( \frac{A_w}{A_c} \right)} \tag{8.18}$$

The radiation heat transfer between the outermost condensing surface ($Q_{r(oc-a)}$ in W) of the distillation chamber and the ambient is estimated by

$$Q_{r(oc-a)} = \sigma \varepsilon_{oc} (T_{oc}^4 - T_a^4) \tag{8.19}$$

The convection heat transfer between the outermost condensing surface ($Q_{c(oc-a)}$ in W) of the distillation chamber and the ambient is estimated by

$$Q_{c(oc-a)} = h_{c(oc-a)} A_{oc} (T_{oc} - T_a) \tag{8.20}$$

where

"$A_{oc}$" is the surface area of outermost condensing surface (m$^2$).
"$T_{oc}$" is the temperature of outermost condensing surface (K).
"$T_a$" is the ambient temperature (K).

"$\varepsilon_w$", "$\varepsilon_c$", and "$\varepsilon_{cc}$" are emissivity of water, condensing surface of interior stage, and outermost condensing cover, respectively.

The convective heat transfer coefficient between the outermost condensing surface and the ambient ($h_{c(oc-a)}$ in W/m$^2$K) is estimated from the Nusselt number ($Nu_D$) relation given for a long horizontal cylinder and the ambient (Chen et al., 2013):

$$Nu_D = \left(0.6 + \frac{0.387(Ra_D)^{1/6}}{\left[1 + (0.559/Pr)^{9/16}\right]^{8/27}}\right)^2 \tag{8.21}$$

The above correlation is valid up to $Ra_D \leq 10^{12}$, where

"$Pr$" is the Prandtl number of air and
"$Ra_D$" is the Rayleigh number estimated using diameter "$D$" of the outermost cover of the tubular stage.

## 8.3 Parameters Influencing the Performance of Multi-effect Tubular Solar Thermal Desalination Systems

### *8.3.1 Number of Effects*

The performance ratio of single-effect, two-effect, and three-effect tubular stills under heat input of 300–400 W to the first effect is about 0.75, 1.40, and 1.70, respectively. Similarly, for a fixed operating temperature of about 80 °C in first effect, distillate yield of two effect still is about 1.71 kg/m$^2$h which is nearly 80.0% higher than the yield of single effect still. Moreover, for the same operating temperature of 80 °C, the yield of three-effect still is about 25.0% higher than the yield of two-effect still. Distillate yield of single-, double-, triple-, and quadruple-effect tubular stills is about 4.51, 7.24, 8.92, and 10.03 kg/m$^2$d, respectively. Yield enhancement of quadruple-effect tubular still is only 9.0% higher than triple-effect still. Hence, triple-effect still can be considered as an optimum configuration for multi-effect tubular solar stills (Alshammari et al., 2021).

### *8.3.2 Operating Pressure*

Tubular multi-effect systems have good pressure-bearing capacity hence they are mostly recommended to be operated under reduced pressure for enhanced performance (Zheng et al., 2013a). Operation of the tubular multi-effect system under negative pressure eliminates non-condensable gases from the system, minimizes convection heat transfer from one effect to the other, and maximizes evaporation heat transfer from one effect to the other which is more beneficial for improved performance. The convection, radiation, and evaporation heat transfer fraction in each effect of the tubular two-effect solar still varies with operating pressure and water temperature. As water temperature increases, convection and radiation energy fraction drops and evaporation energy fraction increase in the two effects. Similarly, as operating pressure is reduced from 101 to 20 kPa for a water temperature of 34.0 °C, the evaporation energy fraction rises from 48.0 to 79.0%. Reduction of operating pressure increases yield and at the saturation pressure, feed water in the troughs starts to boil and gives off a large amount of vapors. Performance ratio increases from 1.7 to 3.8 with increase in the number of effects from 2 to 5 at an operating pressure of 20 kPa (Yan et al., 2019). Xie et al. (2020) suggested 0.40–0.60 kPa to be an optimum operating pressure for three-effect tubular solar still for obtaining enhanced performance. Performance ratio dropped from 1.4 to 0.9 with the reduction in operating pressure from 0.40 to 0.20 kPa and is attributed to the poor yield of the still due to deteriorated outer tubular cover cooling. Moreover, thick outer cover is required to withstand such lower pressure which may also increase system construction cost.

Energy consumption of vacuum pump is nearly similar at all operating temperatures and is dependent only on the required vacuum. Operating temperature of the first effect is higher than the second effect thereby, more vapors are formed in the first effect than in the second effect for a given operating pressure which results in high productivity of the first effect than the second effect (Zheng et al., 2013a). Electric energy consumption of vacuum pump (maintaining 20 kPa vacuum in the still) per kg of distillate production is nearly five times higher at a still operating temperature of 42.0 °C in comparison to the still operating temperature of 62.0 °C. Hence, high water temperature, increased number of effects, and reduced operating pressure are better combinations for achieving high yield and improved performance of multi-effect tubular solar stills (Yan et al., 2019).

### 8.3.3 *Nature of the Gas Media*

The performance of multi-effect tubular solar still can be enhanced by inflating the still with suitable gas media in which the water vapor has high diffusivity. This can be achieved by evacuating the system with the help of vacuum pump and then inflating the system with the chosen gas media. The process of evacuation and inflation needs to be repeated four times such that the system becomes completely occupied with the chosen gas media. The pressure inside the system is maintained a similar to the ambient pressure with the aid of pressure buffer balloon. The mass diffusivity coefficient of water vapor is highest in helium followed by oxygen, air, and carbon dioxide which can be seen in Table 8.1. Moreover, the mass diffusivity values of water vapor in gas media increase with temperature.

The distillate yield of the system with oxygen as gas media (0.58 kg/h at 85 °C) is about 31.8% higher than that of the distillate yield obtained with air as gas media (0.44kg/h at 85 °C). The system with helium as gas media must have given high yield but the highest yield is observed for the system with oxygen as gas media and this can be attributed to the poor ability of helium to mix with water vapor at low operating temperatures. For temperatures above 80.0 °C, the helium molecules turn more active, mix with water vapor, and increase the yield steeply. Type of gas media also affects the condensation phenomena. The vapors evaporated from the water surface push the lighter gas like helium upside and the heavier gas like carbon dioxide downside thereby, vapor condenses more likely on the downside and upside of the tubular still, respectively in these cases. This can be reflected as high upside shell temperature and high downside shell temperature for the system with carbon dioxide and helium as gas media, respectively (Zheng et al., 2013b).

### 8.3.4 Solar Concentrators

Installation of multi-effect tubular solar still directly at the focus of linear concentrators has dual benefit of improved distillate productivity and low specific electric energy consumption per kg of distillate produced due to the elimination of circulation pumps. Moreover, heat-exchanging coils, heat transfer fluids, and storage tanks are also eliminated. Among concentrators, parabolic trough concentrators and Fresnel reflectors are the most suitable options that can be considered for integration with tubular multi-effect solar stills due to their line focusing on incident solar radiation and possibility of sole reflector tracking. The schematic of double effect concentric tubular solar still installed at the focus of manually tracked parabolic trough concentrator is shown in Fig. 8.3.

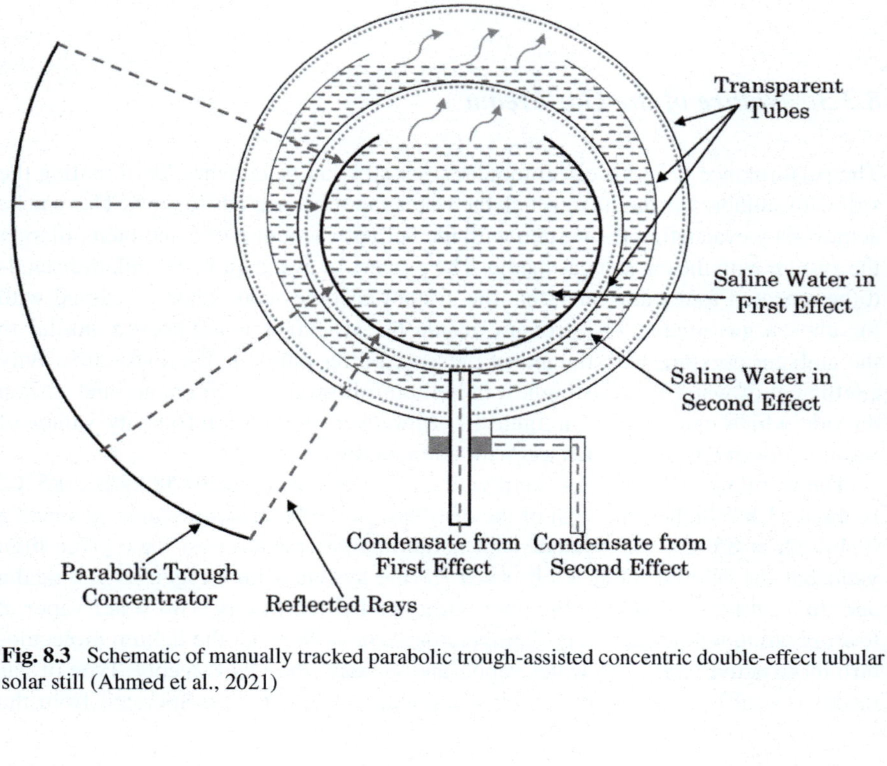


**Fig. 8.3** Schematic of manually tracked parabolic trough-assisted concentric double-effect tubular solar still (Ahmed et al., 2021)

The diameter of the semi-circular tubular central trough, tubular condenser of first effect, tubular trough of second effect, and tubular condenser of first effect is 80 mm, 100 mm, 130 mm, and 160 mm, respectively. The length of the tubular unit, parabolic trough concentrator, and aperture width of the concentrator is about 1.0, 1.0, and 0.87 m, respectively. The distillate productivity of the system is about 5.70 $kg/m^2d$ at a thermal efficiency of 40.4%. The distillate yield of the second effect is about 25.8% of the total yield of the system. Distillate yield of double-effect system is about 31.7% higher than the single-effect system. Moreover, the outer condensing surface temperature of the double-effect system has reduced by about 15.6 °C in comparison to single-effect system whose outermost condensing surface temperature is about 88.6 °C.

### *8.3.5 Falling Film Evaporators*

Multi-effect tubular solar still can be made more compact and less weight by substituting bulk water troughs with tubular thin falling film evaporators. Falling film evaporation and condensation process can be carried out effectively even with low feed water flow rate and low-temperature difference between the evaporating and the condensing surfaces and this is mainly due to the high heat and mass transfer coefficient that can be achieved in the falling film evaporation process. Gang et al. (2019) proposed the concept of falling film evaporator-based multi-effect tubular solar still. The schematic of multi-effect falling film eccentric tubular still integrated with Fresnel reflector and the photograph of multi-effect falling film eccentric tubular still integrated with evacuated tube solar collector is shown in Figs. 8.4 and 8.5a, respectively. The main components of the still are a number of semi-tubular metal plates of increasing diameter stacked one above the other with a necessary gap between them to form crescent-shaped distillation cavities. Lateral sides of the distiller are closely packed to prevent vapor loss. Feed water from the water distributor tube first flows over the 5 cm thick absorbent gauze that is laid over the apex of each semi-tubular metal plate except the last one with the aid of circulation pump. The water absorbed by the gauze then flows slowly and uniformly over the entire outer surface.

The inner surface of the innermost semi-tubular metal plate is exposed to concentrated solar radiation from the Fresnel collector thereby, water film over its outer surface evaporates and condenses over the inner surface of the adjacent semi-tubular metal plate. In the case of the system integrated with evacuated solar collector, the innermost tubular metal plate is filled with hot water as shown in Fig. 8.5a. The released latent heat of condensation of one effect is used as heat source for the occurrence of evaporation in the adjacent effects. The inner surface of the innermost semi-tubular metal plate is black coated to increase its absorptivity and its aperture is covered with transparent glass cover to prevent heat loss to ambient. The axis of each of the semi-tubular metal plates is eccentric to each other hence this system is termed as **multi-effect eccentric tubular solar still**. Two-effect system with first and second effect evaporation areas of 0.36 $m^2$ and 0.53 $m^2$, corresponding feed water

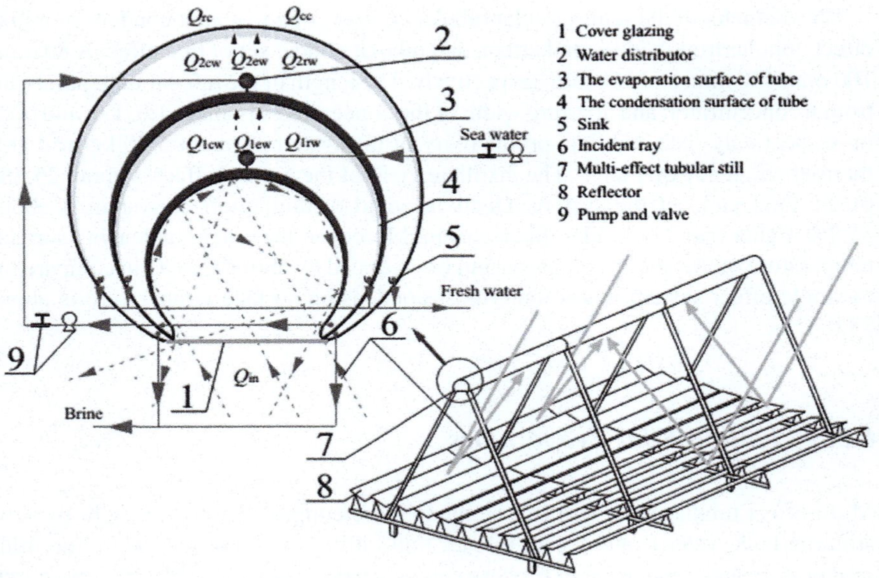


**Fig. 8.4** Schematic of two-effect falling film eccentric tubular still integrated with Fresnel reflector (Gang et al., 2019)

flow rate of 55 kg/h and 40 kg/h, respectively, is capable of producing nearly, 11.35 kg/d of distillate at a performance ratio of 2.88 with the aid of evacuated tube solar collector of 0.92 $m^2$ under a cumulative solar radiation intensity of 22.0 $MJ/m^2d$. For the same performance ratio, the distillate production capacity of the two-effect system is estimated to be about 63.68 kg/d for a direct solar radiation intensity and Fresnel reflector area of 800 $W/m^2$ and 4.8 $m^2$, respectively. The nocturnal distillate production of the system with evacuated tube solar collector is about 38.0% of the total distillate productivity of the system and is due to the presence of bulk hot water inside the innermost tubular system which sustains the evaporation process to continue even after sunset hours which can be seen from Fig. 8.5b.

### 8.3.6 *Immersion Cooling*

Evacuation is a widely recommended technique to enhance the distillate productivity of solar stills. However, vacuum pump used for evacuation consumes about 0.033 kWh of electricity to produce 1 kg of distillate from a solar still operating at a pressure and temperature of about 40 kPa and 62.0 °C, respectively. Evacuation reduces still's operating temperature which is beneficial in terms of fouling and corrosion mitigation. However, it also lowers the condensing surface temperature relative to

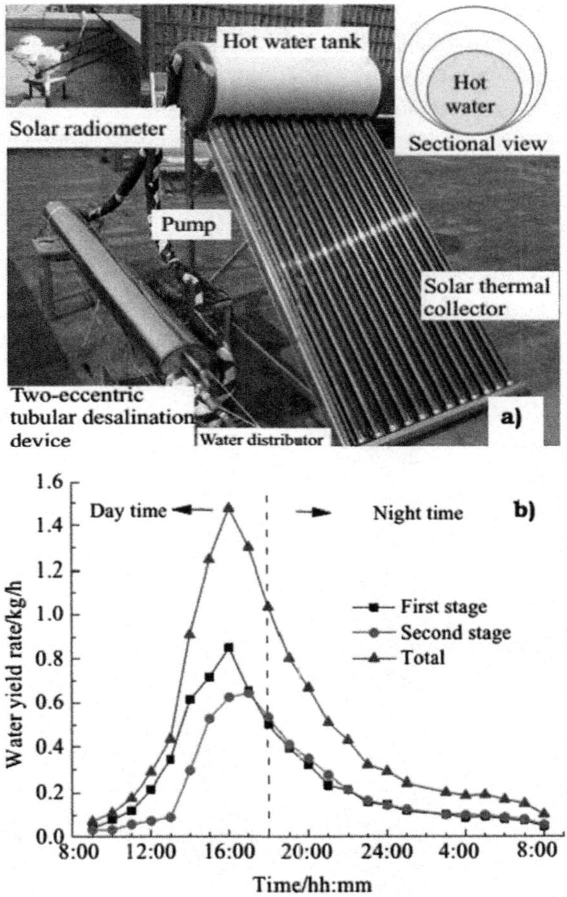


**Fig. 8.5** **a** Photograph of two-effect falling film eccentric tubular still integrated with evacuated tube solar collector **b** Hourly yield of two-effect falling film eccentric tubular still integrated with evacuated tube solar collector (Gang et al., 2019)

the condensing surface temperature noticed during normal pressure operation. This reduces required heat loss to the ambient which leads to vapor accumulation and poor vapor condensation resulting in suction of vapor by vacuum pump and low distillation efficiency of the systems. Hence, lowering condensing surface temperature further by external means is the option available to increase condensation phenomena at very low operating pressures. For low operating pressures, maximum distillate can be obtained at condensing surface temperature of 0 °C, and a further reduction in condensing surface temperature has not resulted in significant yield enhancement. However, achieving this low temperature consumes additional energy and cost.

One of the cheapest and easiest ways to achieve low condensing surface temperature without consuming additional energy is by immersion cooling. In immersion cooling, the multi-effect tubular distillation system is immersed inside a bulk water bath as shown in Fig. 8.6a. The system consists of three effect tubular solar still

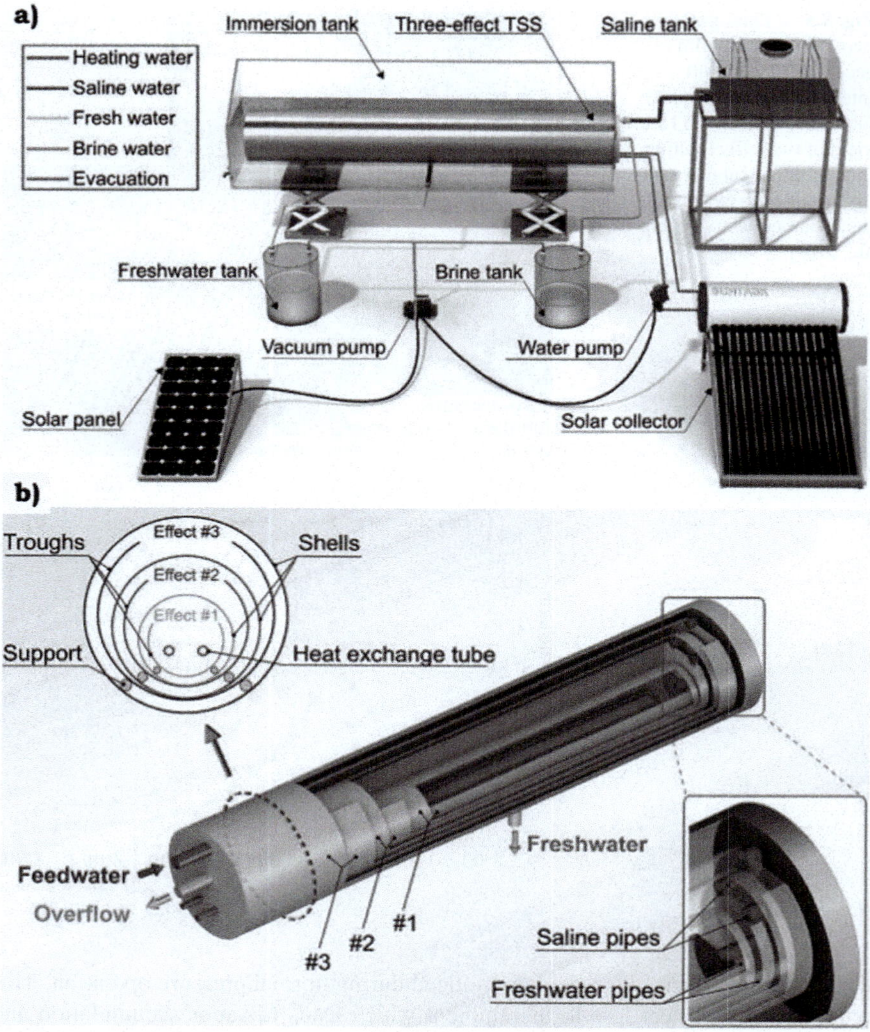


**Fig. 8.6** **a** Schematic of triple-effect tubular solar still under immersion cooling **b** Construction structure of triple-effect tubular solar still (Yan et al., 2021)

immersed inside water bath, solar evacuated tube collector of 1.09 $m^2$, solar PV panel of 30 W capacity, vacuum pump of 8 W capacity, and hot water circulation pump of 10 W capacity. Solar panel is used to power the pumps. Solar collector is used to provide hot heat transfer fluid to the first effect of the three-effect tubular still whose constructional structure is shown in Fig. 8.6b.

Vacuum pump is operated every 1 h for few minutes to maintain the pre-set pressure inside the distillation system. The surface temperature of the natural air-

**Table 8.2** Performance of triple-effect tubular solar still under different cooling methods (Yan et al., 2021)

| Parameters | Cooling method | | | |
|---|---|---|---|---|
| | Natural air | | Immersion cooling | |
| Operating pressure (kPa) | 40.0 | 60.0 | 40.0 | 60.0 |
| Maximum water temperature in Effect 1 (°C) | 83.1 | 86.4 | 74.5 | 79.4 |
| Maximum water temperature in Effect 2 (°C) | 81.5 | 83.7 | 70.1 | 73.1 |
| Maximum water temperature in Effect 3 (°C) | 76.2 | 77.0 | 53.7 | 57.4 |
| Maximum condensing surface temperature of Effect 3 (°C) | 66.3 | 70.7 | 38.5 | 39.3 |
| Total yield (kg/d) | 7.24 | 6.00 | 10.69 | 8.01 |
| Performance ratio | 1.45 | 1.31 | 1.87 | 1.52 |
| Specific energy consumption (kJ/kg of distillate) | 3.70 | 1.40 | 5.20 | 1.70 |

cooled system is higher than the water temperature in the third effect leading to low productivity before noon. The lower surface temperature of immersion-cooled system resulted in enhanced vapor condensation and increased distillate production. The performance of triple effect tubular solar still under natural air cooling and immersion cooling mode for an operating pressure of 40 and 60 kPa is shown in Table 8.2. Operating temperature of the system dropped with reduction in pressure for both the cooling modes. Condensing surface temperature of the third effect in immersion cooled system is about 30.0 °C lower than the condensing surface temperature of the third effect in the natural air-cooled system. Distillate productivity of immersion-cooled system at 40 kPa is about 33.5% higher than the similar system operating at 60 kPa. However, the total electrical energy consumption per kg of distillate produced by both the systems at a operating pressure of 40 kPa is observed to be about three times higher than similar systems operating at a pressure of 60 kPa this may be due to the longer duration of vacuum pump operation for maintaining the required low pressure inside the still every 1 h. Optimum cooling temperature of bulk water for evacuated system is estimated to be about 40 °C. At very low cooling water temperature, the yield drops due to excessive drop in operating temperature of the system by heat losses (Yan et al., 2021).

### *8.3.7 Modular System*

The operation of entire system will be affected if any damage or malfunction occurs at any one of the stages of multi-stage tubular still. In order to overcome this drawback, modular-type system is proposed. Each of the tubular stills in modular-type system works independently and any damage to one of the stills does not stop the operation

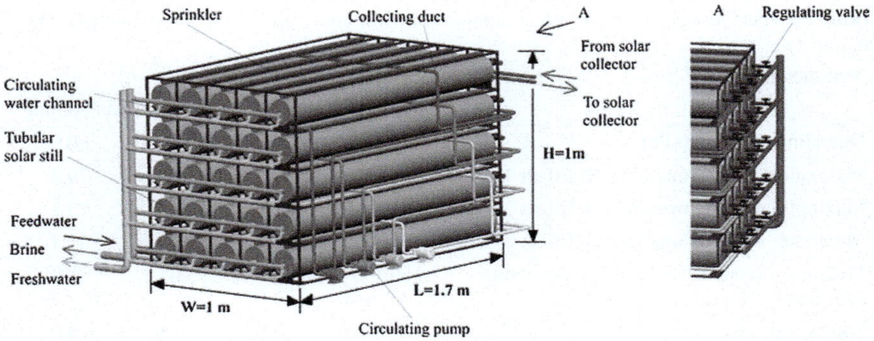


**Fig. 8.7** Conceptual representation of modular multi-effect tubular solar still (Xie et al., 2016)

of the whole plant. The schematic of modular multi-effect tubular solar distillation system is shown in Fig. 8.7. The main components of the system are solar collectors, multiple tubular solar stills, vacuum pumps, sprinklers, regenerators, and circulation pumps. A number of tubular solar stills are stacked in several rows as shown in Fig. 8.7. Inlet saline water gets preheated by passing through regenerators and is then distributed to all the tubular stills. The saline water in the tubular stills of the topmost layers will be heated by the circulating heating fluid from the solar collectors. The hot circulating fluid causes evaporation of saline water whose vapors condense over the inner surface of tubular stills and are collected by suitable provisions.

The condensation process is enhanced by minimizing condensing temperature by spraying water over the outer surface of the tubular stills. The sprayed water gets preheated by absorbing the latent heat of condensation and is then directed through the U-shaped heat exchanger tubes installed inside the tubular solar stills arranged below to heat their saline water to cause evaporation. The preheated sprinkled water in one row is used as a heat source for the next row of tubular solar stills arranged immediately below. A similar kind of water sprinkling, heat recovery, and saline water heating are followed in all the units arranged below. The waste heat energy of reject brine and condensate is recovered using regenerators by circulating feed saline water.

The impact of operating pressure and feed water temperature on distillate yield/m$^2$ of evaporating area of a single tubular solar still is shown in Fig. 8.8. Distillate rate increases with increase in water temperature and reduction in operating pressure. Distillate production rate at vacuum pressure for an operating temperature of 50 °C, 60 °C, 70 °C, and 80 °C is about 95%, 69%, 47%, and 75% higher than the distillate production rate at normal ambient pressure, respectively. The vacuum pump can be actuated every 2.5 min for 2 seconds duration to maintain the vacuum pressure inside the still. Modular system with 25 tubular stills (arranged in 5 rows), 5.0 m$^2$ solar collector (23.0 MJ/m$^2$d), saline water flow rate of 40.0 kg/d in each system, and cooling water flow rate of 1.5 kg/s operating under reduced pressure is capable of producing nearly 17.2 kg of distillate per day per m$^2$ of solar collector area. The

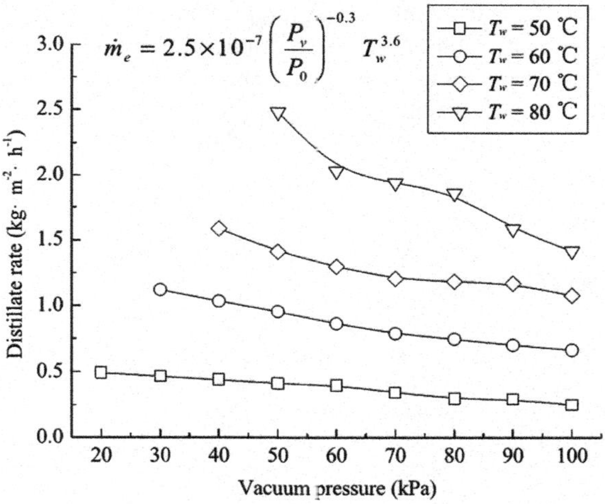


**Fig. 8.8** Impact of operating pressure and feed water temperature on distillate yield of a tubular solar still (Xie et al., 2016)

power requirement of vacuum pump and circulation pump is about 1 kWh/d and 3 kWh/d, respectively. The performance ratio of the modular multi-effect tubular solar still is about 0.81 and is nearly twice that of conventional basin solar still (Xie et al., 2016). The distillate yield correlation derived based on experimental results (shown in Fig. 8.8) has deviation of about −18 to +25% and is valid for water temperature and ambient temperature in range of about 50–80 °C and 20–25 °C, respectively, ($P_V$—vacuum pressure; $P_O$—ambient pressure).

## 8.4 Features of Multi-effect Tubular Solar Thermal Desalination Systems

(a) High condensing surface area than the evaporating surface area.
(b) Major portion of the released latent and sensible heat energy in each effect can be effectively reused for evaporation process.
(c) Tubular structure has good pressure bearing capacity and it can be effectively operated under vacuum.
(d) Easy scaling up for large-scale freshwater production.
(e) Modular tubular solar stills are highly adjustable such that the number of tubular stills in operation can be adjusted based on water demand and weather conditions.
(f) Damage to any one of the modular tubular stills will never lead to shut down of the whole system.
(g) Modular multi-effect tubular still is easy to assemble and low cost compared to other multi-effect systems.
(h) Lower corrosion rate of still components as saline water is in contact only with the troughs.

(i) Eccentric multi-effect tubular still has a larger condensing surface area than its evaporator area thereby condensation rate is enhanced.
(j) Evaporator troughs are enclosed well within the tubular condensing covers leading to low heat losses from it to the ambient thereby, insulation requirement is avoided.

## References

Ahmed, M. M. Z., Alshammari, F., Abdullah, A. S., & Elashmawy, M. (2021). Enhancing tubular solar still performance using double effect with direct sunrays concentration. *Solar Energy Materials & Solar Cells, 230*, 111214.

Alshammari, F., Elashmawy, M., & Ahmed, M. M. Z. (2021). Cleaner production of freshwater using multi-effect tubular solar still. *Journal of Cleaner Production, 281*, 125301.

Chen, Z., Yao, Y., Zheng, Z., Zheng, H., Yang, Y., Hou, L., & Chen, G. (2013). Analysis of the characteristics of heat and mass transfer of a three-effect tubular solar still and experimental research. *Desalination, 330*, 42–48.

Gang, W., Qichang, Y., Hongfei, Z., Yi, Z., Hui, F., & Rihui, J. (2019). Direct utilization of solar linear Fresnel reflector on multi-effect eccentric horizontal tubular still with falling film. *Energy, 170*, 170–184.

Xie, G., Sun, L., Mo, Z., Liu, H., & Du, M. (2016). Conceptual design and experimental investigation involving a modular desalination system composed of arrayed tubular solar stills. *Applied Energy, 179*, 972–984.

Xie, G., Chen, W., Yan, T., Tang, J., Liu, H., & Cao, S. (2020). Three-effect tubular solar desalination system with vacuum operation under actual weather conditions. *Energy Conversion and Management, 205*, 112371.

Yan, T., Xie, G., Sun, L., Du, M., & Liu, H. (2019). Experimental investigation on a two-effect tubular solar still operating under vacuum conditions. *Desalination, 468*, 114057.

Yan, T., Xie, G., Chen, W., Wu, Z., Xu, J., & Liu, Y. (2021). Experimental study on three-effect tubular solar still under vacuum and immersion cooling. *Desalination, 515*, 115211.

Zheng, H., Chang, Z., Chen, Z., Xie, G., & Wang, H. (2013). Experimental investigation and performance analysis on a group of multi-effect tubular solar desalination devices. *Desalination, 311*, 62–68.

Zheng, H., Chang, Z., Zheng, Z., Chen, Z., & Su, Y. (2013). Performance analysis and experimental verification of a multi-sleeve tubular still filled with different gas media. *Desalination, 331*, 56–61.

# Chapter 9
# Vertical Convection Solar Thermal Desalination Systems

## 9.1 System Description and Working

The schematic of passive double-sided vertical convection solar thermal desalination system oriented in the east-west direction is shown in Fig. 9.1. It consists of two vertically oriented tempered glass covers between which a blackened wick-lined absorber plate is installed by maintaining a necessary gap between the glass covers and the absorber plate. The top, bottom, and lateral sides are sealed perfectly to form an air-tight distillation chamber. An overhead tank is provided to supply feed water to the distillation chamber. Upper end of the wick lining of the absorber plate is dipped inside the overhead tank to facilitate wetting by capillary action. Lower end of the wick is left free to allow free drain off excess un-evaporated brine. Top and bottom sides of the distillation chamber are insulated with glass wool to minimize heat energy loss to the ambient. The system oriented in the east-west direction receives solar radiation from the east and the west directions during forenoon and afternoon hours, respectively. Incident solar radiation heats up the wetted wick-lined absorber plate and causes evaporation of feed water. Vapors thus formed move towards the condensing surface (i.e.) glass covers due to convection caused by the partial pressure difference between the heated wick and glass covers. The vapors condense over the inner surface of the glass cover by releasing its latent heat of condensation. The condensate thus generated rolls down the glass cover surface by gravity and is collected in distillate collection troughs. In the meantime, the excess un-evaporated brine in wetted wick leaves the system as rejected through the brine collection trough.



H. Sharon and K. Srinivas Reddy, *Solar Thermal Desalination Technologies for Potable Water*, https://doi.org/10.1007/978-3-031-49155-9_9

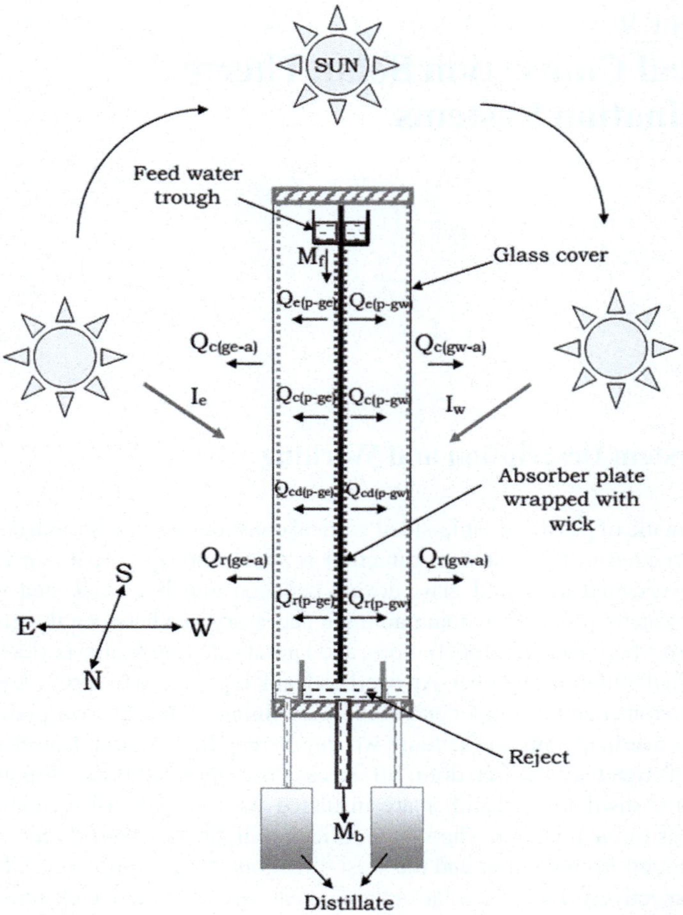


**Fig. 9.1** Schematic of passive double-sided east-west-oriented vertical convection solar still along with the associated energy transport processes

## 9.2 Thermodynamic Modeling of Vertical Convection Solar Thermal Desalination System

Energy transport processes occurring in vertical convection solar still are shown in Fig. 9.1. Assumptions taken into account for formulating energy balance in various components of vertical convection solar still are listed below:

(a) Distiller is free from vapor leakage;
(b) Feed water enters the distiller at ambient temperature and uniformly wets the absorber surface;
(c) Heat capacity of absorber plate and glass cover are considered;

(d) Heat transfer from absorber to glass cover is by convection, radiation, conduction, and evaporation;
(e) Heat loss to ambient is by conduction through insulation and convection & radiation from glass covers;
(f) Reject brine leaves the distiller at absorber temperature;
(g) Vapors reaching the inner surface of top insulation by buoyancy is negligible.

Energy balance in absorber plate

$$\begin{aligned}\begin{pmatrix}\text{rate of change of}\\ \text{absorber temperature}\end{pmatrix} &= \begin{pmatrix}\text{solar radiation absorbed}\\ \text{by the absorber}\end{pmatrix}\\ &+ \begin{pmatrix}\text{heat energy content}\\ \text{of incoming feed water}\end{pmatrix}\\ &- \begin{pmatrix}\text{convection, radiation evaporation \& conduction heat}\\ \text{transfer from absorber plate to east glass cover}\end{pmatrix}\\ &- \begin{pmatrix}\text{convection, radiation evaporation \& conduction heat}\\ \text{transfer from absorber plate to west glass cover}\end{pmatrix}\\ &- \begin{pmatrix}\text{heat energy content}\\ \text{of reject brine}\end{pmatrix} - \begin{pmatrix}\text{heat energy loss from}\\ \text{absorber to ambient}\end{pmatrix}\end{aligned}$$

$$\begin{aligned}(M_pC_p)\frac{dT_p}{dt} &= \tau_g\alpha_pA(I_e + I_w) + m_fc_{pf}(T_f + 273)\\ &- (Q_c + Q_r + Q_{ew} + Q_{cd})_{p-ge} - (Q_c + Q_r + Q_{ew} + Q_{cd})_{p-gw}\\ &- m_bc_{pb}(T_{fo} + 273) - Q_d \end{aligned} \tag{9.1}$$

Energy balance in east-facing glass cover

$$\begin{aligned}\begin{pmatrix}\text{rate of change of east}\\ \text{glass cover temperature}\end{pmatrix} &= \begin{pmatrix}\text{solar radiation absorbed}\\ \text{by the east glass cover}\end{pmatrix}\\ &+ \begin{pmatrix}\text{convection, radiation evaporation \& conduction heat}\\ \text{transfer from absorber plate to east glass cover}\end{pmatrix}\\ &- \begin{pmatrix}\text{convection \& radiation heat loss}\\ \text{from east glass cover to ambient}\end{pmatrix}\end{aligned}$$

$$(M_pC_p)\frac{dT_{ge}}{dt} = \alpha_gA_{ge}I_e + (Q_c + Q_r + Q_{ew} + Q_{cd})_{p-ge} - (Q_c + Q_r)_{ge-a} \tag{9.2}$$

Energy balance in west-facing glass cover

$$\begin{pmatrix}\text{rate of change of west}\\ \text{glass cover temperature}\end{pmatrix} = \begin{pmatrix}\text{solar radiation absorbed}\\ \text{by the west glass cover}\end{pmatrix}$$
$$+ \begin{pmatrix}\text{convection, radiation evaporation \& conduction heat}\\ \text{transfer from absorber plate to west glass cover}\end{pmatrix}$$
$$- \begin{pmatrix}\text{convection \& radiation heat loss}\\ \text{from west glass cover to ambient}\end{pmatrix}$$

$$(M_g C_g)\frac{dT_{gw}}{dt} = \alpha_g A_{gw} I_w + (Q_c + Q_r + Q_{ew} + Q_{cd})_{p-gw} - (Q_c + Q_r)_{gw-a} \tag{9.3}$$

where

$I_w$ solar radiation intensity over the west facing surface (W/m$^2$);
$I_e$ solar radiation intensity over the east facing surface (W/m$^2$);
$A_{ge}$ and $A_{gw}$ surface area of east- and west-facing glass cover (m$^2$), respectively;
$A$ surface area of absorber plate (m$^2$);
$T_p$ temperature of absorber plate (°C);
$T_{ge}$ temperature of east glass cover (°C);
$T_{gw}$ temperature of west glass cover (°C);
$\alpha_g$ absorbtivity of glass cover;
$\alpha_p$ absorbtivity of absorber plate;
$\tau_g$ Transmissivity of glass cover.

Convection heat transfer between absorber plate and east glass cover ($Q_{c(p-ge)}$ in W) is given by

$$Q_{c(p-ge)} = h_{c(p-ge)} A(T_p - T_{ge}) \tag{9.4}$$

For a vertical enclosure of height "$H$ in m" heated one side and cooled other side maintained at a gap "$W$ in m" between the surfaces, the Nusselt number correlation (Nu) is given by Inaba (1984)

$$Nu = 0.271\left(\frac{H}{W}\right)^{-0.21}(Gr\ Pr\ \sin\theta)^{0.250} \tag{9.5}$$

The correlation is valid for

$$5 \le \frac{H}{W} \le 83; \quad 5000 \le Gr Pr \le 1.2 \times 10^6$$

where

$\theta$ tilt angle of the still from horizontal (degree).

The Grashof number (Gr) is estimated by considering a considerable amount of water vapor present inside the still

$$Gr = \frac{g\beta \Delta T W^3}{\upsilon^2} \tag{9.6}$$

$$\Delta T = (T_{es} - T_{cs}) + \frac{(T_e + 273)(P_{vp} - P_{vg})}{\left(\frac{M_a P_t}{M_a - M_v}\right) - P_{vp}} \tag{9.7}$$

where

$\beta$ is the volumetric thermal expansion coefficient of air-vapor mixture (1/K).
$\upsilon$ is the kinematic viscosity of air-vapor mixture ($m^2/s$).
$T_{es}$ is the temperature of evaporating surface (°C).
$T_{cs}$ is the temperature of condensing surface (°C).
$P_t$ is the total or atmospheric pressure (101325 Pa).
$P_{vp}$ is the partial vapor pressure of water vapor at evaporating surface temperature (Pa).
$P_{vg}$ is the partial vapor pressure of water vapor at condensing surface temperature (Pa).
$M_a$ and $M_v$ molar mass of air and water vapor (kg/kmol), respectively.

The convective heat transfer coefficient between the absorber plate and east glass cover ($h_{c(p-ge)}$ in $W/m^2K$) is estimated based on the Nusselt number correlation as

$$h_{c(p-ge)} = \frac{Nu \times K_v}{W} \tag{9.8}$$

where

$K_v$ thermal conductivity of air-vapor mixture (W/mK)

Radiation heat transfer between the absorber plate and east glass cover ($Q_{r(p-ge)}$ in W) is given by

$$Q_{r(p-ge)} = h_{r(p-ge)} A (T_p - T_{ge}) \tag{9.9}$$

Radiation heat transfer coefficient between the absorber plate and east glass cover ($h_{r(p-ge)}$ in $W/m^2K$) is given by

$$h_{r(p-ge)} = \varepsilon_{eff} \sigma \left((T_p + 273)^2 (T_{ge} + 273)^2\right) (T_p + T_{ge} + 546) \tag{9.10}$$

Effective emissivity ($\varepsilon_{eff}$) is represented as

$$\varepsilon_{eff} = \frac{1}{\left[\frac{1}{\varepsilon_w} + \frac{1}{\varepsilon_{ge}} - 1\right]} \tag{9.11}$$

where

$\varepsilon_w$ is the emissivity of water.
$\varepsilon_{ge}$ is the emissivity of east glass cover.

Conduction heat transfer between the absorber plate and east glass cover ($Q_{cd(p-ge)}$ in W) due to the presence/accumulation of air-vapor mixture between them is given by

$$Q_{cd(p-ge)} = K_v A \frac{(T_p - T_{ge})}{W} \tag{9.12}$$

Distilled water production rate of vertical convection solar distiller can be found by modified mass transfer theory proposed by Kiatsiriroat et al. (1986). This model has already been used for estimating vertical solar distiller's distillate production rate with reasonable accuracy. Deviation of the predicted values obtained using this model from the experimental values is only within 4.0–8.0% which has been shown clearly in Table 9.1.

Distillate production rate in one chamber of vertical convection solar still ($m_{de}$ in kg/s) is given as (Kiatsiriroat, 1987)

$$m_{de} = Ag^* \ln\left(1 + \frac{m_{H_2O,cs} - m_{H_2O,es}}{m_{H_2O,es} - 1}\right) \tag{9.13}$$

**Table 9.1** Comparison of experimental distillate production rate in vertical convection still with theoretical data predicted by various mass transfer models (Kiatsiriroat et al., 1986)

| Temperature | | Distillate yield | | | Deviation between experiment values and Dunkle's model | Deviation between experiment values and modified theory |
|---|---|---|---|---|---|---|
| Absorber | Glass cover | Experiment | Dunkle's model | Modified theory | | |
| (°C) | (°C) | (kg/m²h) | | | (%) | (%) |
| 51.6 | 42.2 | 0.168 | 0.276 | 0.182 | 64.29 | 8.33 |
| 53.6 | 44.0 | 0.192 | 0.313 | 0.203 | 63.02 | 5.73 |
| 58.0 | 49.1 | 0.223 | 0.358 | 0.233 | 60.54 | 4.48 |
| 60.8 | 52.0 | 0.252 | 0.407 | 0.264 | 61.51 | 4.76 |
| 62.5 | 53.3 | 0.276 | 0.450 | 0.291 | 63.04 | 5.43 |

where

$m_{H_2O,cs}$ is the mass fraction of water vapor at the condensing surface.
$m_{H_2O,es}$ is the mass fraction of water vapor at the evaporating surface.

$$M_{H_2O,es} = \frac{\left[\frac{0.622P_{es}}{(P_t - P_{es})}\right]}{1 + \left[\frac{0.622P_{es}}{(P_t - P_{es})}\right]} \tag{9.14}$$

$$M_{H_2O,cs} = \frac{\left[\frac{0.622P_{cs}}{(P_t - P_{cs})}\right]}{1 + \left[\frac{0.622P_{cs}}{(P_t - P_{cs})}\right]} \tag{9.15}$$

where

$P_{es}$ is the partial vapor pressure of water vapor at evaporating surface temperature (Pa).
$P_{cs}$ is the partial vapor pressure of water vapor at condensing surface temperature (Pa).
$P_t$ is the total or atmospheric pressure (101325 Pa).

The mass transfer conductance ($g^*$ in kg/m$^2$s) is given as

$$g^* = \frac{0.271(H/W)^{-0.21}(GrSC)^{0.25}}{\left(\frac{W}{(\mu/SC)}\right)} \tag{9.16}$$

where

SC is the Schmidt number.
$\mu$ is the dynamic viscosity of air-vapor mixture estimated at average temperature of evaporating and condensing surface (kg/ms).

Evaporation heat transfer from the absorber plate to the east glass cover ($Q_{ew(p-ge)}$ in W) is given as

$$Q_{cw(p-ge)} = m_{de}h_{fg} \tag{9.17}$$

The evaporative heat transfer coefficient in the east chamber of vertical distiller ($h_{ew(p-ge)}$ in W/m$^2$K) is given as

$$h_{ew(p-ge)} = \frac{m_d h_{fg}}{A(T_p - T_{ge})} \tag{9.18}$$

Convection heat transfer ($Q_{c(ge-a)}$ in W) and radiation heat transfer ($Q_{r(ge-a)}$ in W) from east glass cover to the ambient is estimated by

$$Q_{c(ge-a)} = h_{ca} A_{ge}(T_p - T_{ge}) \tag{9.19}$$

$$Q_{r(ge-a)} = \varepsilon_g \sigma A_{ge}\left((T_{ge} + 273)^4 - (T_a + 273)^4\right) \tag{9.20}$$

where

$T_a$ is the ambient temperature (°C).
$\varepsilon_{ge}$ is the emissivity of east glass cover.

Conduction heat loss from the distiller to the ambient ($Q_d$ in W) is estimated by

$$Q_d = K_{ins} A_{ins} \frac{T_p - T_a}{t_{ins}} \tag{9.21}$$

where

$A_{ins}$ is the surface area of insulation (m$^2$).
$K_{ins}$ is the thermal conductivity of insulation (W/mK).
$t_{ins}$ is the thickness of insulation (m).

Energy transfer and distilled water production rate from the western chamber of the distiller can be found by replacing the subscripts ($ge$) by subscripts ($gw$) and subscripts ($de$) by subscripts ($dw$), respectively, in the above equations.

## 9.3 Parameters Influencing the Performance of Vertical Convection Solar Still

### *9.3.1 Still Configuration*

Vertical convection solar still can be either one-sided or double-sided type. One-sided and double-sided systems have one and two apertures, respectively. One-sided system can be oriented either in sole north, sole south, sole east, sole west, sole northeast, sole northwest, sole southeast, or sole southwest direction. Similarly, double-sided vertical solar still can be oriented either in the east-west or the north-south direction

**Fig. 9.2** Photograph of single sided vertical solar still (Boukar & Harmim, 2007)

(Kiatsiriroat, 1987). The photograph of single-sided and double-sided vertical solar stills is shown in Figs. 9.2 and 9.3, respectively. As orientation of the vertical still changes, the intensity of solar radiation reaching the absorber surface also varies thereby, performance of the system also varies accordingly. Solar radiation intensity reaching the east- and west-oriented vertical absorber plate is shown in Fig. 9.4. East- and west-facing absorber surfaces receive both direct and diffuse radiations during morning and afternoon hours, respectively. However, only diffuse solar radiation is received by the east and west surfaces during afternoon and morning hours, respectively. This observation is due to the movement of sun in the sky from the east to the west. Cumulative solar radiation intensity over the surface facing sole east or sole west will be identical and this leads to similar annual average distillate production for vertical still oriented in sole east or west direction which can be seen from Fig. 9.5. Distillate yield of double-sided vertical still (3.15 kg/m$^2$d) is about two times the distillate yield of single-sided vertical still (1.44 kg/m$^2$d) which can be seen from Fig. 9.5. Boukar and Harmim (2004) obtained a distillate yield of about 0.50–2.30

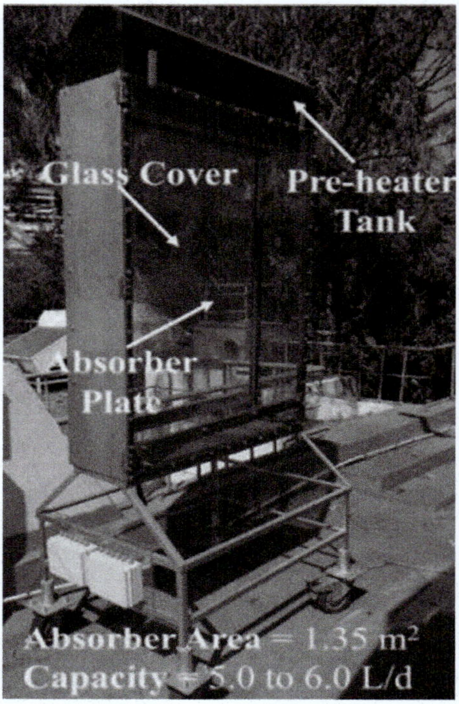


**Fig. 9.3** Photograph of double-sided vertical solar still

kg/m$^2$d for a south-faced single-sided vertical solar still under the climatic conditions of Algerian desert at an energy efficiency of about 7.85–21.19% (Boukar & Harmim, 2005).

### *9.3.2 Feed Water Preheating*

Evaporation rate in solar stills can be increased by preheating feed water. Feed water can be preheated by circulating it through solar collectors. (Hou et al., 2018a) proposed a single-effect concentric tubular vertical solar still whose schematic is shown in Fig. 9.6. The inner concentric tube is filled with water and its outer surface is lined with water absorption material which acts as evaporating surface. Brackish water from feed water tank is distributed over the evaporating surface with the aid of gravity using perforated silicone tube. The inner concentric tube is surrounded by concentric outer tube whose inner surface acts as condensing surface and its outside is exposed to the ambient. The water inside the inner concentric tube is heated with the aid of solar collector and this hot water act as heating medium for initiating water evaporation from the wick surrounding the inner concentric tube. The generated vapors condense over the inner surface of the outer concentric tube. The schematic

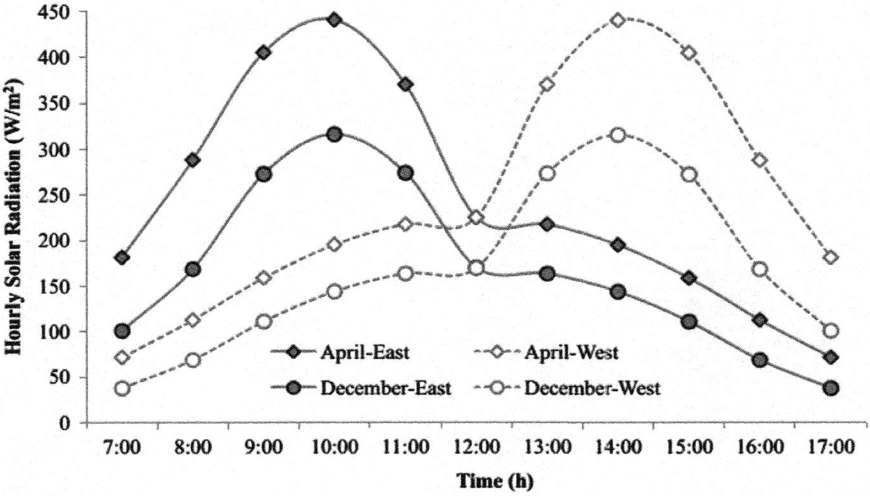


**Fig. 9.4** Hourly variation of solar radiation intensity incident over the vertical aperture oriented in sole east and sole west directions (Sharon, 2020)

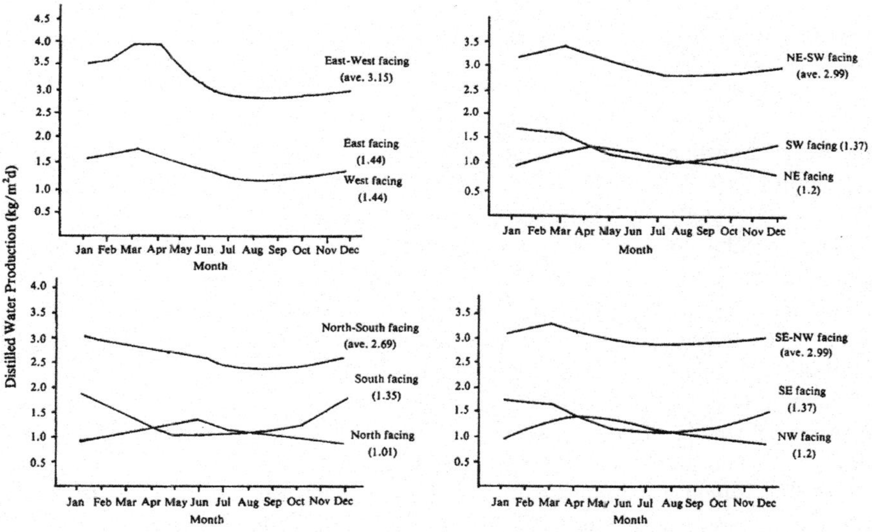


**Fig. 9.5** Monthly distillate productivity of vertical solar still for different orientations (Kiatsiriroat, 1987)

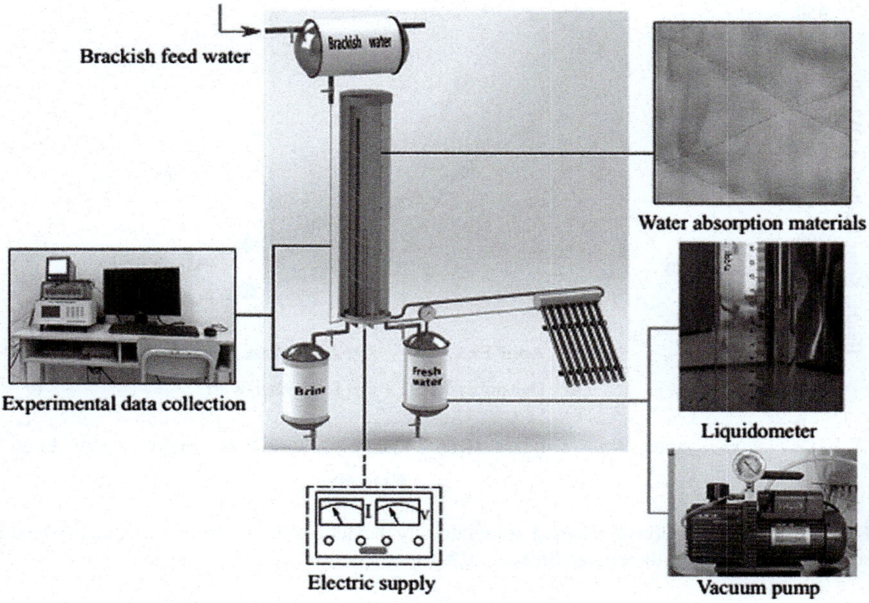


**Fig. 9.6** Schematic of vertical tubular convection-based single-effect solar still operating under reduced pressure (Hou et al., 2018a)

of energy transport process occurring in vertical tubular convection solar still is shown in Fig. 9.7. The distillate yield of this system is about 55, 100, 145, and 210 g/h for feed water temperatures of 50, 60, 70, and 80 °C, respectively.

Use of solar collectors may require installation of circulation pump and consumption of additional power which increases the capital, operation, and maintenance cost. Instead of using solar collectors and circulation pumps, overhead feed water tank of vertical still itself can be modified into a preheating unit such that the modified arrangement ignores pump, separate solar collector, and additional space (Sharon & Reddy, 2015). Moreover, the feed water flow is also achieved by gravity. The schematic of vertical solar still with a modified preheating unit is shown in Fig. 9.8. The overhead tank can be modified into single-basin and double-basin stills. The vertical still with these corresponding preheating units is termed cascaded vertical-single-slope still (CVSS) and cascaded vertical-double-slope still (CVDS), respectively. Since the feed water is preheated exterior to the vertical still the system can also be termed **active vertical solar still**.

The modified overhead tank causes shadow over the absorber of vertical still and this may affect the performance of the cascaded desalination system. However, the absorber plate temperature of shaded active vertical still will be still higher than the absorber plate temperature of passive vertical still without pre-heater and shading. Hence, the modification of overhead tank into preheater seems to be a better and positive option for enhancing vertical solar still's yield. The CVDS system has

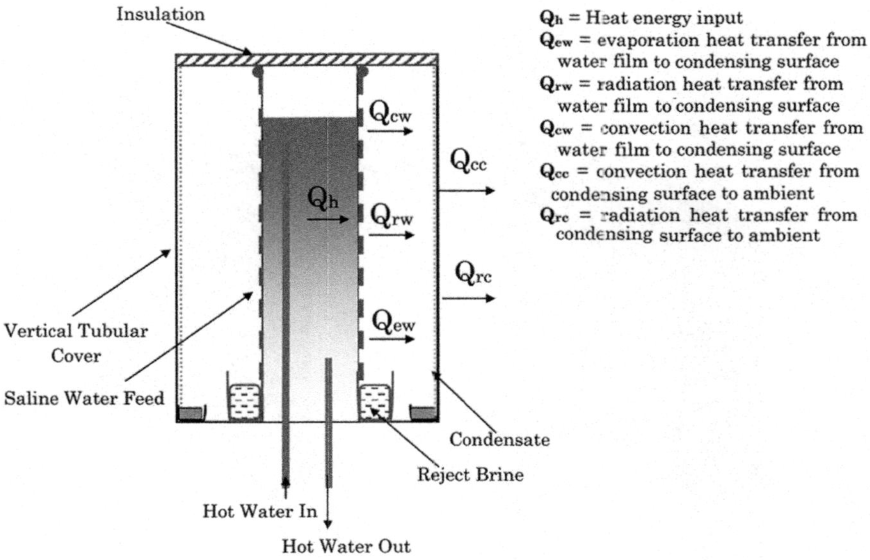


**Fig. 9.7** Schematic representing energy transport in vertical tubular convection solar still (Hou et al., 2018a)

higher productivity than the CVSS system due to the high cumulative solar irradiance over the double-slope still in comparison to single-slope still based pre-heater. The distillate yield of geometrically optimized CVDS system is around 6.00 kg/m$^2$d, while, for passive vertical solar still of same dimension, it is only about 4.25 kg/m$^2$d (Sharon & Reddy, 2015).

### 9.3.3 *Absorber Area*

Absorber area is an important component of vertical solar desalination system. Distillate yield increases with increase in absorber area of the still up to 4.0 m$^2$. Further increase in absorber area beyond 4.0 m$^2$ has shown no significant improvement in distillate yield. This observation can be attributed to the fact that as surface area increases, amount of solar radiation incident over it increases and, at the same time, heat losses from it also increases, thereby above 4.0 m$^2$ the absorber area becomes ineffective (El-Sebaii, 1998; Sharon & Reddy, 2015). In addition for fixed absorber area, CVDS and CVSS systems with maximum breadth has about 51.95 and 51.73% higher distillate yield, respectively, than the system with maximum height (Sharon & Reddy, 2015). This observation is mainly due to the fact that the Nusselt number of air-vapor mixture within the vertical still drops with increase in height due to the collision of cold air from the condensing surface with the heated air-vapor

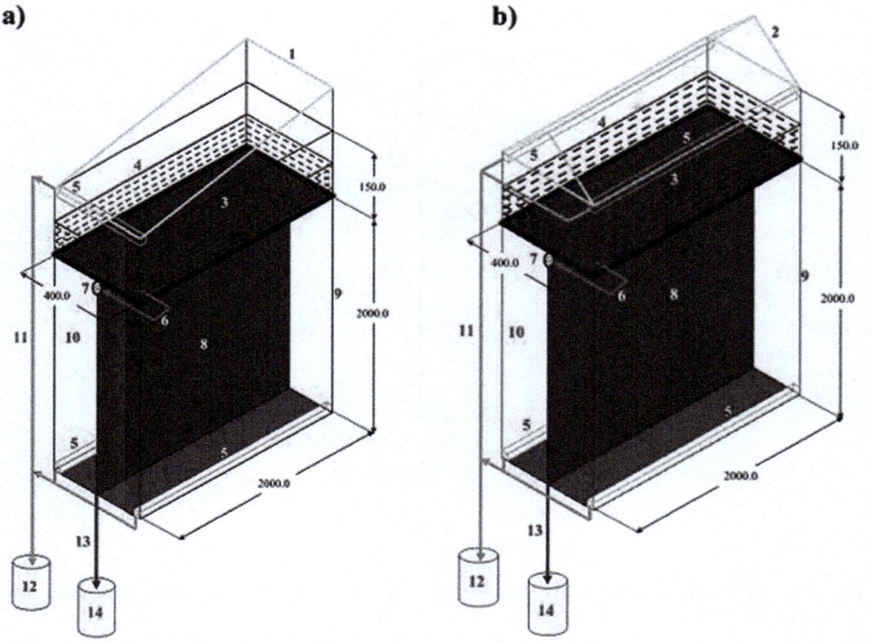


All Dimensions are in mm
(Not to Scale)

| | | | |
|---|---|---|---|
| 1 | Glass Cover of Single Slope Still | 8 | Absorber Plate of Vertical Still |
| 2 | Glass Cover of Double Slope Still | 9 | East Glass Cover of Vertical Still |
| 3 | Absorber Plate of Basin Type Still | 10 | West Glass Cover of Vertical Still |
| 4 | Water Mass in Basin Type Still | 11 | Distillate Carrying Tube |
| 5 | Distillate Collection Trough | 12 | Distillate Collection Tank |
| 6 | Pipe Connecting Feed Water Tank and Feed Water Distributor | 13 | Brine Carrying Tube |
| 7 | Feed Water Distributor | 14 | Brine Collection Tank |

**Fig. 9.8** Schematic of **a** cascaded vertical-single-slope still (CVSS) and **b** cascaded vertical-double-slope still (CVDS) (Sharon & Reddy, 2015)

mixture from the evaporating surface thereby the mass transfer reduces which can be understood from Eqs. 9.5, 9.11, and 9.12.

## 9.3.4 Gap Between Absorber Plate and Glass Cover

In the case of passive vertical still (Fig. 9.1), the distillate yield drops with increase in gap. However, in the case of active vertical still (Fig. 9.5), the yield increases with increase in gap between the evaporating and condensing surface from 0.05 m to 0.20 m. Further increase in gap beyond 0.20 m results in a significant drop in distillate yield. When the gap increases in active vertical still, the breadth of the basin still kept above the vertical still also increases thereby more solar radiation is collected and the feed water is preheated to high temperature thereby more evaporation is expected. However, increased gap also indicates increased air space and increased resistance for movement of vapors from the evaporating to condensing surface. Up to 0.20 m, preheating effect dominates, and beyond 0.20 m, air resistance dominates (Sharon & Reddy, 2015).

### *9.3.5 Feed Water Flow Rate, Thickness, and Salinity*

Distillate yield of passive double-sided vertical solar still is only about 0.75 kg/m$^2$d for a feed water mass flow rate of 0.005 kg/s (Kiatsiriroat et al., 1987a). One of the simplest and easiest ways to boost the distillate yield without much investment is to reduce the feed water mass flow rate and its thickness. Distillate yield of passive vertical still increases from 3.17 kg/m$^2$d to 7.00 kg/m$^2$d with increase in feed water flow rate from 10.0 kg/d to 30.0 kg/d. Further increase in feed water flow rate beyond 30.0 kg/d decreases the yield. Drop in yield beyond 30.0 kg/d is due to increased heat capacity of feed water which indicates the requirement of additional energy for initiating evaporation. Lower yield observed for flow rate below 30.0 kg/d is due to the non-availability of enough feed water for evaporation against the system's actual production capacity. Moreover, the reduced yield at very low mass flow rate can also be attributed to the formation of dry patches over the evaporating surface/wick which hinders the capillary effect (Sharon, 2020).

In general, the recommended minimal feed water flow rate for safe and enhanced performance of wick-based solar evaporation systems is about 1.30–2.00 times the system's evaporation rate. Optimum feed water flow rate in vertical solar still can be fixed anywhere between 30.0 to and.0 kg/d because within this range the feed water flow rate is about 1.47–2.03 times the evaporation rate. Moreover, the yield of the system drops by about 2.50% for every 1.50 wt% increase in salinity of the feed water. This observation can be attributed to the increased boiling point elevation with salinity which reduces the partial pressure of vapor near the evaporating surface. The reduced partial pressure difference between the evaporating and condensing surfaces leads to reduced mass transfer (Sharon, 2020). Reducing the film thickness from 0.01 m to 0.002 m has also enhanced distillate yield of passive vertical solar still by about 7.70%. This enhancement can be attributed to the reduced heat capacity and increased solar radiation transmissivity of thin feed water film in comparison to thick feed water film (El-Sebaii, 1998).

### 9.3.6 Utilization of Latent Heat of Condensation

Performance of vertical solar still can be improved by recovering and utilizing its latent heat of condensation which is wasted/rejected to the ambient via the condensing cover. This idea has led to the development of multi-effect vertical solar still which is more sophisticated, high productive but little bit expensive. The schematic of solar collector integrated multi-effect vertical convection solar still is shown in Fig. 9.9. Heat exchanger fluid heated in solar collector is circulated at a mass flow rate of 30.0 kg/h through the serpentine tube installed at the back surface of vertical multi-effect

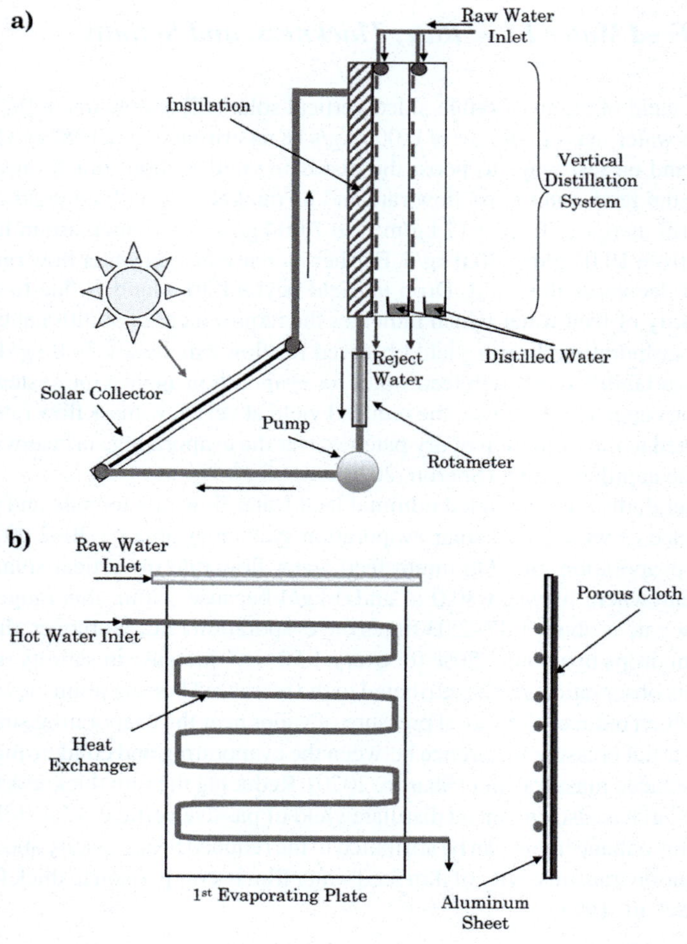


**Fig. 9.9** Schematic of **a** solar collector integrated multi-effect vertical convection solar still **b** Serpentine heat exchanger at the back side (outer surface) first aluminum plate **c** Porous cloth lined inner surface of first aluminum plate (Kiatsiriroat et al., 1987b)

still's first plate to transfer the collected heat energy to the distillation system. The heat energy thus transferred initiates feed water evaporation in first plate's wetted wick-lined inner surface. The vapors thus generated condense over the outer surface of adjacent (second) plate by releasing its latent heat. This released latent heat is utilized for feed water evaporation in the second plate and the similar process repeats in all other plates. Each plate with an outer condensing surface and an inner evaporating surface is called an effect. The outer surface of the last plate is exposed to the ambient air (Kiatsiriroat et al., 1987b).

Distillate yield increases from 1.80 kg/m$^2$d to 4.0 kg/m$^2$d with increase in number of plates/effects from one to five. This yield enhancement indicates effective utilization of latent heat of condensation for enhanced evaporation. Further, increase in the number of effects beyond five has not resulted in significant yield improvement and it may be attributed to the increased thermal resistance offered by the additional plates and air space. Lining the outer surface of the last plate by wetted wick has also been adopted to enhance condensation phenomenon through evaporative cooling. Evaporative cooling of the last plate has increased the yield by about 35.0% which can be seen in Fig. 9.10a (Kiatsiriroat et al., 1987b). Effect of the ratio of evaporating surface area to the solar collector area on distillate yield is shown in Fig. 9.10b. It can be observed that the optimum ratio is about five, beyond which the yield enhancement is nearly negligible. This indicates that the solar energy collected by the solar collector of 1 m$^2$ area can be effectively used for water distillation in multi-effect system with cumulative evaporating surface area nearly five times the solar energy collection area.

Reddy and Sharon (2016) pre-heated feed water in a solar collector and distributed it directly to each effect of vertical multi-effect still as shown in Fig. 9.11. Moreover, in this system, one surface of the bulk feed water storage tank is used as a condensing plate of the last effect thereby, its temperature is nearly maintained at ambient condition and thus enhanced condensation phenomena is achieved. The optimum number of effects and convection gap between the effects is about 5 and 0.05 m, respectively. The yearly average yield of the optimized system is about 6.8 kg/m$^2$d for a feed water mass flow rate of about 0.002 kg/s.

Chang et al. (2015) investigated the performance of multi-effect concentric vertical tubular still shown in Fig. 9.12. The yield and gained output ratio (GOR) of the system increased with increase in heat input, number of effects, and operating temperature. The yield of the single-effect, double-effect, and triple-effect systems is about 0.56, 1.08, and 1.40 kg/h, respectively, for a fixed operating temperature of 80 °C. The GOR of single-effect, double-effect, and triple-effect units is about 0.66, 1.21, and 1.89, respectively, for a heating temperature of 60–80 °C.

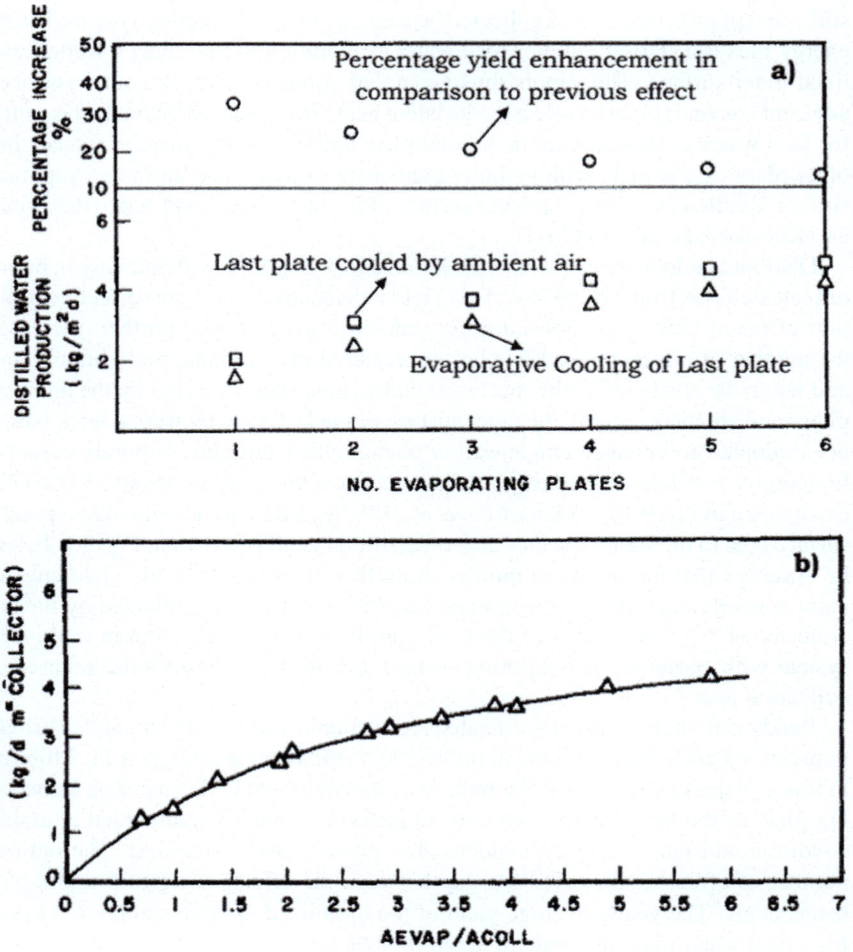


**Fig. 9.10** **a** Variation of distillate yield with number of plates or effects **b** Impact of ratio of evaporating surface to solar collector area on distillate yield (Kiatsiriroat et al., 1987b)

## 9.3.7 Operating Pressure

Reduced mass flow rate of feed water and high feed water temperature around 90 °C can increase distillate productivity and gained output ratio (GOR) but it also leads to increase scale formation and reduced efficiency of the water absorbing material in vertical tubular desalination systems. Hence, to achieve high yield at optimum flow rate and low heat input evacuation or operation of the system under reduced pressure is highly recommended. Evacuation helps to remove non-condensable gases and reduces mass transfer resistance of vapor from the evaporating to condensing surface.

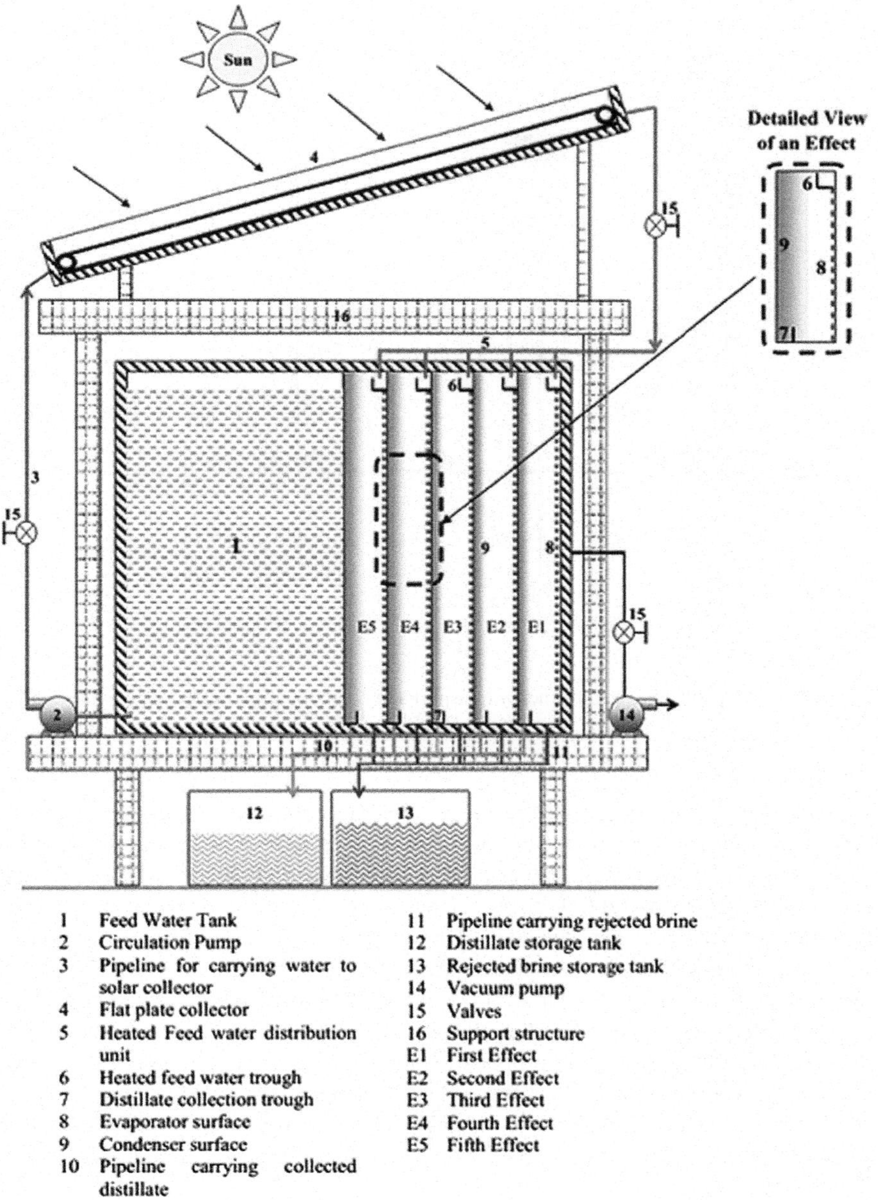


**Fig. 9.11** Schematic of active vertical multi-effect solar still (Reddy & Sharon, 2016)

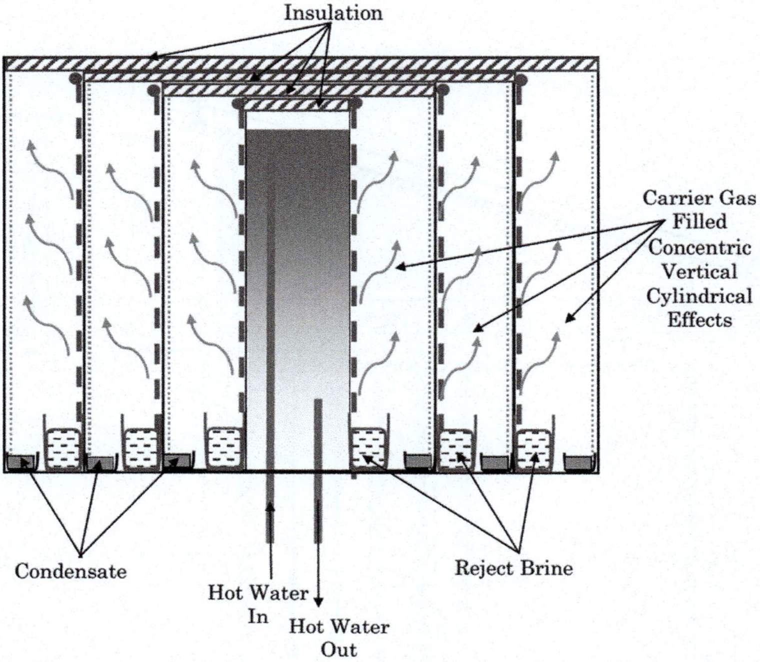


**Fig. 9.12** Schematic representing energy transport and components in triple effect concentric vertical still (Hou et al., 2018b)

In addition, lowering operating pressure reduces saturation temperature thereby, more vapors are generated from the feed water at a given temperature, and the distillate yield increases. Lower temperature difference is observed between the evaporating and condensing surface at low-pressure operation condition and is mainly due to the reduced thermal resistance. The yield of the vertical tubular single effect desalination system is about 0.25 kg/h at an operating temperature and pressure of 80 °C and 75 kPa, respectively which is nearly 24.0% higher than the yield obtained at ambient pressure. The outer tube has to bear the external pressure from the ambient during low-pressure operation hence proper thickness of the outer tube must be selected to bear the stress (Hou et al., 2018a). The yearly average yield of the optimized active vertical multi-effect solar still under low-pressure operation mode is about 21.3 kg/m$^2$d for a feed water mass flow rate of about 0.002 kg/s (Reddy & Sharon, 2016).

**Table 9.2** Mass diffusion coefficient of water vapor in different carrier gases at different temperatures (Hou et al., 2018b)

| Mass diffusion coefficient of vapor in Different carrier gases ($10^{-7}$ $m^2$/s) | | | | | | |
|---|---|---|---|---|---|---|
| Air | Carbon di-oxide | Helium | Nitrogen | Oxygen | Argon | Temperature (°C) |
| 290.38 | 238.62 | 962.54 | 295.04 | 300.17 | 289.55 | 50 |
| 306.29 | 310.38 | 1018.44 | 310.38 | 315.81 | 305.09 | 60 |
| 321.80 | 327.34 | 1072.37 | 327.34 | 333.01 | 321.58 | 70 |
| 340.52 | 345.25 | 1121.74 | 345.25 | 352.12 | 340.29 | 80 |

### *9.3.8 Carrier Gases*

The thermophysical and transport properties of carrier gas within the distillation system affect distillate yield of the desalination system. The mass diffusion coefficient of water vapor in different carrier gases at different temperatures is shown in Table 9.2. The mass diffusion coefficient of vapor in helium is highest which is then followed by oxygen, nitrogen, argon, carbon dioxide, and air. Helium is lighter than water vapor while carbon dioxide is heavier than water vapor and this affects the evaporation and condensation phenomena due to the variation in diffusion and buoyancy force on water vapor.

The solar still must be first evacuated using vacuum pump and then inflated with the corresponding carrier gas. For all the carrier gases, distillate yield of the triple-effect concentric tubular vertical solar still increases with temperature due to the rapid rise in vapor diffusion coefficient with temperature which can be seen in Table 9.2. For an operating temperature of 80 °C, the still with lighter carrier gas (helium) has a yield of about 1.19 kg/h which is about 30.8, 25.2, 20.9, 20.3, and 18.1% higher than the still with air, carbon dioxide, argon, nitrogen, and oxygen, respectively, as carrier gas. Hence, helium is a better alternative for air to obtain an improved yield from the still (Hou et al., 2018b).

### *9.3.9 Wickless Evaporating Surface*

Evaporation rate in solar stills has been increased by incorporating thin porous wick as evaporating surface and it is an essential component in vertical-type solar stills. However, these porous wicks suffer from gradual peeling off after a certain operation time, pore blockage due to salt/slit deposition, and accidental contact with condensing surface due to bulging thereby results in distillate contamination. Moreover, evaporation temperature and feed water flow rate cannot go beyond a certain range in wicked evaporating surfaces to prevent slit deposition and bubble formation. These

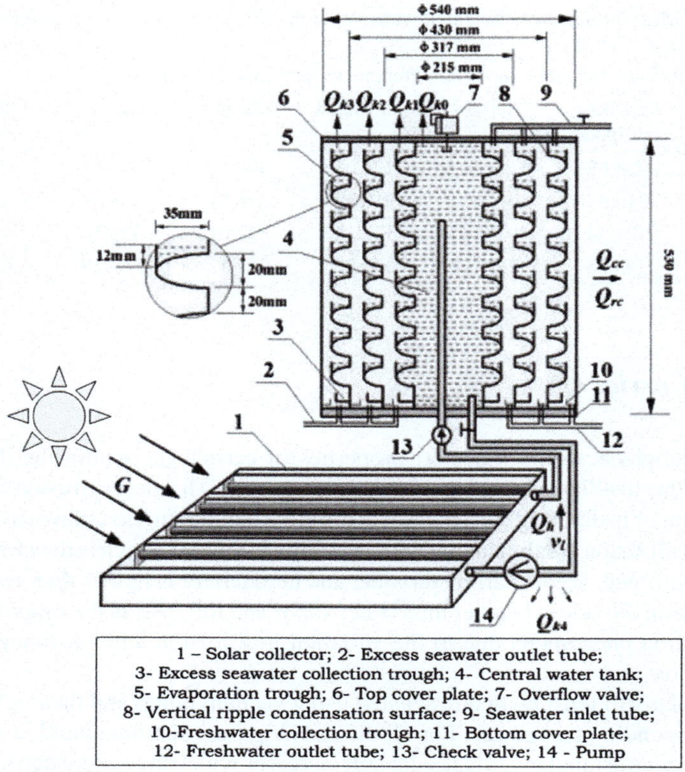


**Fig. 9.13** Schematic of wickless multi-effect vertical tubular still (Xie et al., 2015)

drawbacks can be avoided and the reliability of vertical solar stills can be increased by developing wick-free evaporating surfaces.

Xie et al. (2015) developed a vertical tubular triple-effect wickless solar still whose schematic is shown in Fig. 9.13. The wick-based evaporating surfaces are replaced by corrugated tubular vertical hollow cylinders of 1 mm thickness whose outer surfaces are provided with multiple water troughs. Yield of each stage of the system increased with increase in heat input. The first stage has a higher yield than other stages and the difference in yield among the stages diminishes with increase in heat input. The highest overall performance ratio of the system is noticed to be about 1.81 for a heat input of 600 W, which is comparatively higher than other triple-effect stills. The cumulative yield of the system is about 5.88 kg/m$^2$d at a performance ratio of 1.35 for cumulative solar radiation intensity of 22.5 MJ/m$^2$d. The characteristics and performance of the system under cumulative solar radiation intensity of 22.5 MJ/m$^2$d is shown in Table 9.3. Distillate yield of the system is about 10.3 kg/m$^2$d and 3.63 kg/m$^2$d for a cumulative solar radiation intensity of 26.1

**Table 9.3** Characteristics and performance of vertical tubular wickless multi-effect solar still (Xie et al., 2015)

| | | | |
|---|---|---|---|
| Shape | Cylindrical outside, corrugated barrel inside | | |
| Overall dimensions | Diameter 540 mm × Height 530 mm | | |
| Effective area of solar collector | 1.48 $m^2$ | | |
| Incident solar radiation | 22.5 $MJ/m^2d$ | | |
| Efficiency of solar collector | 45.6% | | |
| Actual heat available (100%) | 10.27 $MJ/m^2d$ (15.2 MJ/d) | | |
| Length of connection pipeline | 2.5 m | | |
| Solar still material | 304 Stainless steel | | |
| | First stage | Second stage | Third stage |
| Convection gap (mm) | 50 | 50 | 38 |
| Seawater capacity (kg) | 3.1 | 5.0 | 8.0 |
| Evaporation area ($m^2$) | 0.27 | 0.39 | 0.51 |
| Condensation area ($m^2$) | 1.15 | 1.51 | 0.85 |
| Convection heat transfer in the cavity (MJ) | 1.66 | 1.72 | 1.85 |
| Radiation heat transfer in the cavity (MJ) | 1.13 | 1.20 | 0.77 |
| Evaporation heat transfer in the cavity (MJ) | 7.84 | 6.46 | 5.64 |
| Heat Transferred in each stage (MJ) | 10.63 | 9.38 | 8.26 |
| Percentage of heat transferred from the available heat (%) | 69.9 | 61.7 | 54.3 |
| Yield ($kg/m^2d$) | 2.23 | 1.76 | 1.89 |

$MJ/m^2d$ (summer) and 17.3 $MJ/m^2d$ (winter), respectively. Performance ratio can be increased by preventing/minimizing heat losses to the ambient by incorporating insulation with optimum thickness.

## 9.4 Features of Vertical Convection Solar Thermal Desalination Systems

(a) Simple, lightweight, and easy to fabricate.
(b) Nearly no chances of distillate contamination by incoming feed water.
(c) Distillate productivity per $m^2$ of absorber area is nearly closer to basin-type solar stills.
(d) Occupies less ground area and is more suitable for rooftop installations.
(e) Passive system requires shade-free area for installation and efficient operation.

(f) Passive systems are not effective in utilizing the solar radiation potential at and around solar noon hours.
(g) Multi-effect solar collector-assisted evacuated systems are attractive and more suitable for meeting the drinking water demand of a single family.

## References

Boukar, M., & Harmim, A. (2004). Parametric study of a vertical solar still under desert climatic conditions. *Desalination, 168*, 21–28.

Boukar, M., & Harmim, A. (2005). Performance evaluation of a one-sided vertical solar still tested in the Desert of Algeria. *Desalination, 183*, 113–126.

Boukar, M., & Harmim, A. (2007). Design parameters and preliminary experimental investigation of an indirect vertical solar still. *Desalination, 203*, 444–454.

Chang, Z., Zheng, Y., Chen, Z., Zheng, H., Zhao, M., Su, Y., & Mao, J. (2015). Performance analysis and experimental comparison of three operational modes of a triple-effect vertical concentric tubular solar desalination device. *Desalination, 375*, 10–20.

El-Sebaii, A. A. (1998). Parametric study of a vertical solar still. *Energy Conversion and Management, 39*, 1303–1315.

Hou, J., Yang, J., Chang, Z., Zheng, H., & Su, Y. (2018). The mass transfer coefficient assessment and productivity enhancement of a vertical tubular solar brackish water still. *Applied Thermal Engineering, 128*, 1446–1455.

Hou, J., Yang, J., Chang, Z., Zheng, H., & Su, Y. (2018). Effect of different carrier gases on productivity enhancement of a novel multi-effect vertical concentric tubular solar brackish water desalination device. *Desalination, 432*, 72–80.

Inaba, H. (1984). Experimental study of natural convection in an inclined air layer. *International Journal of Heat and Mass Transfer, 27*(8), 1127–1139.

Kiatsiriroat, T., Bhattacharya, S. C., & Wibulswas, P. (1986). Prediction of mass transfer rates in solar stills. *Energy, 11*(9), 881–886.

Kiatsiriroat, T. (1987). Review of research and development on vertical solar stills. *The ASEAN Journal on Science & Technology for Development, 6*(1), 15.

Kiatsiriroat, T., Bhattacharya, S. C., & Wibulswas, P. (1987). Transient simulation of vertical solar still. *Energy Conversion and Management, 27*(2), 247–252.

Kiatsiriroat, T., Bhattacharya, S. C., & Wibulswas, P. (1987). Performance analysis of multiple effect vertical still with a fat plate solar collector. *Sol Wind Technol, 4*, 451–457.

Reddy, K. S., & Sharon, H. (2016). Active multi-effect vertical solar still: Mathematical modeling, performance investigation and enviro-economic analyses. *Desalination, 395*, 99–120.

Sharon, H., & Reddy, K. S. (2015). Performance investigation and enviro-economic analysis of active vertical solar distillation units. *Energy, 84*, 794–807.

Sharon, H. (2020). Energy, exergy and enviro-economic assessment of productivity enhanced passive double sided vertical convection solar distiller for fresh water production. *Sustainable Energy Technologies and Assessments, 42*, 100846.

Xie, G., Xiong, J., Liu, H., Xu, B., Zheng, H., & Yang, Y. (2015). Experimental and numerical investigation on a novel solar still with vertical ripple surface. *Energy Conversion and Management, 98*, 151–160.

# Chapter 10
# Vertical Diffusion Solar Thermal Desalination Systems

## 10.1 Concept of Diffusion Solar Still

The productivity of solar stills can be enhanced significantly by reducing the gap between the evaporating and condensing surface. However, this is not possible in basin-type solar stills as there are huge chances of contamination of distillate with the incoming feed water. This drawback can be avoided or nearly made nil in vertical solar thermal desalination systems. Vertical solar thermal desalination system in which the gap between evaporating and condensing surface is reduced such that humid air convection gets arrested and conduction/diffusion dominates is termed as **vertical diffusion solar thermal desalination systems**. These systems are also called vertical diffusion solar stills. Convection gets suppressed when the Grashoff number drops below 2000. The recommended maximum limit for gap up to which convection between parallel plates gets suppressed is below 0.012 m. Multiple effect systems are more common configurations of vertical diffusion solar stills.

## 10.2 Thermodynamic Modeling of Vertical Multi-effect Diffusion Solar Still (VMEDS)

Necessary assumptions considered for developing an appropriate thermodynamic model for vertical multi-effect diffusion still are listed below:

(a) System is air tight, leakage-free, and perfectly insulated (Elsayed, 1986).
(b) Metallic plates and wick are assumed to be very thin such that temperature gradients across their thickness are neglected (Tanaka, 2016b).
(c) Feed water enters the system at ambient temperature and leaves the system at evaporating surface temperature.
(d) Gap between the condensing and evaporating surface is small such that convection is arrested (Elsayed, 1986; Tanaka, 2016b).



H. Sharon and K. Srinivas Reddy, *Solar Thermal Desalination Technologies for Potable Water*, https://doi.org/10.1007/978-3-031-49155-9_10

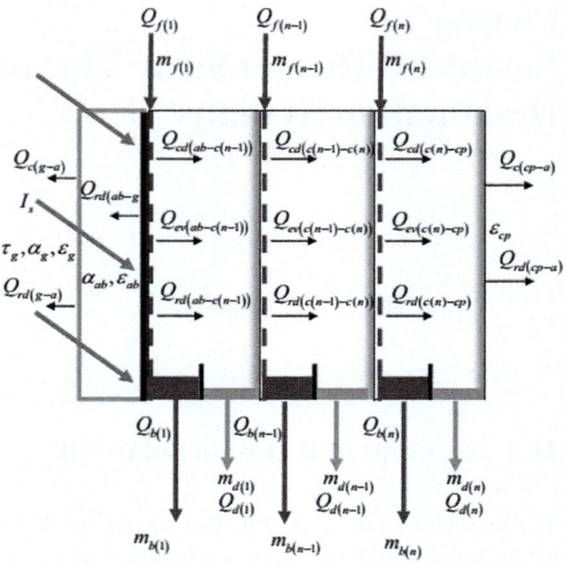

**Fig. 10.1** Energy and mass transport process in passive multi-effect vertical diffusion solar still (Sharon et al., 2020)

(e) Conduction, radiation, and evaporation heat transfer occurs between the evaporating and condensing surface (Lim et al., 2018; Tanaka & Nakatake, 2004).
(f) Radiation heat loss occurs from the absorber plate to the glass cover.
(g) Heat loss from the desalination system to the ambient is by convection and by radiation from the glass cover and last condensing surface.
(h) Feed water flow rate is twice the distillate productivity rate in each diffusion chamber (Elsayed, 1986; Tanaka, 2016b).

Energy and mass transport occurring in passive multi-effect vertical diffusion still is presented in Fig. 10.1. Transient energy balance equations for each component of '$n$' diffusion effect solar still are shown below:

Energy transport in glass cover:

$$\begin{pmatrix}\text{rate of change of}\\ \text{glass cover temperature}\end{pmatrix} = \begin{pmatrix}\text{solar radiation absorbed}\\ \text{by the glass cover}\end{pmatrix} + \begin{pmatrix}\text{radiation heat transfer from}\\ \text{absorber to glass cover}\end{pmatrix} - \begin{pmatrix}\text{convection heat transfer}\\ \text{from glass cover to ambient}\end{pmatrix} - \begin{pmatrix}\text{radiation heat transfer}\\ \text{from glass cover to ambient}\end{pmatrix}$$

$$(mc)_g \frac{dT_g}{dt} = Q_{s(g)} - Q_{rd(ab-g)} - Q_{c(g-a)} - Q_{rd(g-a)} \tag{10.1}$$

Energy transport in absorber/evaporator of 1st diffusion effect:

$$\begin{pmatrix}\text{rate of change of}\\ \text{absorber temperature}\end{pmatrix} = \begin{pmatrix}\text{solar radiation}\\ \text{absorbed by absorber}\end{pmatrix} + \begin{pmatrix}\text{energy content of}\\ \text{flowing feed water}\end{pmatrix} - \begin{pmatrix}\text{radiation heat transfer from}\\ \text{absorber to glass cover}\end{pmatrix} - \begin{pmatrix}\text{convection, radiation\&evaporation heat}\\ \text{transfer from absorber to condenser of 1st effect}\end{pmatrix} - \begin{pmatrix}\text{energy content}\\ \text{of reject brine}\end{pmatrix}$$

$$(mc)_{ab}\frac{dT_{ab}}{dt} = Q_{s(ab)} + Q_{f(1)} - Q_{rd(ab-g)} - (Q_{cd} + Q_{rd} + Q_{ev})_{(ab-c1)} - Q_{b(1)} \tag{10.2}$$

Energy transport in the evaporator of the second diffusion effect/condenser of first diffusion effect:

$$\begin{pmatrix}\text{rate of change of}\\ \text{temperature of 2nd effect}\end{pmatrix} = \begin{pmatrix}\text{convection, radiation \& evaporation}\\ \text{heat transfer from 1st effect to 2nd effect}\end{pmatrix} + \begin{pmatrix}\text{energy content of}\\ \text{flowing feed water}\end{pmatrix} + \begin{pmatrix}\text{sensible heat lost by the distillate}\\ \text{of 1st effect to the 2nd effect}\end{pmatrix} - \begin{pmatrix}\text{convection, radiation \& evaporation}\\ \text{heat transfer from 2nd effect to 3rd effect}\end{pmatrix} - \begin{pmatrix}\text{energy content}\\ \text{of reject brine}\end{pmatrix}$$

$$\begin{aligned}(mc)_{c1}\frac{dT_{c1}}{dt} =& (Q_{cd} + Q_{rd} + Q_{ev})_{(ab-c1)} + Q_{f(2)} + Q_{d(ab-c1)}\\ &- (Q_{cd} + Q_{rd} + Q_{ev})_{(c1-c2)} - Q_{b(2)}\end{aligned} \tag{10.3}$$

Energy transport in the evaporator of the $n$th diffusion effect/condenser of $(n - 1)$th diffusion effect:

$$\begin{pmatrix}\text{rate of change of}\\ \text{temperature of } n\text{th effect}\end{pmatrix} = \begin{pmatrix}\text{convection, radiation \& evaporation heat}\\ \text{transfer from } (n-1)\text{th effect to } n\text{th effect}\end{pmatrix} + \begin{pmatrix}\text{energy content of}\\ \text{flowing feed water}\end{pmatrix} + \begin{pmatrix}\text{sensible heat lost by the distillate of}\\ (n-1)\text{th effect to the } n\text{th effect}\end{pmatrix} - \begin{pmatrix}\text{convection, radiation \& evaporation heat}\\ \text{transfer from } n\text{th effect to last condenser plate}\end{pmatrix} - \begin{pmatrix}\text{energy content}\\ \text{of reject brine}\end{pmatrix}$$

$$\begin{aligned}(mc)_{c(n)}\frac{dT_{c(n)}}{dt} =& (Q_{cd} + Q_{rd} + Q_{ev})_{(c(n-1)-c(n))} + Q_{f(c(n))}\\ &+ Q_{d(c(n-1)-c(n))} - (Q_{cd} + Q_{rd} + Q_{ev})_{(c(n)-cp)} - Q_{b(c(n))}\end{aligned} \tag{10.4}$$

where $n = 2, 3, 4 \ldots$.

Energy transport in the condenser of the $n$th diffusion effect:

$$\begin{pmatrix}\text{rate of change of temperature}\\ \text{of last condenser plate}\end{pmatrix}$$
$$= \begin{pmatrix}\text{convection, radiation \& evaporation heat transfer}\\ \text{from } (n)\text{th effect to the last condenset plate}\end{pmatrix}$$
$$+ \begin{pmatrix}\text{sensible heat lost by the distillate of}\\ (n)\text{th effect to the last condenser plate}\end{pmatrix}$$
$$- \begin{pmatrix}\text{convection heat transfer from last}\\ \text{condenser plate to the ambient}\end{pmatrix}$$
$$- \begin{pmatrix}\text{radiation heat transfer from last}\\ \text{condenser plate to the ambient}\end{pmatrix}$$

$$(mc)_{cp}\frac{dT_{cp}}{dt} = (Q_{cd} + Q_{rd} + Q_{ev})_{(c(n)-cp)} + Q_{d(c(n)-cp)} + Q_{c(cp-a)} - Q_{rd(cp-a)} \quad (10.5)$$

Solar radiation absorbed by glass cover ($Q_{s(g)}$ in W) is given as

$$Q_{s(g)} = I_s A_g \alpha_g \quad (10.6)$$

where

$I_s$—solar radiation intensity over the glass cover (W/m$^2$).
$A_g$—area of the glass cover (m$^2$).
$\alpha_g$—absorbtivity of the glass cover.

Solar radiation absorbed by the absorber plate ($Q_{s(ab)}$ in W) is given as

$$Q_{s(ab)} = I_s A_{ab} \tau_g \alpha_{ab} \quad (10.7)$$

where

$A_{ab}$—area of the absorber (m$^2$).
$\alpha_{ab}$—absorbtivity of the absorber.
$\tau_g$—transmissivity of the glass cover.

Energy content of feed water entering into each diffusion effect ($Q_{f(n)}$ in W) is given as

$$Q_{f(n)} = m_{f(n)} c_{pw} (T_a + 273.15) \quad (10.8)$$

where

$m_{f(n)}$—Mass flow rate of feed water entering nth effect (kg/s).
$c_{pw}$—specific heat capacity of feed water (J/kgK).
$T_a$—ambient temperature (°C).

Energy content of brine leaving each diffusion effect ($Q_{b(n)}$ in W) is given as

$$Q_{b(n)} = m_{b(n)} c_{pb} (T_{e(n)} + 273.15) \tag{10.9}$$

where

$m_{b(n)}$—Mass flow rate of brine leaving $n$th effect (kg/s).
$c_{pb}$—specific heat capacity of brine (J/kgK).
$T_{e(n)}$—evaporating surface temperature of $n$th effect (°C).

Radiation heat transfer between absorber plate and glass cover ($Q_{rd(ab-g)}$ in W) is given by

$$Q_{rd(ab-g)} = \frac{\sigma A_{ab} \left(T_{ab}^4 - T_g^4\right)}{(1/\varepsilon_{ab}) + (1/\varepsilon_g) - 1} \tag{10.10}$$

where

$T_{ab}$—temperature of the absorber plate (K).
$T_g$—temperature of the glass cover (K).
$\varepsilon_{ab}$—Emissivity of absorber plate.
$\varepsilon_g$—Emissivity of glass cover.

Convection heat loss ($Q_{c(g-a)}$ in W) and radiation heat loss ($Q_{r(g-a)}$ in W) from the glass cover of diffusion still to the ambient is given by

$$Q_{c(g-a)} = h_{ca} A_g (T_g - T_a) \tag{10.11}$$

$$Q_{r(g-a)} = \sigma A_g \varepsilon_g (T_g^4 - T_a^4) \tag{10.12}$$

where

$h_{ca}$—convective heat transfer coefficient between glass cover and ambient (W/m$^2$K).

Conduction heat transfer between evaporator and condenser of ($n$ − 1)th diffusion effect ($Q_{cd(c(n-1)-c(n))}$ in W) is estimated by

$$Q_{cd(c(n-1)-c(n))} = K_{av} A_e \frac{T_{c(n-1)} - T_{c(n)}}{W} \tag{10.13}$$

where

$K_{av}$—thermal conductivity of air-vapor mixture (W/mK).
$A_e$—area of the evaporating surface ($m^2$).
$T_{c(n-1)}$—temperature of the $(n-1)$th effect (°C).
$T_{c(n)}$—temperature of the $(n)$th effect (°C).
$W$—Diffusion gap (m).

Radiation heat transfer between evaporator and condenser of $(n-1)$th diffusion effect ($Q_{rd(c(n-1)-c(n))}$ in W) is estimated by (Lim et al., 2018)

$$Q_{rd(c(n-1)-c(n))} = \frac{\sigma A_e (T_{c(n-1)}^4 - T_{c(n)}^4)}{(1/\varepsilon_w) + (1/\varepsilon_p) - 1} \tag{10.14}$$

where

$\varepsilon_w$—Emissivity of water.
$\varepsilon_p$—Emissivity of effect (metal partition).

Evaporation heat transfer between evaporator and condenser of $(n-1)$th diffusion effect ($Q_{ev(c(n-1)-c(n))}$ in W) is estimated by (Tanaka, 2016b)

$$Q_{ev(c(n-1)-c(n))} = \left[\frac{D_{av} P_t}{R_{wv} W T_{av}} \ln\left(\frac{P_t - P_{c(n)}}{P_t - P_{c(n-1)}}\right)\right] A_e h_{fg} \tag{10.15}$$

where

$D_{av}$—diffusivity of water vapor in air ($m^2/s$).
$R_{wv}$—specific gas constant of water vapor (J/kgK).
$T_{av}$—mean temperature of the corresponding diffusion chamber (°C).
$P_{c(n)}$—partial pressure of water vapor corresponding to the condenser temperature of $(n-1)$th effect (Pa).
$P_{c(n-1)}$—partial pressure of water vapor corresponding to the evaporator temperature of $(n-1)$th effect (Pa).
$P_t$—total pressure or atmospheric pressure (101325 Pa).
$h_{fg}$—latent heat of evaporation (J/kg).

Convection heat loss ($Q_{c(cp-a)}$ in W) and radiation heat loss ($Q_{r(cp-a)}$ in W) from last condensing surface of diffusion still to ambient is given by

$$Q_{c(p-a)} = h_{ca} A_{cp} (T_{cp} - T_a) \tag{10.16}$$

$$Q_{r(cp-a)} = \sigma A_{cp} \varepsilon_{cp} (T_{cp}^4 - T_a^4) \tag{10.17}$$

where

$h_{ca} = 2.8 + 3.0(V)$ for $V \le 5.0$ m/s.
$h_{ca} = 6.15(V)^{0.8}$ for V > 5.0 m/s.

$V$—Wind speed (m/s).
$A_{cp}$—surface area of last condensing plate ($m^2$).
$T_{cp}$—temperature of last condensing plate (K).
$\varepsilon_{cp}$—Emissivity of last condensing plate.

## 10.3 Types of Vertical Multiple Effect Diffusion Solar Stills (VMEDS)

(a) Heat pipe-integrated vertical multiple effect diffusion solar still.
(b) Basin-type vertical multiple effect diffusion solar still.
(c) Tilted wick still-integrated vertical multiple effect diffusion solar still.
(d) Flat reflector-integrated vertical multiple effect diffusion solar still.
(e) Passive multi-effect vertical diffusion solar still.

### *10.3.1 Basin-Type Vertical Multiple Effect Diffusion Solar Still*

The schematic of basin-type vertical diffusion multiple effect solar still is shown in Fig. 10.2. It is a hybrid system formed by integrating basin still and multiple effect vertical diffusion still. In this system, the vertical back wall of basin solar still acts as the absorber plate for vertical diffusion still. A part of the solar radiation transmitted through the glass cover of basin still reaches the vertical back wall of the still and gets absorbed. The back wall also acts as a condensing surface of basin still thereby some portion of humid air inside the basin condenses over the vertical wall by releasing its latent heat of condensation. Vertical absorber of vertical multiple effect solar diffusion still integrated with basin still utilizes the absorbed solar radiation and latent heat of condensation to start evaporation in its first diffusion cell. Latent heat of condensation released by the first cell/effect is utilized by the next effect for evaporation and the process continues in the same manner in all other adjacent effects. The gap between the partition plates forming each cell is less than 0.01 m. Each of the metal partitions is lined with wick over its inner surface to facilitate the effective distribution of feed water over its evaporating surface. The fibers of wick after absorbing feed water become heavier and get suppressed by gravity thereby, contact of wicks with the condensing surface is avoided (Tanaka et al., 2000a). As productivity of vertical diffusion still is higher than basin still, double glass cover can be used in basin still to prevent heat losses from it thereby, this heat energy can be used as additional heat source to produce more distillate in the integrated vertical diffusion system.

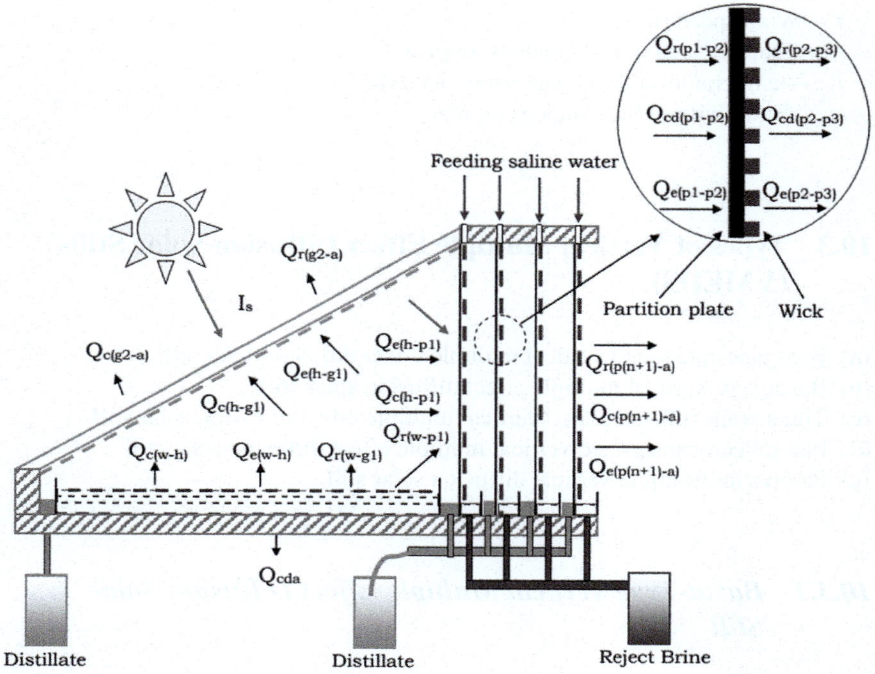


**Fig. 10.2** Basin-type vertical multiple effect diffusion solar still (Tanaka et al., 2000b)

### *10.3.2 Basin-Type Vertical Multiple Effect Diffusion Solar Still with Solar Pond*

Diurnal and nocturnal distillate production of basin-type vertical four effect diffusion still can be enhanced by utilizing multiple floating cotton wicks in basin and integrating the system with mini solar pond, respectively (Dhindsha & Mittal, 2018). The Schematic of such a system is shown in Fig. 10.3. Basin section of distillation system (1.15 $m^2$) is provided with double glass cover (1.32 $m^2$) to reduce thermal losses to the ambient and to increase the available latent heat energy for utilization by the vertical four effect diffusion still. Trapezoidal mini solar pond of top and bottom areas 2.65 and 0.50 $m^2$ and depth 0.90 m incorporated with 2.65 $m^2$ reflector (aluminum sheet) and copper heat exchanger is used to supply heat energy to the basin type VMEDS using a 50 W circulation pump. The total distillate production rate of this system is about 56.92% higher in comparison to the system without solar pond and floating wicks. Incorporating floating wicks and solar pond has drastically increased diurnal and nocturnal productivity by about 50.0% and 71.0%, respectively. The thermal efficiency of the system with solar pond is about 80.29% and is higher than the efficiency of the bare system by about 21.0%. Thermal efficiency of the distillation system integrated with solar pond can be estimated by

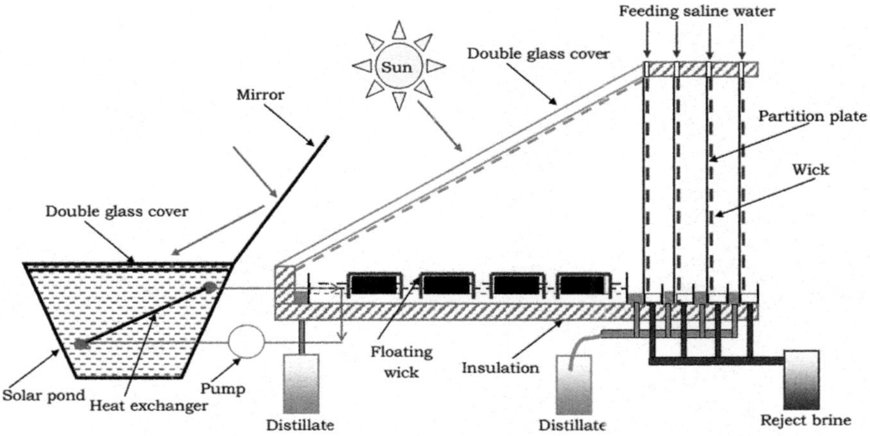


**Fig. 10.3** Trapezoidal solar pond integrated floating wick basin-type-vertical multiple effect diffusion solar still (VMEDS) (Dhindsha & Mittal, 2018)

$$\eta_{th} = \frac{m_d \times h_{fg}}{\underbrace{(I_t \times A_g)}_{\text{incident solar energy}} + \underbrace{(mc_p\Delta T)_{(9:00 to 9:30 pm)} + (mc_p\Delta T)_{(3:00 to 3:30 am)}}_{\text{Energy supplied by solar pond}} + \underbrace{W_p}_{\text{pump work}}} \tag{10.18}$$

### 10.3.3 *Basin-Type Vertical Multiple Effect Diffusion Solar Still with Portable Electric Generators*

Waste heat utilization to produce useful work is always recommended as it increases efficiency and minimizes pollution and fossil fuel consumption. Park et al. (2014) stressed that diffusion stills with back up heating system could make them more economically feasible and acceptable among the public. The schematic of basin-type vertical two effect diffusion still of effective aperture area 0.77 m$^2$ tested by supplying waste exhaust gas at 283 °C from a four-stroke portable gasoline generator (5.6 KVA) is shown in Fig. 10.4. Basin section is provided with a serpentine-type stainless heat exchanger of 0.025 m diameter and 6.0 m length to transfer waste heat energy from the exhaust gas to the basin water. Glass cover of the basin system is covered to prevent direct exposure to the sun and wind during experimentation. Distillate production increased from 14.1 kg/d to 21 kg/d in a linear fashion with increase in heat input from 3000 to 5800 kJ/h. Amount of distillate generated from the basin section is larger than that generated in the two-effect diffusion part of the system.

A similar system with eleven diffusion effects supplied with heat energy from solar radiation and waste exhaust gas has been tested under the climatic condition

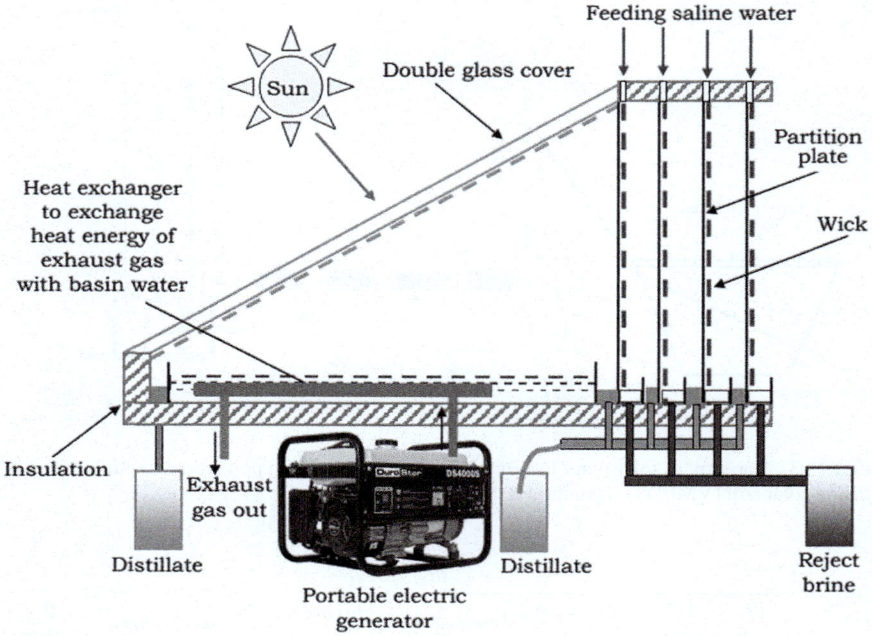


**Fig. 10.4** Basin-type-VMEDS utilizing waste exhaust heat from portable generator (Park et al., 2016)

of Daejeon, South Korea (Park et al., 2016). Sea water temperature in basin section and first metallic partition of diffusion still reached a steady-state value of about 72 °C in 2 h and 63 °C in 4 h, respectively, for heat input of 22.37 MJ. Low amount of distillate is obtained from the basin section for a minimum water depth (0.01 m) due to the partial submergence of heat exchanger tube (0.025 m) in basin water. High productivity is observed in the second effect of vertical diffusion system and it dropped with a further increase in the number of effects. Productivity from vertical eleven effect diffusion section is about 5.54 times larger than basin section for a heat input of 22.37 MJ. Performance ratio of the system is about 13.4–16.7 and the system requires at least 3.0 MJ/d of heat input to produce distillate.

Cumulative distillate productivity ($M_d$ in kg/m$^2$-d) of this desalination system is correlated to heat input ($Q_{in}$ in MJ/d) by the following relation (Park et al., 2016):

$$M_d = -2.0 + 0.89(Q_{in}) \tag{10.19}$$

Yeo et al. (2019) investigated the performance of similar system by adding reflecting fins to the heat exchanger surface and by using various heat sources like solar energy, waste heat from generator (5.6 kVA), and electric heater (1.4 kW). Schematic of the system is shown in Fig. 10.5. For the same heat input, productivity and perfor-

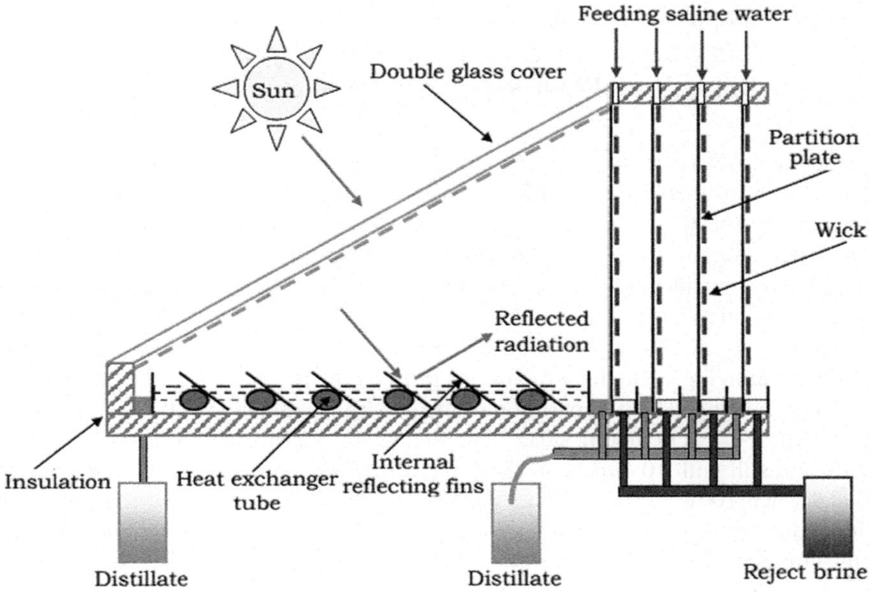


**Fig. 10.5** Basin-type-VMEDS system with heat exchanger and reflecting fins (Yeo et al., 2019)

mance ratio of the system is higher for electric heater and waste heat-based energy sources rather than solar energy. Energy from electric heater and waste heat from exhaust gas are directly absorbed by basin water, while solar radiation has to be transmitted through double glass cover and some of it will be reflected away by the reflecting fins leading to reduced performance of the system under solar-only condition.

Insulating side glass covers of the system has helped to enhance the productivity of solar-only system by about 16.7% as it prevents heat lost to the ambient. Depth of feed water needs to be a minimum (0.005 mm) for solar-only condition and it has to be a maximum (0.015 mm) for waste heat-only mode as submergence of heat exchanger in basin water is required for enhanced heat transfer from the exhaust gas tube to the basin water. Incorporating reflecting fins is found to be a bad option in these types of stills as it drops the production capacity of the system due to resistance caused by improper contact with heat exchanger surface and improper orientation. Performance ratio of the system is lower for solar-only mode (0.71) and higher for waste heat-only mode (2.32).

### 10.3.4 Basin-Type Vertical Multiple Effect Diffusion Solar Still with External Heat Exchangers for Recovering Waste Heat From Reject Hot Brine

Temperature of reject brine from vertical diffusion stills is high in comparison to the ambient or inlet feed water temperature. Utilizing the waste thermal energy of reject brine for preheating feed water could increase distillate productivity, system efficiency and minimize energy requirement. However, magnitude of performance improvement achieved by incorporating this technique depicts the acceptability of this modification.

Nosoko et al. (2005) theoretically investigated the performance increment of VMEDS utilizing waste heat recovered from its distillate and reject brine for feed water preheating. The first partition of the system is maintained at a fixed temperature of 100 °C by circulating steam. Coefficient of performance (COP) of the vertical diffusion still with 10 effects and 0.005 m diffusion gap increased rapidly from 4.6 to 7.4 when 100% of the distillate and reject brine heat energy were recovered and utilized for feed water preheating. The rapid increase in COP is due to reduced steam consumption of the system because of feed water preheating. However, only a 3.82% increment in distillate production was noticed. COP is mathematically expressed as

$$\text{COP} = \frac{\text{Total heat of evaporation from all effects}}{\text{Enthalphy of steam supplied}} \tag{10.20}$$

Kaushal et al. (2017a) experimentally investigated the performance of basin-type vertical four effect diffusion still supplied with feed water preheated by waste heat energy of its reject brine. The schematic of this configuration is shown in Fig. 10.6. Thermal energy of hot reject brine is recovered using tube in counter-flow tube heat exchanger and is used to preheat the incoming feed water. Flow rate of preheated feed water is maintained at 2.5 times higher than the distillate production rate and is distributed equally to all the diffusion effects. Circulation of feed water through the heat exchanger has been achieved by gravity using an overhead tank. Around 10 °C increment in feed water temperature is noticed throughout the day due to preheating caused by reject brine. On partial cloudy day (18.5 MJ/m$^2$-d cumulative solar intensity) with high ambient temperature (37.7–40.8 °C) distillate productivity improvement of the modified system is about 14% higher than the reference system without feed water preheating and floating wick. On a clear sunny day (21.1 MJ/m$^2$-d cumulative solar intensity) with low ambient temperature (33.0 °C), distillate productivity improvement of the modified system is about 21.2% higher than the reference system without feed water preheating and floating wick. Low distillate production enhancement percentage observed for the day with high ambient temperature is due to the fact of reduced heat losses from the condensing surface which in turn reduced the temperature difference between the vertical plates resulting in low cumulative distillate production rate of the system.

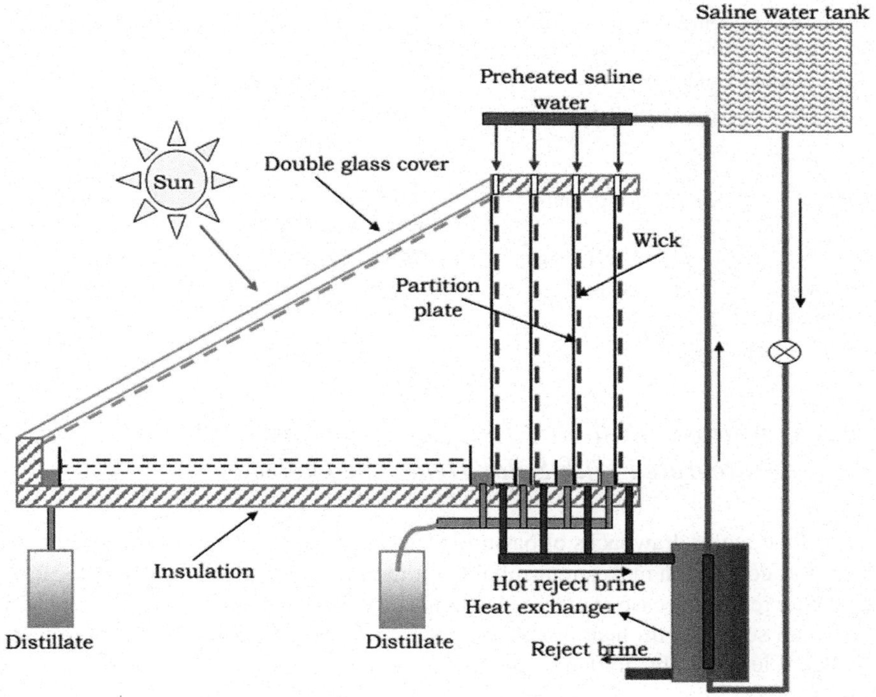


**Fig. 10.6** Basin-type-VMEDS utilizing waste heat of reject for preheating feed water (Kaushal et al., 2017b)

The system performed very well without any maintenance problem for more than 1 year under the climatic condition of Patiala (30.3 °N 76.3 °E), India. Daily distillate productivity ($M_d$ in kg/m$^2$d) of basin-type vertical multi-effect diffusion still supplied with preheated feed water can be related to cumulative solar radiation intensity ($G_t$ in MJ/m$^2$d), daily average ambient temperature ($T_a$ in °C), number of effects ($n$), diffusion gap ($\delta_p$ in mm), daily average feed water flow rate ($m_f$ in g/m$^2$s) and daily average feed water temperature ($T_f$ in °C). The relation can be expressed in terms of nonlinear power equation based on experimental results obtained by Kaushal et al. (2017b)

$$M_d = 0.0004(G_t)^{0.730}(T_a)^{1.585}(T_f)^{0.377}(n)^{0.401}(\delta_p)^{-0.068}(m_f)^{-0.102} \quad (10.21)$$

The above equation is able to predict values with a deviation of about −17% to +14% from the experimental values and can be used under the conditions displayed in Table 10.1.

**Table 10.1** System and ambient parameters considered for development of correlation for estimation of daily distillate production rate (Kaushal et al., 2017b)

| System design conditions | | Weather parameters | |
|---|---|---|---|
| Number of effects ($n$) | 2–7 | Cumulative solar radiation intensity ($G_t$) | 16.3–24.6 MJ/m$^2$-d |
| Diffusion gap ($\delta_p$) | 10–16 mm | Average ambient temperature (Ta) | 33–40 °C |
| Feed water rate ($m_f$) | 0.19–0.38 g/m$^2$-s | Wind velocity (V) | 0.3–7.2 m/s |
| Feed water temperature ($T_f$) | 34–49 °C | | |

### 10.3.5 Vertical Multiple Effect Diffusion Solar Still Integrated With Solar Heat Pipe Collector

One of the major drawbacks of basin-type VMEDS is its large volume, weight, and transport cost which makes it unsuitable for mobile applications in remote locations and other regions in case of emergency. This drawback has been rectified by replacing the basin section with heat pipes and tilted wick still and this modification makes VMEDS less weight, foldable, and easily portable to any location without much shipping cost.

The schematic of heat pipe solar collector integrated VMEDS is shown in Fig. 10.7. Copper tubes attached below the absorber plate of solar collector represent the evaporator section of heat pipe and are connected to the copper tubes attached to the front side of the first metal plate of VMEDS which represents the condenser section of heat pipe thereby, the tube arrangement becomes a closed loop. The heat pipe tubes were filled partially with working fluid and evacuated to facilitate ease of evaporation of the working fluid. Working fluid gets vaporized by absorbing heat energy from the absorber plate of solar collector and moves towards the condenser section where it gets condensed by releasing heat energy to the first partition of vertical multiple effect diffusion still thereby, distillation process starts in this VMEDS. The front side of first metal plate along with condenser section is covered by insulation to prevent heat loss to the ambient.

Tanaka & Nakatake (2004) theoretically investigated VMEDS integrated with ethanol heat pipe solar collector under the climatic conditions of Okinawa (26.1 °N latitude), Japan. Aperture area of solar heat pipe collector and evaporation area of each diffusion effect is 2.0 m$^2$. Vertical ten effect diffusion still with 0.003 m and 0.005 m diffusion gap is able to produce around 21.8 and 19.2 kg/m$^2$-d of distillate, respectively, for a cumulative incident solar radiation intensity of about 24.4 MJ/m$^2$-d. Heat pipe-integrated VMEDS produced 13.0% higher distillate than basin-type VMEDS. Solar heat pipe collector tilt angle lower than the latitude of the test location is needed during summer solstice day and tilt angle twice the latitude of the location is required during winter solstice day for maximum distillate production

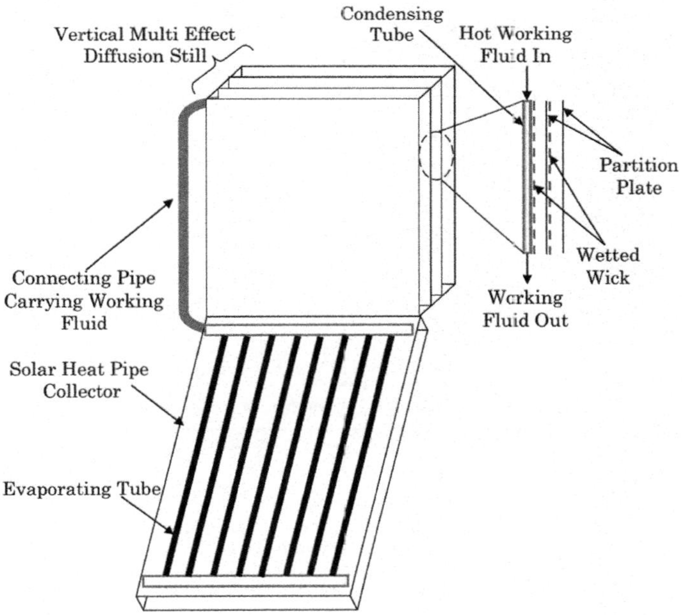


**Fig. 10.7** VMEDS integrated with solar heat pipe collector (Tanaka et al., 2005)

of VMEDS. For spring, autumn equinox day and overall year-round performance, tilt angle of heat pipe collector equal to the latitude of the concerned location is recommended. Ratio of solar heat pipe collector area to the first partition area of vertical diffusion still must be 1 to have maximum distillate productivity. However, the ratio of 2 is recommended for practical purposes to eliminate requirement of heavy feed water tanks and pumps for feed water tank re-filling (Tanaka et al., 2004).

### *10.3.6 Vertical Multiple Effect Diffusion Solar Still Integrated With Heat Pipe Utilizing Waste Heat from Portable Generators and Biomass Stoves as Heat Source*

Most of the water-scarce regions lack both electric power and gas supply for cooking. In these zones, mostly portable generators and biomass stoves are used to meet the power and cooking needs, respectively. In both these activities, a lot of heat energy is lost to the ambient through hot exhaust gas. These high-temperature exhaust gases can be used to generate potable water from waste or saline feed water using diffusion stills. Direct exposure of distillation system to hot exhaust gas could damage the system which may increase the maintenance and capital cost. However, heat pipes

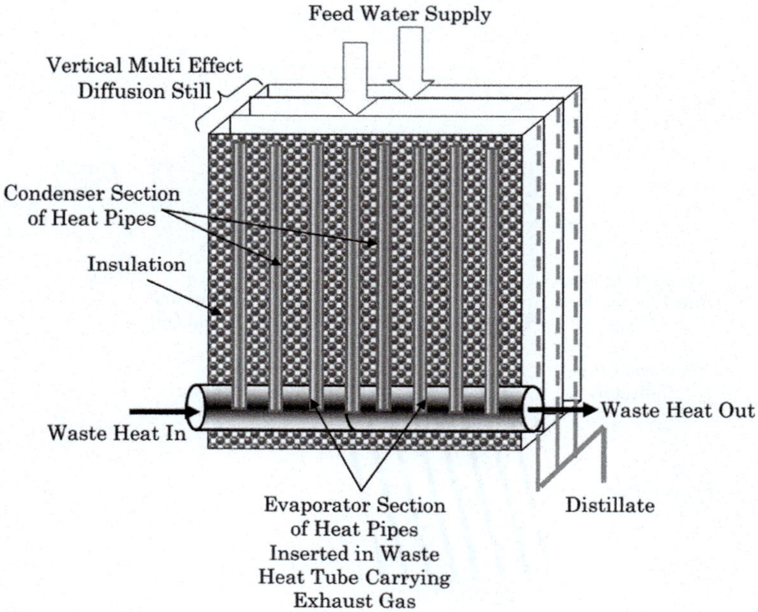


**Fig. 10.8** Schematic of VMEDS with heat pipe utilizing heat energy from exhaust gas of generator (Tanaka & Park, 2010)

can be used to transport heat energy from the exhaust gas to the evaporator section of the desalination system.

Tanaka & Park (2010) proposed and theoretically investigated the performance aspects of VMEDS with heat pipe utilizing waste heat from exhaust of portable generator. A schematic representation of the system is shown in Fig. 10.8. Vertical heat pipes (0.01 m diameter, 1.0 m length) with constant mass of distilled water as working fluid are attached to the back side of the first metal plate of diffusion still (1.0 $m^2$) which is then insulated to prevent heat loss to the ambient. Lower end of heat pipes is immersed inside horizontal waste gas pipe through which hot exhaust gas (250 °C) from portable generator flows. Energy analysis indicated the loss of nearly 52.5% of exhaust gas heat energy to the ambient and is due to the high temperature of the exhaust gas (140 °C) leaving the waste gas pipe. Better utilization of the exhaust gas can be achieved by incorporating fins and increasing the residence time of exhaust gas in waste gas pipe but this may increase back pressure and affect electric generator's performance negatively which is not acceptable. Increasing the number of diffusion effects increases distillate production rate in a rapid manner for high mass flow of hot exhaust gas. COP is above 1.0 for the distiller with number of partitions above 4 due to effective reuse of latent heat of condensation. Approximately, 20.0 kg of distillate can be produced in 3 h by vertical 10-effect diffusion system utilizing 40.0 $m^3$/h of exhaust gas at 250 °C.

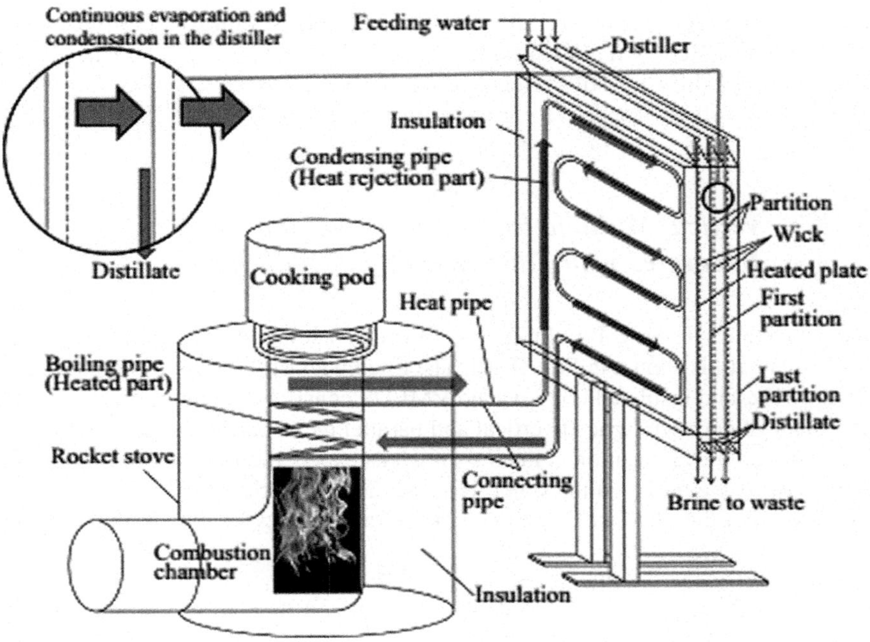


**Fig. 10.9** Schematic of VMEDS integrated with rock stove via heat pipe (Tanaka, 2016b)

Tanaka & Park (2012) validated the theoretical results from a previous study by carrying out experiments in heat pipe-integrated single-effect vertical diffusion still utilizing exhaust gas from portable generator as heat source. Temperature of the distiller components and its distillate production rate reached steady state in 20 min and 60 min, respectively. About, 65–85% of the heat energy transported by the heat pipe has been utilized for distillate production effectively in the diffusion still. COP of single effect system is around 0.35 and it increased to 2.5 for 10 effect system. However, COP gets affected negatively by increased diffusion gap.

Tanaka (2016a) developed and experimented single-effect and four-effect vertical diffusion stills with 0.010 m diffusion gap integrated with distilled water-based heat pipe. Hot exhaust gas from highly efficient rocket elbow stove is used as heat source for the heat pipe. The schematic of this system is shown in Fig. 10.9. Evaporator section of heat pipe has three-turn helical coil made of a 0.012 m diameter copper tube and is placed over the combustion chamber of the stove. Condenser section of heat pipe is made from pipe of similar diameter and is attached to the first metal plate (0.23 $m^2$) of vertical diffusion still. Two connecting pipes each of length 0.5 m are used to connect evaporator and condenser section of heat pipe. Side of the first metal plate with condenser section and connecting pipes was insulated to prevent heat loss to the ambient. Dried Japanese cedar wood having a gross heating value of 15.0 MJ/kg is used as feed for stove. Temperature of combustion chamber and first metal plate reached 800 °C and 90 °C, respectively, within 20 min from the start of experiment. Required cooking time is generally less than 2 h. The single-effect and four-effect diffusion stills integrated with rocket elbow stove produced 0.80 and 1.35 kg of distillate, respectively, during this time period.

### 10.3.7 *Vertical Multiple Effect Diffusion Solar Still Integrated with Tilted Wick Solar Still*

The schematic of vertical multiple effect diffusion solar still integrated with tilted wick solar still is shown in Fig. 10.10. The tilted wick still consists of double-glass cover, wick, and insulated bottom. The VMEDS consists of double-glass cover and a number of narrowly spaced vertical and parallel metal partitions. Inner surface of each vertical metal partition is lined with wicks expect the last partition. Wicks are continuously wetted by saline water. Wick of the first vertical partition is connected to the wick of the tilted still such that concentrate leaving the first vertical partition

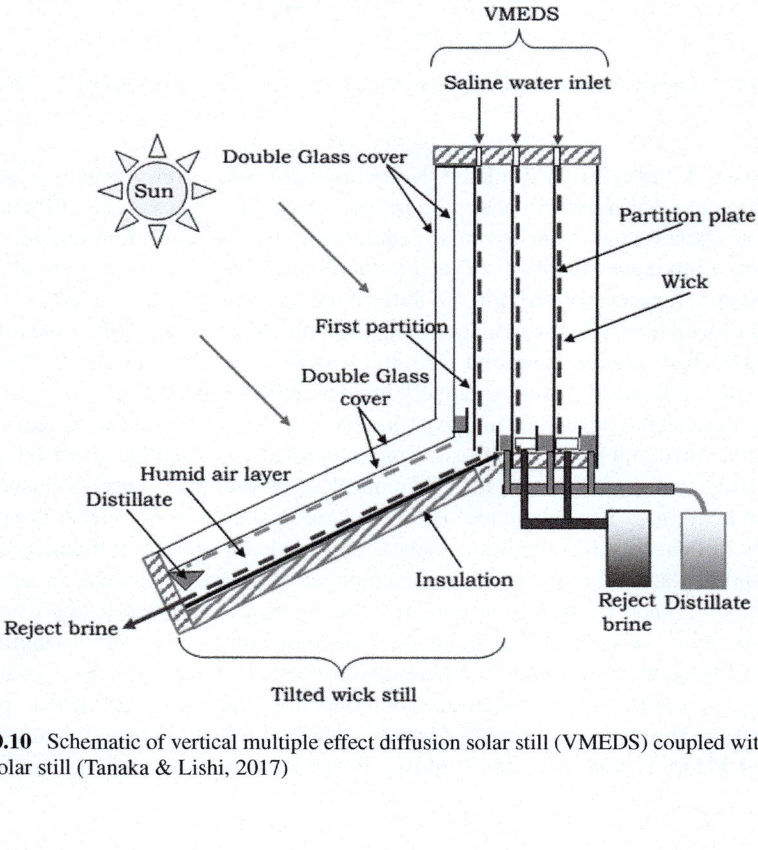


**Fig. 10.10** Schematic of vertical multiple effect diffusion solar still (VMEDS) coupled with tilted wick solar still (Tanaka & Lishi, 2017)

is fed to the tilted solar still. The humid air layer of tilted solar still is in connection with the gap between the first vertical metal partition and vertical glass cover such that part of the vapor generated inside the tilted still moves up by natural convection and condenses over the outer surface of first vertical metal partition and inner surface of vertical glass cover. Solar radiation is the only heat source for tilted solar still, while solar radiation in combination with latent heat of condensation of vapors condensing over the outer surface of first vertical partition act as heat source for VMEDS system. Latent heat of condensation released from first partition is used for feed water evaporation in the next partition. In this manner, heat energy supplied to the first partition is effectively recycled in multiple effect system to achieve increased distillate productivity.

The major advantages of this arrangement are as follows:

(a) Inclination angle of tilted wick still can be adjusted according to seasons to tap more solar radiation.
(b) Cost of tilted solar still is much lower than solar collector.
(c) Flat plate mirror integrated VMEDS needs to be rotated according to sun's position in sky to absorb solar radiation effectively while this avoided in tilted still-integrated VMEDS.
(d) Size of this system is smaller than basin-type VMEDS thereby, reduced construction, transportation, and maintenance cost can be expected.

### *10.3.8 Vertical Multiple Effect Diffusion Solar Still Integrated With Reflectors*

Heat pipe-integrated VMEDS is highly productive and portable but one of its major drawbacks is the requirement of very tight and perfect sealing arrangement for proper operation of heat pipe. This hinders its fabrication by unskilled local workers in developing or under privileged countries. This drawback leads to the replacement of heat pipe collector in VMEDS with reflecting mirror. A schematic of reflector-integrated VMEDS is shown in Fig. 10.11. It consists of VMEDS, adjustable flat mirrors, and casters for manual azimuth tracking of the whole system. Front surface of first partition of VMEDS receives solar radiation reflected by the flat mirror apart from the global solar radiation received by it because of its vertical orientation. Glass cover is also provided in front of the first partition to prevent/minimize heat losses to the ambient.

Theoretical investigation on the performance of the system in equator (0° Latitude) was carried out by Tanaka & Nakatake (2005a). Path of the sun in the sky for different days in equator and 40°N latitude along with azimuth angle ($\Upsilon$) of the still in different directions is shown in Fig. 10.12. It could be inferred that the cumulative solar radiation absorbed by the first partition of VMEDS would be equal in both spring and autumn equinox days. Similarly, still oriented towards the south in winter solstice and towards the north in summer solstice days would receive same solar radiation

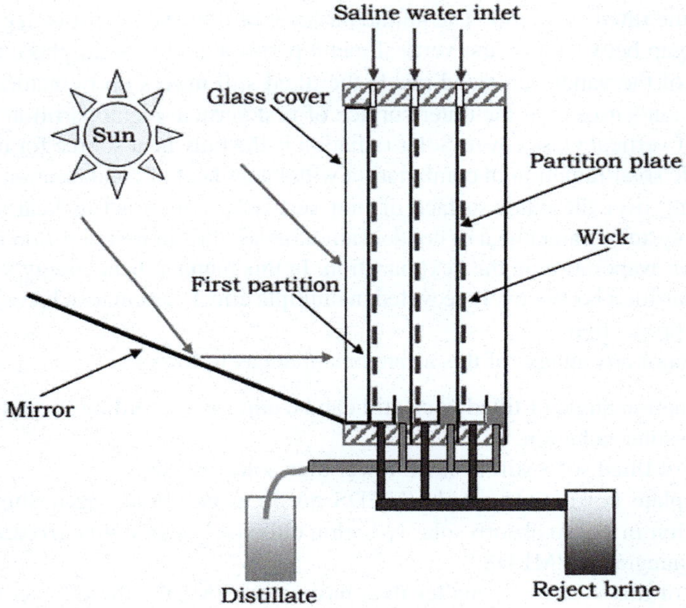


**Fig. 10.11** Schematic of adjustable VMEDS unit integrated with flat plate reflector (Tanaka, 2009)

intensity when rotated (east to west) clockwise and counter clockwise, respectively. Optimum azimuth angle for operation of the system in spring/autumn equinox day is −90° (towards east) during morning and +90° (towards west) during evening. Optimum azimuth angle for operation of the system in summer/winter solstice day is −45° (towards north/southeast) during morning and +45 (towards north/southwest) during evening hours. Optimum flat mirror orientation angle ($\theta$) from horizontal is about 15 ° on spring/autumn equinox day and 8 ° on winter/summer solstice day. The rotation of the system to adjust its azimuth position is done manually once in a day.

Table 10.2. presents the cumulative solar radiation reaching the first partition of VMEDS with, without mirror and rotation under optimum conditions. Cumulative solar radiation intensity received by the system is higher in winter and summer solstice days in comparison to spring/ autumn equinox days. Flat mirror (1.0 $m^2$) integrated vertical ten effect diffusion still (0.010 m diffusion gap; 1.0 $m^2$ aperture area) can generate at least 29.2 and 34.5 kg/$m^2$-d of distillate during spring equinox and winter solstice days, respectively, in 0 ° Latitude location.

Tanaka & Nakatake (2007a) recommended to fix flat reflector at 10 ° and 0 ° tilt angle for optimum performance of the flat mirror integrated VMEDS during summer and winter seasons in high latitude locations (above 10 °N). Month-wise recommended tilt angle of flat mirror, rotation angle of distiller along with its daily distillation production rate for various locations based on latitude is presented in Table 10.3.

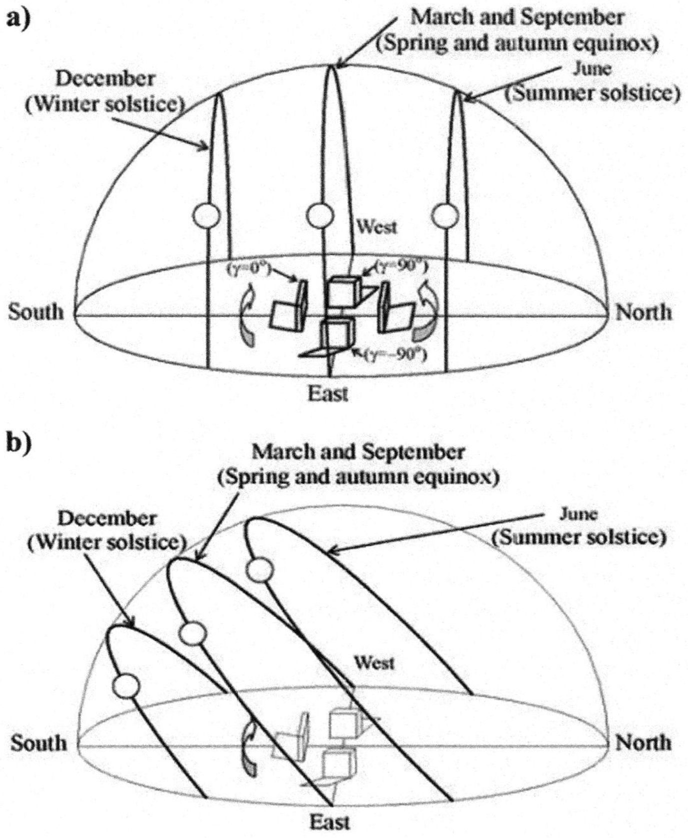


**Fig. 10.12** Path of the sun in sky at **a** 0° Latitude and **b** 40°N Latitude (Tanaka & Nakatake, 2007a)

Tanaka & Nakatake (2007b) conducted experiments on single effect vertical diffusion still (0.010 m diffusion gap) integrated with flat mirror finished metal plate under the climatic conditions of Kurume (33.2 N latitude), Japan with tap water as feed. Area of first partition of distillation system receiving solar radiation and flat plate reflector are kept equal (0.097 $m^2$). Absorptivity of first partition, emissivity of first partition, and reflectivity of reflector are 0.92, 0.32, and 0.95, respectively. Summary of experimental results on single effect vertical diffusion still integrated with flat reflector are tabulated in Table 10.4. Low productivity of the system during July with high global horizontal solar radiation intensity is due to the fact that in summer most of the rays would be parallel to vertical surface facing south near noon as a result large portion of the radiation would not be effectively utilized by the system for distillate production. Productivity of the system is reasonably low in winter. These drawbacks can be easily overcome by utilizing multiple effect sys-

**Table 10.2** Cumulative solar radiation intensity reaching the first partition of VMEDS under different conditions in 0° Latitude (Equator) (Tanaka & Nakatake, 2005a)

| Type of day | Orientation | Mirror | Rotation | Cumulative Solar Radiation Intensity ($MJ/m^2$-d) |
|---|---|---|---|---|
| Spring/autumn equinox | Towards south | No | No | 7.5 |
| Spring/autumn equinox | Towards south | Yes | No | 13.5 |
| Spring/autumn equinox | Towards south | Yes | East to West | 24.0 |
| Winter solstice | Towards south | No | No | 9.5 |
| Winter solstice | Towards south | Yes | No | 18.0 |
| Winter solstice | Towards south | Yes | Southeast to Southwest | 25.6 |

tem instead of single-effect system. Tanaka (2009) obtained distillate yield of about 5 times higher than the single-effect system with four-effect vertical diffusion still integrated with flat reflector.

Occurrences of air bubbles on wick surface have been observed under high solar radiation intensity due to evolution of dissolved gases from the feed water at high temperatures. This problem can be rectified by attaching the wick over the metal partition using thin thermal melting film. High productivity of the flat mirror-based VMEDS justifies it to be a suitable distiller for desalination in remote and coastal villages (Tanaka & Nakatake, 2007a).

## 10.4 Parameters Influencing the Performance of Vertical Multiple Effect Diffusion Solar Still (VMEDS)

Distillate production capacity of vertical diffusion stills is influenced by system design and operational parameters. Design parameters include tilt angle, material properties, diffusion gap thickness, and number of diffusion effects. Operational parameters include heat energy input, ambient temperature, wind speed, basin water depth. and saline feed water flow rate.

### *10.4.1 Diffusion Gap Thickness*

Diffusion gap thickness refers to the horizontal space between the evaporating and condensing surfaces in vertical diffusion stills. The gap thickness is so small such

**Table 10.3** Optimum tilt angle of mirror, orientation distiller, and daily production rate of distiller for various months and locations (Tanaka & Nakatake, 2007a)

| Location | 0°N | | | 10°N | | | 20°N | | | 30°N | | | 40°N | | |
|---|---|---|---|---|---|---|---|---|---|---|---|---|---|---|---|
| Month | $\theta^*$ | $\Upsilon^*$ | $M_d$ | $\theta^*$ | $\Upsilon^*$ | $M_d$ | $\theta^*$ | $\Upsilon^*$ | $M_d$ | $\theta^*$ | $\Upsilon^*$ | $M_d$ | $\theta^*$ | $\Upsilon^*$ | $M_d$ |
| January | 10 | ±45 | 32.5 | 10 | ±45 | 38.0 | 0 | ±22.5 | 38.0 | 0 | ±22.5 | 35.0 | 0 | ±22.5 | 27.5 |
| February | 10 | ±67.5 | 31.5 | 10 | ±45 | 32.5 | 10 | ±45 | 35.0 | 0 | ±45 | 38.0 | 0 | ±22.5 | 35.0 |
| March | 10 | ±90 | 30.0 | 10 | ±67.5 | 30.0 | 10 | ±67.5 | 32.0 | 10 | ±45 | 35.0 | 0 | ±45 | 38.0 |
| April | 10 | ±112.5 | 31.5 | 10 | ±90 | 30.0 | 10 | ±90 | 30.0 | 10 | ±67.5 | 32.5 | 10 | ±45 | 35.0 |
| May | 10 | ±135 | 32.5 | 10 | ±112.5 | 31.0 | 10 | ±90 | 30.0 | 10 | ±90 | 32.5 | 10 | ±67.5 | 35.0 |
| June | 10 | ±135 | 33.0 | 10 | ±112.5 | 31.5 | 10 | ±112.5 | 30.0 | 10 | ±90 | 32.5 | 10 | ±67.5 | 35.0 |
| July | 10 | ±135 | 32.5 | 10 | ±112.5 | 31.0 | 10 | ±90 | 30.0 | 10 | ±90 | 32.0 | 10 | ±67.5 | 35.0 |
| August | 10 | ±112.5 | 31.5 | 10 | ±90 | 30.0 | 10 | ±90 | 29.0 | 10 | ±67.5 | 32.5 | 10 | ±45 | 35.0 |
| September | 10 | ±90 | 30.0 | 10 | ±67.5 | 30.0 | 10 | ±67.5 | 31.5 | 10 | ±45 | 35.0 | 0 | ±45 | 38.0 |
| October | 10 | ±67.5 | 31.0 | 10 | ±45 | 32.5 | 0 | ±45 | 35.0 | 0 | ±45 | 38.0 | 0 | ±22.5 | 32.5 |
| November | 10 | ±45 | 32.5 | 10 | ±45 | 35.0 | 0 | ±22.5 | 36.0 | 0 | ±22.5 | 35.0 | 0 | ±22.5 | 27.5 |
| December | 10 | ±45 | 35.0 | 0 | ±45 | 38.0 | 0 | ±22.5 | 38.0 | 0 | ±22.5 | 32.5 | 0 | ±22.5 | 22.5 |

$\theta^*$–optimum mirror tilt angle; $\Upsilon^*$–optimum azimuth orientation of still; $\pm\Upsilon$ represents $-\Upsilon$ (east) to $+\Upsilon$ (west)
$M_d$—daily distillate production rate of still (kg/m$^2$d)

**Table 10.4** Summary of experimental results obtained from single effect vertical diffusion still integrated with flat reflector (Tanaka & Nakatake, 2007b)

| Month | $\theta$ | $\Upsilon$ | $\sum I_h$ (MJ/m$^2$d) | $\sum I_t$ (MJ/m$^2$d) | $M_{d,exp}$ (kg/m$^2$d) | $M_{d,cal}$ (kg/m$^2$d) | $(M_{d,exp}/M_{d,cal}) - 1$ (%) |
|---|---|---|---|---|---|---|---|
| July | 0° | ±90° | 25.0 | 15.5 | 2.75 | 2.95 | −6.8 |
| October | 0° | ±22.5° | 16.5 | 16.6 | 4.19 | 4.10 | +2.2 |
| November | 0° | ±22.5° | 8.1 | 7.8 | 1.06 | 1.51 | −28.2 |

$\sum I_h$—global solar radiation horizontal surface; $\sum I_t$—solar radiation absorbed by first partition $M_{d,exp}$—daily distillate obtained from experiments; $M_{d,cal}$—daily distillate predicted theoretically

that free convection is suppressed and conduction occurs. Distillate productivity is indirectly proportional to diffusion gap such that it decreases exponentially with increase in diffusion gap thickness. Increased diffusion gap indicates increased air space which leads to large thermal resistance between the evaporating and condensing surface resulting in reduced productivity of the distiller (Tanaka & Nakatake, 2004). Diffusion gaps in vertical diffusion stills can be maintained lower than the gap in basin stills because fibers from wick are suppressed by gravity hence chance of contamination of distillate by incoming brine is reduced/avoided (Tanaka et al., 2000a). Overall distillate productivity of ethanol heat pipe integrated VMEDS increased by 15.0% with reduction in diffusion gap from 0.005 to 0.003 m (Tanaka & Nakatake, 2004).

Productivity enhancement by about 31.0% and 78.0% are observed on basin still integrated VMEDS with reduction in gap from 0.01 to 0.005 m and 0.01 to 0.002 m, respectively (Tanaka et al., 2000b). Bending of metal partitions by gravity, thermal expansion, and wind loads can cause contact of evaporating wick with condensing surface leading to distillate contamination (Tanaka et al., 2004). Too small gap can also cause re-absorption of distillate by the evaporating wick. This can be avoided if spacers are used between the evaporating and condensing surfaces (Tanaka & Nakatake, 2005b). For practical purposes, diffusion gap thickness of 0.01 m is highly recommended but this will be accompanied by a penalty of 16.5% drop in daily distillate productivity.

### 10.4.2 Number of Diffusion Effects

Number of diffusion effects refers to number of vertical metallic partitions used in VMEDS. In additional diffusion effects, latent heat of condensation released by previous effects/first effect is utilized for evaporation of feed saline water to increase distillate productivity. Magnitude of distillate production in each effect drops from first effect to the last effect because the temperature difference between the adjacent effects drops as moved from first to last (Tanaka et al., 2000a). Daily productivity increased rapidly for system with lower number of effects and it reached a saturation point for higher number of effects. Tanaka et al. (2000b) observed saturation for effects larger than 20. The occurrence of saturation is mainly due to increased thermal resistance and reduced heat energy entry through diffusion effects after a particular number of effects.

It is always recommended to select optimum number of effects for VMEDS for enhanced and economic distillate production. Optimum number of effects also depends on the gap and heat input. Tanaka & Nakatake (2004) recommended 9 and 11 effects for heat pipe integrated VMEDS with 0.003 and 0.005 m diffusion gap based on economic analysis. Tanaka & Park (2010) suggested selection of optimum number of effects for portable electric generator integrated VMEDS to be based on capacity of generator, drinking water demand and capital cost of distiller. Lim et al. (2018) found that the optimum number of diffusion effects is season-based due to variation in solar radiation intensity with season. However, 11 numbers of effects are recommended for year-round economic performance of the distiller.

### 10.4.3 Tilt Angle

Overall distillate productivity of basin-type VMEDS increases with increase in tilt angle between glass cover and horizontal basin on winter solstice day due to increase in solar radiation reaching the glass cover and incident over the first partition of VMEDS. On summer solstice day, distillate productivity decreased with increase in tilt angle because of low radiation reaching the glass cover of basin and first partition of diffusion still. On spring and autumn equinox days, distillate productivity first increased with tilt angle up to certain value (twice the value of local latitude) and then remains nearly constant (Tanaka et al., 2000b).

Impact of variation in metal partition tilt angle from 10 ° to 90 ° on tiltable MEDS productivity was investigated by Lim et al. (2018) for various days. Tilt angle of about 40 ° and 60 ° is recommended for enhanced productivity on spring/autumn equinox days and winter solstice days, respectively. Productivity increases in general with reduction in tilt angle of MEDS but this also causes bending of metal partitions, contamination of distillate by incoming feed saline water and increased ground area occupancy of the distiller. Ground area occupied by MEDS installed at 90 ° tilt angle is about 10 times lower than the ground area occupied by MEDS installed at 10 ° tilt angle (Lim et al., 2018). Tanaka & Nakatake (2007a) recommended tilt angle of about 10 ° from horizontal for flat mirror attached to VMEDS for optimum year-round performance of the mirror integrated VMEDS.

### 10.4.4 Material Properties

Material properties considered in this discussion are limited to emissivity of front surface of first partition of VMEDS and reflectivity of flat mirror integrated with VMEDS. Absorbtivity of blackened front partition of VMEDS is an important property and is generally considered to be around 0.90–0.92 in most of the studies (Tanaka et al., 2000a; Tanaka & Nakatake, 2007a; Tanaka, 2009; Lim et al., 2018). Selective black coating with low emissivity is recommended for front surface of first partition of VMEDS and basin liner. High cost and less availability of such coatings hinders its applicability in underdeveloped and remote regions. Hence, ordinary blackboard paints are normally used in VMEDS for practical purposes. Cumulative distillate productivity of VMEDS system with blackboard paint coating is about 8 to 10% lower than VMEDS system with selective black coating due to high radiative heat loss from the first partition of former system to the ambient via glass cover (Tanaka & Nakatake, 2005).

Highly polished mirror-finished reflecting metal plates are recommended for increasing productivity of flat mirror integrated VMEDS. Low-cost and easily available ordinary glass mirrors with reflectivity of 0.80 are suitable for practical applications in remote regions. Daily cleaning is just sufficient to maintain the reflectivity of glass mirrors. Accumulation of dust and scratches formed during cleaning may be a problem for mirror-finished metal plates to maintain its reflectivity (0.95). Replacing mirror-finished metal plates (0.95 reflectivity) with glass mirrors (0.80 reflectivity) can cause drop in productivity by about 10–11% (Tanaka & Nakatake, 2005b). Transmissivity of the glass cover ($\tau_g$) is related to incidence angle ($\theta_i$ in degree) and is given as (Tanaka & Nakatake, 2007a)

$$\tau_g = 2.642 \cos\theta_i - 2.163 \cos^2\theta_i - 0.320 \cos^3\theta_i + 0.719 \cos^4\theta_i \tag{10.22}$$

## 10.4.5 Heat Energy Input and Ambient Temperature

Energy required for evaporation process in VMEDS system is provided by solar radiation and/or waste exhaust gas. It is a well-known fact that the daily productivity of still increases with input energy. Experiments conducted by (Tanaka et al., 2000a; Tanaka et al., 2005) on vertical single diffusion still showed linear relation between solar intensity and daily distillate production rate. Solar radiation over the glass cover of the basin-type-VMEDS is about 22.7 MJ/m$^2$-d and 20.2 MJ/m$^2$-d in winter and summer, respectively. In spite of high cumulative solar radiation intensity daily productivity in winter is around 8.50% lower than in summer due to lower ambient temperature in winter (20 °C) than summer (33 °C). Combination of high solar radiation intensity (23.7 MJ/m$^2$-d) and high ambient temperature (30 °C) have lead to high daily distillate productivity in autumn in comparison to summer (Tanaka et al., 2000b). Lower ambient temperature indicates low feed water inlet temperature to basin and VMEDS section which also indicates higher thermal energy requirement for raising temperature of feed water in each effect (Tanaka et al., 2004). High ambient temperature also favors high nocturnal production in basin-type-VMEDS system (Kaushal et al., 2017a). Impact of solar radiation intensity and ambient temperature on daily productivity of VMEDS is displayed in Table 10.5.

**Table 10.5** Daily productivity of VMEDS system with solar radiation intensity and ambient temperature

| Type of diffusion still | Diffusion gap (m) | Number of effects | Ambient temperature (°C) | Cumulative solar intensity $MJ/m^2d$ | Distillate production $Kg/m^2d$ |
|---|---|---|---|---|---|
| Basin-type-VMEDS (Tanaka et al., 2002) | 0.005 | 11 | 19.0–25.0 | 15.2–19.6 | 9.8–16.3 |
| | | | 19.0–30.0 | 20.7–22.4 | 14.8–18.7 |
| Basin-type-VMEDS (Kaushal et al., 2016) | 0.010 | 1 | 29.0–39.0 | 13.47 | 2.78 |
| | | | 31.0–38.0 | 16.22 | 3.47 |
| Heat pipe-coupled VMEDS (Tanaka et al., 2004) | 0.005 | 11 | 33.0 | 27.4 | 20.2 |
| | | | 20.0 | 21.8 | 13.3 |
| Flat mirror-coupled VMEDS (Tanaka et al., 2005) | 0.010 | 10 | 30.0 | 25.6 | 34.5 |
| | | | 30.0 | 24.0 | 29.2 |
| Flat mirror-coupled VMEDS (Tanaka et al., 2009) | 0.005 | 6 | 10.0–25.5 | 20.2 | 13.5 |
| Basin-type-VMEDS with heat recovery (Kaushal et al., 2017a) | 0.010 | 3 | 40.8 | 18.5 | 8.12 |
| | | | 37.7 | 18.5 | 6.79 |

Impact of thermal energy input from various heat sources on productivity of basin type-VMEDS with fin heat exchanger is shown in Fig. 10.13a. The slope of productivity curve is higher for system operated with waste heat and electric heater compared to solar energy input. Low productivity with solar energy input is mainly attributed to the reduced available radiation reaching the basin liner after reflection and transmission losses. In addition, improper orientation of reflecting fins in heat exchanger also reflects the incident radiation away from the first partition (Yeo et al., 2019). Increasing heat input enhances daily productivity in VMEDS section to a large extent in comparison to the basin section in basin type-VMEDS system (Park et al., 2016) which can be seen in Fig. 10.13b.

### *10.4.6 Wind Speed*

Impact of wind speed on distillate productivity of VMEDS has not been reported in literatures so far. However, Lim et al. (2018) have investigated wind speed impact on

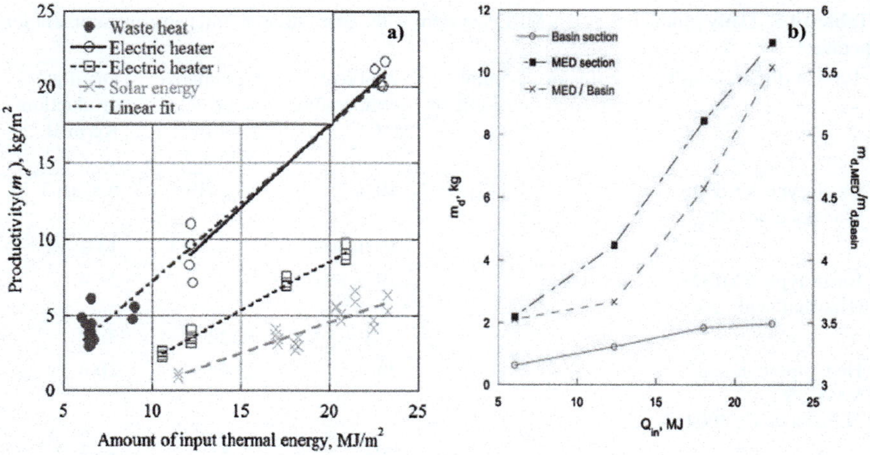


**Fig. 10.13** **a** Impact of different heat sources on basin-type-VMEDS productivity (Yeo et al., 2019) **b** impact of heat energy on productivity of basin section and VMED section of basin-type-VMEDS (Park et al., 2016)

tiltable MEDS. In general, wind speed has positive impact on distillate productivity of conventional solar stills as it reduces temperature of condensing surface/glass cover. In case of basin type-VMEDS system, heat loss from the glass cover of basin system is reduced by using double glass cover this implies negative effect of wind speed on productivity of VMEDS (Tanaka et al., 2000a; Park et al., 2016; Kaushal et al., 2016).

Productivity of tiltable MEDS drops from 13.0 kg/m$^2$-d to 5.5 kg/m$^2$-d with increase in wind speed from 1.0 m/s to 5.0 m/s over the black wick lined glass cover (Fig. 10.14). Enhanced distillate production in the range of 0.40 to 0.73 kg/m$^2$-d is noticed for increased wind speed over the last partition of tiltable MEDS as it enhances the condensation process. Compared to the enhancement obtained by high wind speed, drop in productivity is observed to be higher. Hence, installation of VMEDS system must be in locations with low wind speed or the system must be provided with double glass cover to prevent heat loss (Lim et al., 2018).

### *10.4.7 Basin Water Depth*

Basin water depth is an important parameter which affects the distillate productivity of basin type-VMEDS significantly. As water mass/depth in basin increased, productivity dropped considerably due to increased heat capacity which resists the immediate temperature rise of basin water. Time delay observed between peak solar radiation and peak distillate production of the system justifies the role of heat capacity in reduced distillate production of the system. Around 35 and 103 min time delay

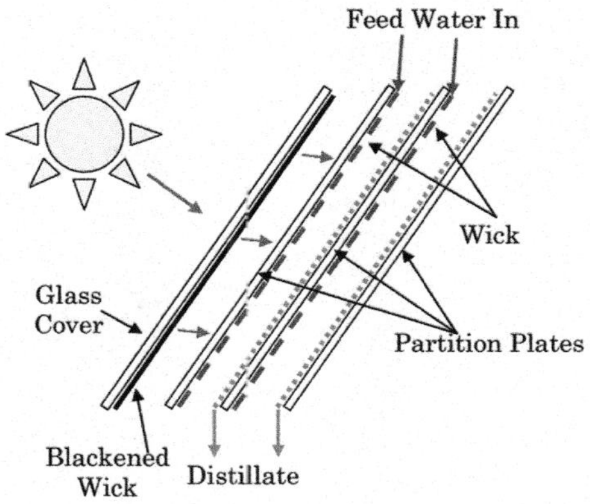


**Fig. 10.14** Schematic of tiltable MEDS (Lim et al., 2018)

is observed between peak distillate production and peak solar radiation for basin water of 10 and 30 kg/m$^2$ of basin area (Tanaka et al., 2000b). Kaushal et al. (2017a) suggested incorporation of floating wicks in basin to reduce heat capacity of basin water to cause immediate water evaporation after incidence of solar radiation.

Productivity of electric heater integrated basin type-VMEDS drops with increasing water depth and reduced heat input. Lower daily productivity at 0.010 m basin water depth for 23 MJ heat input compared to 0.020 m basin water depth indicates the condition of dried out evaporation and stressed the addition of feed water to maintain required water level for proper operation of the system. For very low water levels, the electric heating coil or waste heat exchanger in basin directly exchanges heat energy with humid air trapped in basin, which in turn increases the first partition temperature of integrated VMEDS rapidly resulting in enhanced productivity of VMEDS and overall distillate productivity of basin type-VMEDS system (Park et al., 2016). For water depth of 0.030 m, productivity of basin system is higher while productivity of VMEDS system is lower. On contrary for water depth of 0.010 m, productivity of VMEDS system is higher than productivity of basin system.

### *10.4.8 Saline Feed Water Flow Rate*

Overall distillate productivity of VMEDS is highly dependent on saline feed water flow rate in each effect. Generally, feed water flow rate ($m_f$) for each effect of VMEDS is fixed by estimating the evaporation rate ($m_e$) during peak hours on those considered seasons. Feed water flow rate is kept nearly twice the evaporation rate in each effect (i.e., $m_f/m_e = 2.0$). Feed water flow rate decreases from first effect to last effect gradually as evaporation rate in each effect also decreases in this order (Tanaka & Nakatake, 2005). Daily distillate productivity decreases with increase in $m_f/m_e$ ratio as more fraction of supplied heat energy will be utilized to raise feed water temperature which leads to reduced evaporation rate from each effect (Tanaka

**Fig. 10.15** Salt deposition over first partition of basin type-VMEDS (Kaushal et al., 2017a)

et al., 2004). Reduced $m_f/m_e$ ratio can increase distillate productivity initially but it has serious impacts like dry patch formation over the wick surface, corrosion of partition plate and low productivity after few hours of operation (Tanaka et al., 2004).

Photograph of salt deposition over the first partition of basin type-VMEDS system is shown in Fig. 10.15. No dry patch formation is observed by Tanaka et al. (2000a) while experimenting on single effect vertical diffusion still for $m_f/m_e$ ratio around 2.0. Even though the optimum $m_f/m_e$ ratio is 2.0, for practical purposes $m_f/m_e$ ratio of 3.0 is highly recommended. Increasing $m_f/m_e$ ratio from 2.0 to 3.0 could reduce the productivity of VMEDS by about 15.0% (Tanaka & Nakatake, 2005b). Increasing $m_f/m_e$ ratio above 2.0 can prevent dry patch formation and also helps to flush out salts from the wick surfaces (Dhindsa & Mittal, 2018).

Feed water supplying unit for VMEDS system is somewhat complicated as flow rate varies with each effect in a single system (Park et al., 2014). Tanaka & Nakatake (2005) recommended capillary tubes of varying length and varying water head in overhead tanks for supplying feed water to VMEDS which is shown in Fig. 10.16a. Park et al. (2016) suggested increased thickness for feed water channels to prevent bending and recommended incorporation of two wicks from single pocket for ease fabrication of feed water distribution channels. Schematic of feed water channel recommended by Park et al. (2016) is shown in Fig. 10.16b. Dhindsa & Mittal (2018) used adjustable needle valve of 0.0025 m inner diameter to control feed water flow rate in VMEDS. Kaushal et al. (2016) utilized common header pipe of 0.075 m diameter attached with a number of 0.003 m diameter capillary tubes to supply feed water to each effect. Photograph of the feed water flow arrangement by Kaushal et al. (2016) is shown in Fig. 10.16c. Tanaka & Nakatake (2005a) suggested use of easily available medical syringe needles for feed water distribution in VMEDS system.

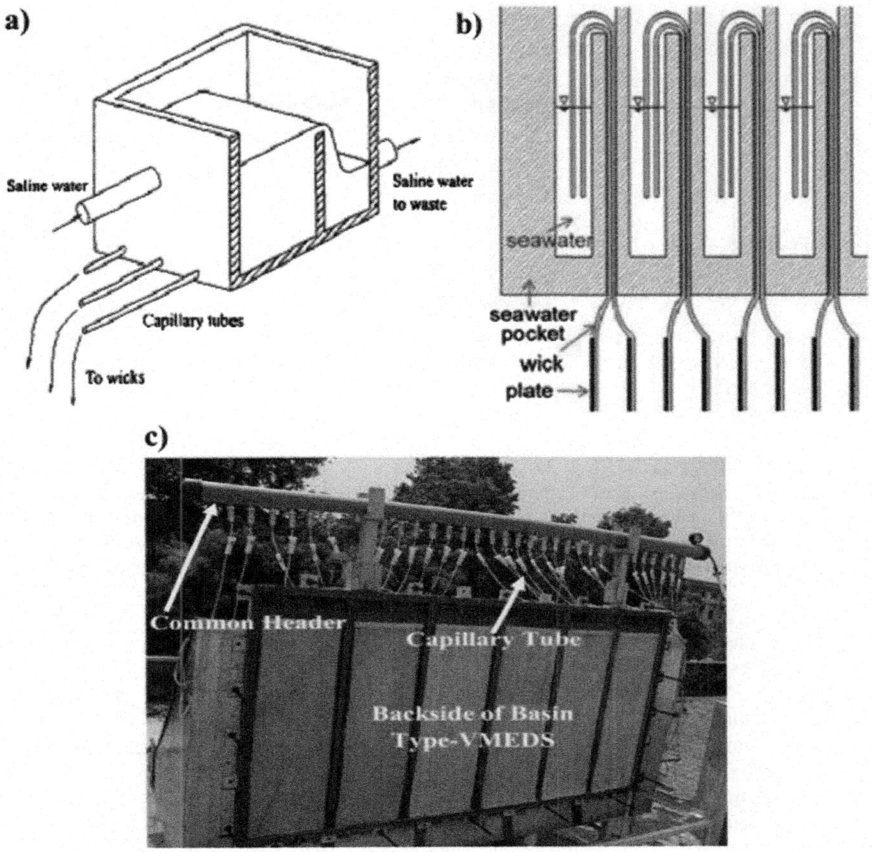


**Fig. 10.16** **a** Feed channel with varying length capillary tubes (Tanaka et al., 2002) **b** Simplified feed channel (Park et al., 2016) **c** feed channel with common header (Kaushal et al., 2016)

## 10.5 Features of Vertical Multiple Effect Diffusion Solar Still

(a) More compact, portable, and less ground area occupancy.
(b) High distillate productivity.
(c) Waste heat energy from generators, stoves, etc., can be effectively used as heat source.
(d) Complex feed water distribution unit.
(e) More chances of salt deposition over the wick as required feedwater flow rate is very low.
(f) Requires high-quality metal partitions to prevent bending of metal partitions due to thermal stress.

(g) Spacers between the partitions are necessary to prevent contact of evaporating surface wick with condensing surface.
(h) Additional arrangements like mirrors and heat pipe solar collectors are required to utilize solar radiation, especially at noon.

## References


Chang-Dae, P., Byung-Ju, L., Kyung-Yul, C., & LeeSung-Soo, K.-M. (2016). Experimental evaluation of hybrid solar still using waste heat. *Desalination, 379*, 1–9.

Dhindsa, G. S., & Mittal, M. K. (2018). Experimental study of basin type vertical multiple effect diffusion solar still integrated with mini solar pond to generate nocturnal distillate. *Energy Conversion and Management, 165*, 669–680.

Elsayed, M. M. (1986). Effects of parametric conditions on the performance of an ideal diffusion still. *Applied Energy, 22*, 187–203.

Kaushal, A. K., Mittal, M. K., & Gangacharyulu, D. (2016). Development and experimental study of an improved basin type vertical single distillation cell solar still. *Desalination, 398*, 121–132.

Kaushal, A. K., Mittal, M. K., & Ganagacharyulu, D. (2017). An experimental study of floating wick basin type vertical multiple effect diffusion solar still with waste heat recovery. *Desalination, 414*, 35–45.

Kaushal, A. K., Mittal, M. K., & Ganagacharyulu, D. (2017). Productivity correlation and economic analysis of floating wick basin type vertical multiple effect diffusion solar still with waste heat recovery. *Desalination, 423*, 95–103.

Lim Byung-Ju, Y., Sang-Seok, C.K.-Y., & Chang-Dae, P. (2018). Numerical analysis of the performance of a tiltable multi-effect solar distiller. *Desalination, 435*, 23–34.

Nosoko, T., Kinjo, T., & Park, C. D. (2005). Theoretical analysis of a multiple-effect diffusion still producing highly concentrated seawater. *Desalination, 180*, 33–45.

Park, C. D., Lim, B. J., Noh, Y. D., Lee, S. S., & Chung, K. Y. (2014). Parametric performance test of distiller utilizing solar and waste heat. *Desalination and Water Treatment*, 1–7.

Se-Dong, Y., Byung-Ju, L., Ga-Ram, L., & Chang-Dae, P. (2019). Experimental study of effects of different heat sources on the performance of the hybrid multiple-effect diffusion solar still. *Solar Energy, 193*, 324–334.

Sharon, H., Reddy, K. S., & Gorjian, S. (2020). Parametric investigation and year round performance of a novel passive multi-chamber vertical solar diffusion still: Energy, exergy and enviro-economic aspects. *Solar Energy, 211*, 831–846.

Tanaka, H. (2009). Experimental study of vertical multiple effect diffusion solar still coupled with a flat plate reflector. *Desalination, 249*, 34–40.

Tanaka, H. (2016). Theoretical analysis of a vertical-multiple-effect diffusion solar still coupled with a tilted wick still. *Desalination, 377*, 65–72.

Tanaka, H. (2016). Thermal distillation system utilizing biomass energy burned in stove by means of heat pipe. *Alexandria Engineering Journal, 55*, 2203–2208.

Tanaka, H., & Chang-Dae, Park. (2010). Distillation utilizing waste heat from a portable electric generator. *Desalination, 258*, 136–142.

Tanaka, H., & Chang-Dae, Park. (2012). Experimental study of distiller with heat pipe utilizing waste heat from a portable electric generator. *Desalination, 302*, 43–49.

Tanaka, H., & Lishi, K. (2017). Experimental study of a vertical single-effect diffusion solar still coupled with a tilted wick still. *Desalination, 402*, 19–24.

Tanaka, H., & Nakatake, Y. (2004). A vertical multiple-effect diffusion type solar still coupled with a heat pipe solar collector. *Desalination, 160*, 195–205.

Tanaka, H., & Nakatake, Y. (2005). A simple and highly productive solar still: a vertical multiple effect diffusion type solar still coupled with a flat plate mirror. *Desalination, 173*, 287–300.

Tanaka, H., & Nakatake, Y. (2005). Factors influencing the productivity of a multiple effect diffusion type solar still coupled with a flat plate reflector. *Desalination, 186*, 299–310.

Tanaka, H., & Nakatake, Y. (2007). Numerical analysis of the vertical multiple effect diffusion solar still coupled with a flat plate reflector: optimum reflector angle and optimum orientation of the still at various seasons and locations. *Desalination, 207*, 167–178.

Tanaka, H., & Nakatake, Y. (2007). Outdoor experiments of a vertical diffusion solar still coupled with a flat plate reflector. *Desalination, 214*, 70–82.

Tanaka, H., Nosoko, T., & Nagata, T. (2000). A highly productive basin-type-multiple-effect coupled solar still. *Desalination, 130*, 279–293.

Tanaka, H., Nosoko, T., & Nagata, T. (2000). Parametric investigation of a basin-type-multiple-effect coupled solar still. *Desalination, 130*, 295–304.

Tanaka, H., Nosoko, T., & Nagata, T. (2002). Experimental study of basin-type, multiple effect, diffusion coupled solar still. *Desalination, 150*, 131–144.

Tanaka, H., Nakatake, Y., & Tanaka, M. (2005). Indoor experiments of the vertical multiple-effect diffusion-type solar still coupled with a heat-pipe solar collector. *Desalination, 177*, 291–302.

# Chapter 11
# Hybrid Solar Chimney-Thermal Desalination Systems

## 11.1 System Description and Operation

The solar chimney is also called the solar updraft tower and is developed to generate power. It consists of a fixed central chimney, a wind turbine coupled generator, and a transparent cover. The transparent cover serves as solar collector and is opened on all sides as shown in Fig. 11.1. The transparent cover is made of polyvinyl chloride or polyvinyl fluoride sheet of 0.1 mm thickness (Zuo et al., 2020). The solar radiation passing through the transparent cover heats up the ground which in turn heats up the air trapped between the cover and the ground. The hot air rises and flows through the central chimney due to a drop in air density. The kinetic energy of flowing hot air is utilized by the turbine attached at the bottom of the central chimney to generate electricity. The heat energy absorbed by the ground will be able to operate the system even at night. Moreover, the expected lifetime of the system is about 100 years. The solar chimney power plant installed in Spain has a chimney of 10.16 m diameter, 194.6 m height, solar collector of 45000 m$^2$ area (radius about 122 m), and centrally installed turbine of about 50 kW capacity (Maia et al., 2019).
The major advantages and limitations of solar chimney power plants are:

(a) Less maintenance cost, no fuel costs, no requirement of concentrators, and trackers;
(b) Construction materials are cheap and are locally available even in developing countries;
(c) Low heat to electrical energy conversion efficiency (only 0.50–10%);
(d) Requirement of tall structures;
(e) Large ground space occupancy.



H. Sharon and K. Srinivas Reddy, *Solar Thermal Desalination Technologies for Potable Water*, https://doi.org/10.1007/978-3-031-49155-9_11

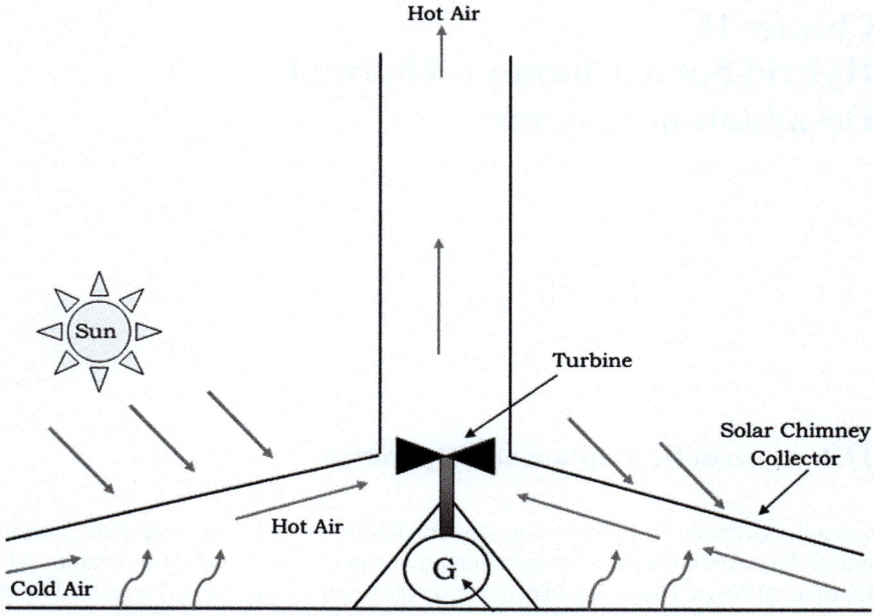


**Fig. 11.1** Schematic of solar chimney power plant (Rahbar & Riasi, 2019)

## 11.2 Hybrid Solar Chimney-Desalination/Water Production Systems

The vast ground area below the transparent cover of solar chimney power plant can also be utilized effectively for drying and desalination purposes. In hybrid solar chimney-desalination/water production systems, both power and potable water are produced. A number of hybrid configurations have been proposed and investigated theoretically by researchers. A few of those hybrid configurations are listed below:

(a) Solar chimney-desalination system with feed water sprayers.
(b) Hybrid solar chimney-basin desalination system.
(c) Hybrid solar chimney with seawater absorber bed.
(d) Hybrid solar chimney with seawater and ground absorber bed.
(e) Wind supercharger-assisted hybrid solar chimney-desalination system.
(f) Hybrid solar chimney-PV module integrated desalination system.
(g) Hybrid solar chimney-based atmospheric water extraction system.

### *11.2.1 Solar Chimney-Desalination System with Feed Water Sprayers*

In this configuration, the feed water sprayers are installed at the inlet of the central tower of the solar chimney system as shown in Fig. 11.2. The sole purpose of this system is to desalinate saline water and therefore it has no turbine. Feed water sprayers are used to spray fine droplets (30 μm diameter) of seawater directly into the hot air which are heated by the solar collector section of the solar chimney. These droplets when come in contact with the hot air evaporate and mix with the air thereby the humidity of air increases. The water vapor in the humid air condenses over the inner wall of central tower. A separate condenser can also be used to condense the vapor. The condensed vapor after post-treatment can be utilized for human consumption. The central tower height, radius, solar collector radius, and gap between collector cover & ground are 200.0 m, 5.0 m, 120 m, and 2.0 m, respectively.

The air temperature at the inlet of central tower (0 m) will be about 309 K and it increases rapidly up to the stagnant zone. After crossing the stagnant zone, temperature of the flowing air drops slowly along the central tower height and finally reaches the ambient temperature at exit (210 m). The air temperature is highly dependent

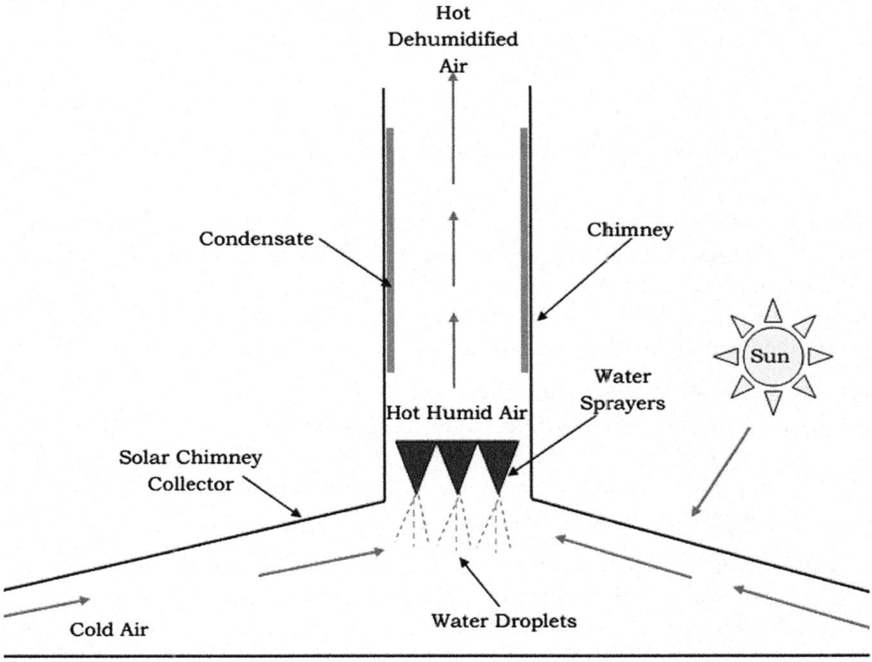


**Fig. 11.2** Schematic of solar chimney-desalination system with feed water sprayers (Ming et al., 2017a)

on the feed water spray rate and it decreases with increase in water spray rate. This observation indicates that most of the heat energy of air has been utilized to evaporate the sprayed feed water. The average airflow velocity is nearly similar for the system with feed water spray rates of 0, 0.1, and 0.3 kg/s. The velocity of flowing air is high at the central tower inlet and it drops rapidly up to a central tower height of 60 m. Beyond 60 m, the airflow velocity slightly increases and reaches the lowest value at the exit. The air velocity distribution along the height of central tower is found to be low for the system with a water spray rate of 0.5 kg/s which clearly indicates that the addition of moisture to the flowing air degrades the velocity of air (i.e.) power output of the turbine in solar chimney power plant.

The relative humidity of air at central chimney inlet is only 32.5% due to high temperature of air at this location (as it is heated by the solar collector). With complete evaporation of the sprayed water, the relative humidity of air rises rapidly up to 15 m and then slightly decreases and reaches ambient relative humidity at the exit for a water spray rate of 0.1, 0.3, and 0.4 kg/s. However, for a water spray rate of 0.5 kg/s, the relative humidity of air is around 100% at the central tower height of 15 m. Therefore, in this situation, the vapor will be able to condense at lower height. Moreover, the condensation efficiency also increases from 0.0 to 79.4% with increase in water spray rate from 0.4 to 0.8 kg/s. The results indicate that there is always a tradeoff between the power generation and distillate production capacity of solar chimney-desalination systems. Hence, it becomes necessary to identify the optimum configuration of the system to have maximum utilization factor.

### *11.2.2 Hybrid Solar Chimney-Basin Desalination System*

In the case of basin solar stills, the latent heat of condensation is wasted to the ambient. A huge ground area is also required to produce distillate using solar stills for a community. Solar chimney power plant also occupies huge land area but the space below its collector is unoccupied. The following reasons have been cited to justify the integration of basin solar still within solar chimney:

(a) The sea water is an excellent heat energy storage medium.
(b) Latent heat energy released from the still can be utilized to heat the air inside the solar chimney.
(c) Additional area requirement for the still is eliminated as the area below the transparent cover of solar chimney is effectively utilized.

The schematic of hybrid solar chimney-basin desalination system is shown in Fig. 11.3. The basin still installed within the solar chimney is annular, shallow, and has black-painted bottom. The bottom and sides of the annular still are insulated to prevent heat losses to the ground. Insulation is mandatory to prevent drastic performance drop of the solar chimney system. Circular transparent roof of the solar chimney is elevated from the still's glass cover using brackets and its height gradually increases towards the center of the chimney as shown in Fig. 11.3. Solar radiation

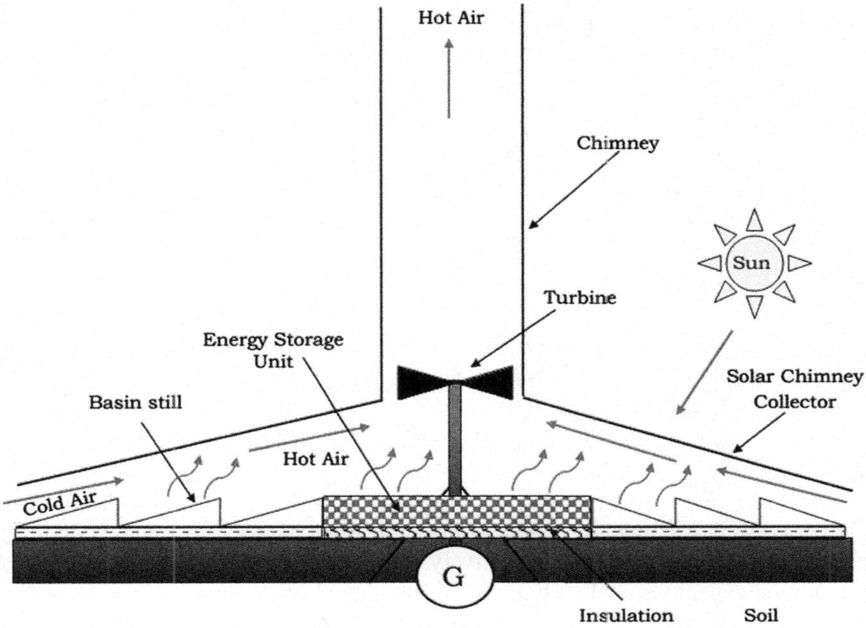


**Fig. 11.3** Schematic of hybrid solar chimney-basin desalination system (Zuo et al., 2020)

reaching the black-painted surface of the annular still initiate saline water evaporation in basin. The formed vapors condense over the inner surface of basin still's glass cover by releasing latent heat energy. In addition, convection and radiation heat exchange also take between the basin water and the glass cover. The heat energy gained by the glass cover due to these three modes of heat transfer heats up the air between the solar still's glass cover and transparent solar chimney cover by convection heat transfer. The hot air due to its low density rises and passes through the central tower by buoyancy effect. The kinetic energy of flowing hot air is converted into mechanical energy by the turbine installed in the airflow path. The cold ambient air surrounding the lower edge of the transparent cover enters the collector and fills the vacant place and the above process repeats.

The annual freshwater production rate and annual power generation rate of this system with a central tower height of 200 m and diameter of 10 m, collector diameter of 250 m, and height of 2.0 m is about 283.0 MWh and 69.5 kilo tons, respectively, for an average ambient temperature of 27.2 °C and average solar radiation intensity of 800 W/m$^2$ for six hour operation per day. Energy storage medium of about 40 m diameter is also placed centrally and is mandatory to reduce/avoid temperature differences in airflow between the peak and valley value during day and night thereby, power fluctuation is prevented and the quality of power generated is also improved (Zuo et al., 2012) investigated a small-scale prototype of the system and found that the temperature rise of air occurs only up to the first 1/3rd section of the solar chimney

**Table 11.1** Performance comparison of various solar chimney systems

| Performance parameters | Simple solar chimney | Hybrid solar chimney basin desalination system | Hybrid solar chimney basin desalination system (optimum Basin still location) |
|---|---|---|---|
| Power generation (kW) | 22.0 | 11.0 | 18.5 |
| Freshwater yield (kg/s) | 0.0 | 10.4 | 6.5 |
| Utilization factor (%) | 0.74 | 55.4 | NA |

collector. Moreover, the temperature rise of flowing air in conventional solar chimney is higher than in hybrid solar chimney-basin desalination system. This is mainly due to the utilization of part of the incident solar energy for desalination in hybrid solar chimney integrated basin desalination system, while, in the case of conventional solar chimney, entire solar radiation is utilized only for heating the air.

The daily utilization efficiency of conventional solar chimney and hybrid solar chimney-basin desalination system of the same dimensions and climatic conditions is about 0.74% and 55.4%, respectively (Zuo et al., 2011). However, the power production capacity of solar chimney decreases with the incorporation of desalination system within it. The effective performance of the hybrid solar chimney-basin desalination system can be achieved only if part of the ground is covered with distillation system rather than covering it fully. The recommended location for covering the ground below the solar collector of solar chimney with desalination system is between 85 m and 125 m radius from the center (Asayesh et al., 2017). The performance comparison of various solar chimney systems is tabulated in Table 11.1.

### 11.2.3 Hybrid Solar Chimney-Desalination System with Sea Water Absorber Bed

Basin solar stills inside hybrid solar chimney-desalination system can be replaced by open shallow sea water absorber bed to eliminate complex construction (Zhou et al., 2010). In this arrangement, the dry ambient air entering the collector of solar chimney gets humidified by the vapors of the evaporating seawater directly. The water vapors in the hot humid air are recovered with the aid of condenser installed at the top of the central tower as shown in Fig. 11.4. The hot humid air while passing over the turbine at the central tower inlet generates power. As the vapor directly mixes with air, the velocity and temperature of the air drop thereby, power generated by the turbine in this arrangement is observed to be lower than the power generated from conventional solar chimney of similar dimensions.

The condensate from the condenser is also allowed to pass over the water turbine installed at the lower end of the exit pipeline of condenser to generate additional power. The modified system with a collector diameter, central tower height, and

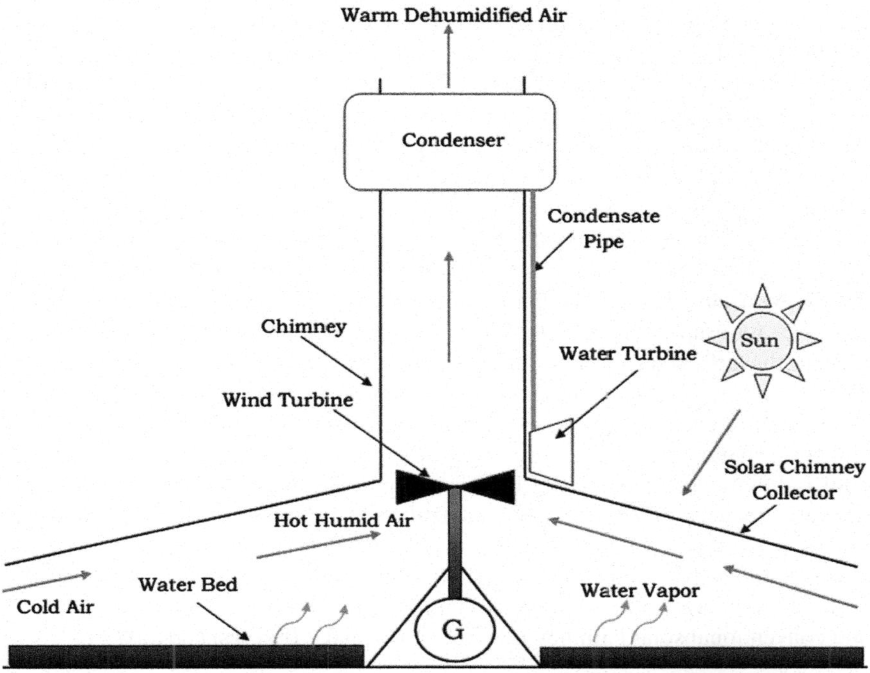


**Fig. 11.4** Schematic of hybrid solar chimney-desalination system with sea water absorber bed (Zhou et al., 2010)

radius of 6000 m, 500 m, and 80 m, respectively, is capable of producing 3807 kg/s of condensate, wind turbine power output of 39.2 MW, and water turbine power output of 16.1 MW for an indicient solar radiation intensity of 1000 W/m$^2$. The power output of water turbine is only 50% of the power output of the wind turbine. The utilization efficiency of the modified system and conventional system is only 0.46% and 0.20%, respectively.

## *11.2.4 Hybrid Solar Chimney-Desalination System with Ground and Sea Water Absorber Bed*

Performance of hybrid solar chimney-desalination system with sole seawater absorber bed is lower than the performance of conventional solar chimney (Zhou et al., 2010). Kiwan et al. (2018) addressed this issue by dividing the absorber area of solar chimney into two sections. The first section is bare ground and the second section is water bed. The water bed inside the solar chimney is of doughnut shape and the water level in the bed is always maintained constant by necessary controllers. The schematic of hybrid solar chimney-desalination system with two absorber sections is shown in Fig. 11.5a–b. In the first section, the air is heated by the solar energy absorbed by the ground, and in the second section, the air gets humidified by the water bed. The generated hot humid air passes over the wind turbine which is installed in the airflow path through the central tower to generate power. The water vapor in the hot humid air condenses over the inner surface of the chimney by releasing its latent heat energy.

The temperature of air inside the solar chimney collector increases in the bare ground section and once it reaches the water bed section it lowers slightly and then remains constant. This can be attributed to the utilization of the heat energy content of hot air for evaporating water in the water bed. Increasing radius of the water bed reduces the airflow rate, power output of turbine, and distilled water production rate. The reduced temperature of air with increased water bed radius indicates reduced pressure difference of air between central tower inlet and outlet thereby airflow rate and power output drops. The reduced distillate production rate with increased water pool radius is attributed to the low air temperature because low air temperature indicates reduced ability to collect moisture. The optimum water depth in water bed of outer diameter 35.0 m is about 4.5 cm. At lower water depth, the water available for evaporation is decreased and at high water depth more energy is needed to initiate evaporation of the water. The utilization efficiency of the modified system and conventional solar chimney is about 71.0% and 1.0%, respectively.

### *11.2.5 Wind Supercharger-Assisted Hybrid Solar Chimney-Basin Desalination System*

Hybrid solar chimney-desalination system suffers reduced power generation in comparison to sole solar chimney power plant. This reduced power generation is due to reduced pressure difference between the system air and ambient air. This reduced pressure difference is attributed to the utilization of part of the solar energy collected by the solar chimney collector for desalination in hybrid systems. Hence, in order to enhance the power generation and distillate yield further, a wind pressure ventilator (wind supercharger) is proposed by Zuo et al. (2018a).

The schematic of hybrid solar chimney-desalination system with a wind pressure ventilator is shown in Fig. 11.6. The wind pressure ventilator consists of a central shaft to which ventilator blades are attached. The free ends of two ventilator blades in line are connected by the H-type vertical axis wind wheel (4° blade installation angle, 10 m diameter, and 15 m height) structure. It is recommended to maintain height to diameter ratio between 0.5 and 3.0 for H-type vertical wind wheel. The negative pressure below the wind pressure ventilator increases to a value of −68.22 Pa from −61.5 Pa with increase in blade installation angle from 0 to 3°. Further increase in installation angle decreases the negative pressure. Blade installation angle is a crucial factor and a small change in it may affect the performance of wind pressure

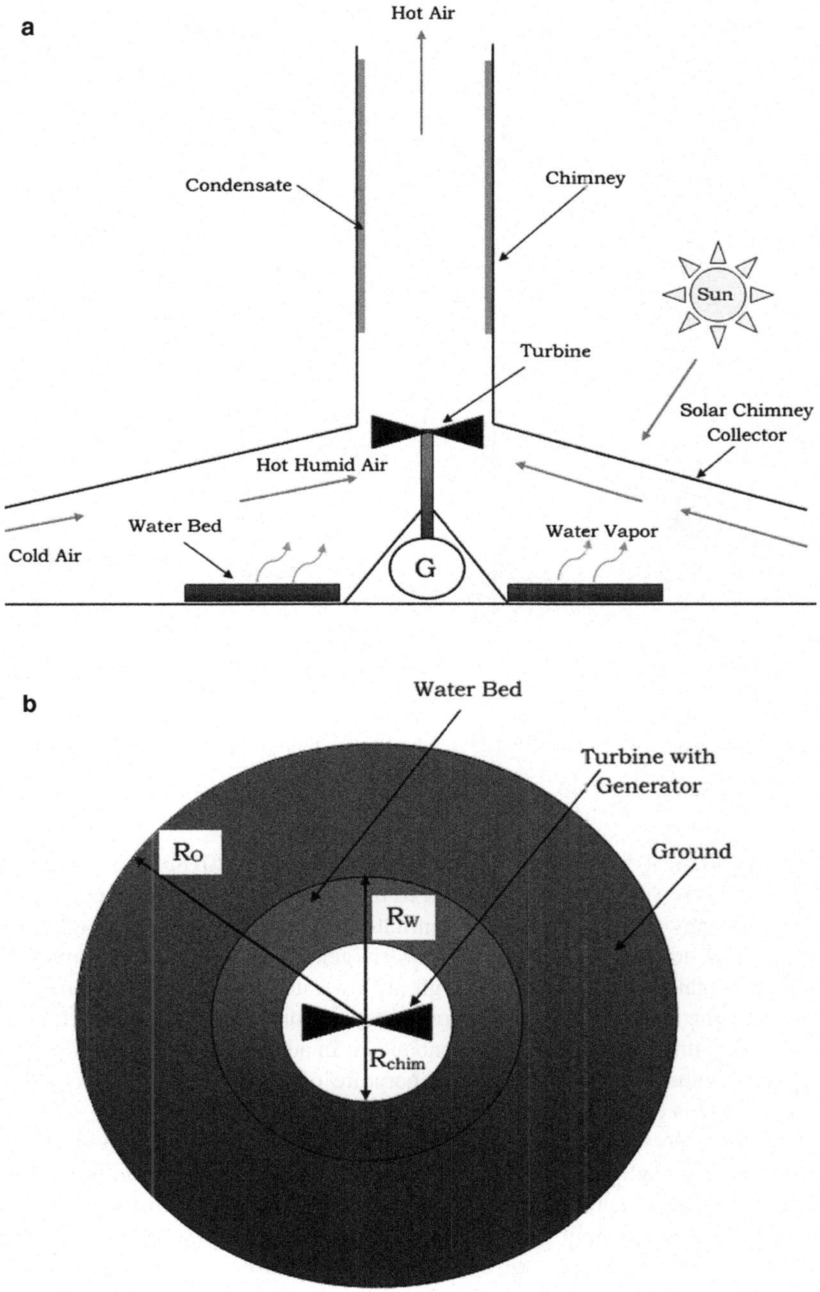


**Fig. 11.5** **a** Schematic of hybrid solar chimney-desalination system with two absorber sections (Chimney height 200 m; chimney diameter 10 m; collector diameter 250 m; collector entrance height 6 m; solar radiation intensity 800 W/m$^2$; ambient temperature 27.2 °C) (Kiwan et al., 2018)

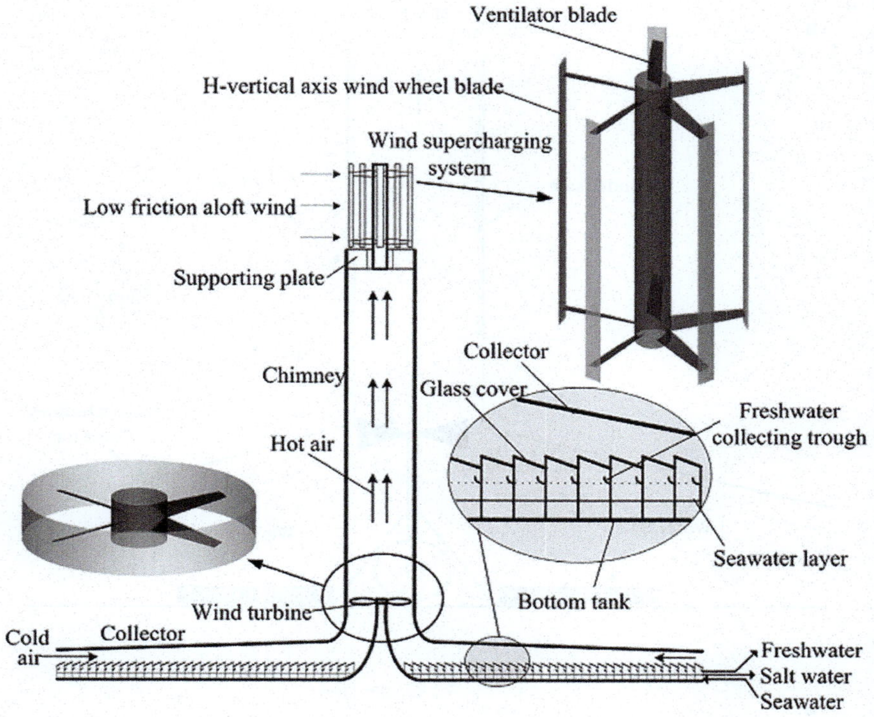


**Fig. 11.6** Schematic of wind-supercharged hybrid solar chimney basin desalination system (Zuo et al., 2018a)

ventilator and the hybrid solar chimney system (Zuo et al., 2018b). The central shaft of wind pressure ventilator is connected to the supporting plate at the central tower exit via bearings. The wind pressure ventilator rotates by utilizing the high-altitude wind thereby additional negative pressure is generated within the chimney. This additional negative pressure causes high airflow rate over the wind turbine blades installed at the bottom thereby the power generation capacity of the hybrid solar chimney-desalination system will be enhanced. In addition, this increased flow rate of air also reduces the glass cover temperature of basin solar stills thereby high distillate yield is expected.

The relative pressure at collector inlet is zero and the pressure variation within the collector is also negligible hence the heat addition can be assumed to be a constant pressure heat addition process. The relative pressure in the **transition section** is about 0 to −80 Pa and this high-pressure drop increases airflow speed in the transition section. The obstruction caused by the turbine increases the relative pressure by about +40 to +80 Pa at the turbine inlet. The relative pressure then drops rapidly to −440 pa just above turbine blade and then increases up to the value of relative pressure (−80 Pa) at central tower outlet. The variation of shaft power output and

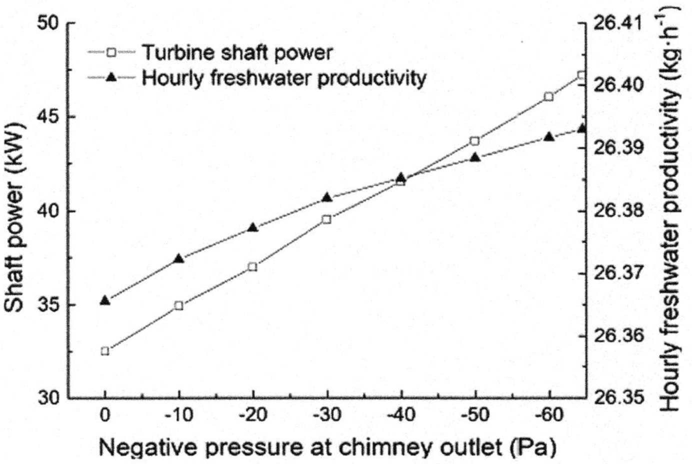


**Fig. 11.7** Variation of shaft power and freshwater yield with negative pressure at the chimney outlet (Zuo et al., 2018a)

freshwater yield of hybrid solar chimney-desalination system of dimension similar to the prototype installed in Spain for different negative pressure at chimney outlet for ambient temperature of 25 °C and solar radiation intensity of 800 W/m$^2$ is shown in Fig. 11.7. It can be seen clearly that the yield and power output increases by 30.0 g/h and 14.7 kW, respectively, with the drop in relative chimney outlet pressure from 0 Pa to −64.5 Pa. Thus, incorporation of wind pressure ventilator to increase both power and yield output is justified (Zuo et al., 2018a).

### *11.2.6 Wind Supercharged Hybrid Solar Chimney-Desalination System with Special Condenser*

The schematic of wind supercharged hybrid solar chimney-desalination system with a special condenser is shown in Fig. 11.8. A special type of condenser is installed within the system to improve the distillate yield and power output of wind-supercharged hybrid solar chimney-desalination system. The condenser enhances condensation rate and increases feed water temperature by utilizing the latent heat of the condensing vapors. The condenser consists of a bundle of tubes placed across the air path connecting solar collector exit and chimney inlet. Cold seawater is pumped to the upper header of the heat exchanger through which it gets distributed to the individual tubes. The water vapor in the hot humid air entering the condenser from the collector gets condensed over the external surface of the tubes and the sea water gets preheated by utilizing this released latent heat. The preheated sea water reaches the desalination tank of the hybrid system to replenish sea water and to maintain a constant water

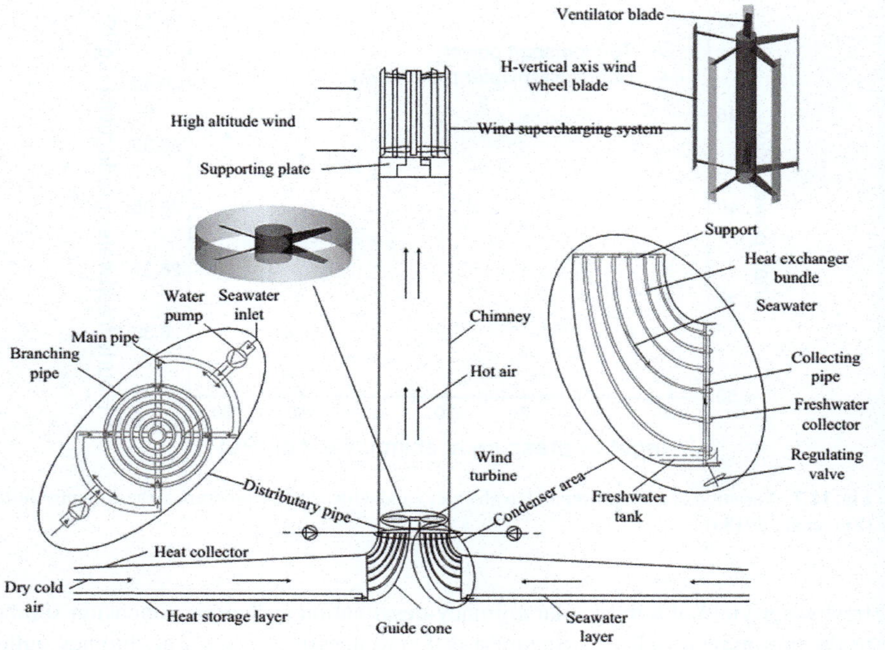


**Fig. 11.8** Schematic of wind-supercharged solar chimney desalination system coupled with a special condenser (Zuo et al., 2019a)

level of 0.03 m in the absorber section. The hot dehumidified air after crossing the heat exchanger passes through the wind turbine placed above the heat exchanger to generate power. Wind supercharger at the chimney outlet increases pressure drop in chimney and enhances power output (Zuo et al., 2019a).

The temperature of air at collector inlet is about 298.15 K and it rises to 312.0 K at collector outlet by absorbing solar radiation and by utilizing heat energy convected from the water absorber bed. The air entering the collector is in direct contact with the seawater absorber bed thereby its moisture content increases along the radial distance from collector inlet to outlet. The temperature of moist air drops rapidly in the condenser section due to the exchange of heat energy with cold seawater. The relative pressure inside the collector is nearly constant and is about 0.0 Pa. Due to flow area contraction in the condenser section, the relative pressure declines in condenser section and the flow velocity increases. The relative pressure at turbine inlet increases due to flow resistance caused by turbine blades. The relative pressure of air after crossing the turbine is low (Zuo et al., 2019a).

The power output of conventional solar chimney, hybrid solar chimney desalination system with a special condenser and wind-supercharged hybrid solar chimney-desalination system with a special condenser is about 36.0 kW, 33.9 kW, and 45.6 kW, respectively, at an optimum turbine rotation speed of 100 rpm. The distillate yield of hybrid solar chimney-desalination system with special condenser and wind supercharged hybrid solar chimney-desalination system with special condenser is about 10.16 ton/h and 9.81 ton/h, respectively. The solar energy utilization factor of hybrid solar chimney-desalination system with a special condenser and wind-supercharged hybrid solar chimney-desalination system with a special condenser is nearly 16.65 and 16.73% higher than that of conventional solar chimney system (Zuo et al., 2019b).

### *11.2.7 Thermal Power Plant Coupled Hybrid Solar Chimney-Basin Desalination System*

The heat energy stored in the seawater of hybrid solar chimney desalination system may not be sufficient enough to operate the system effectively during non-sunshine or low solar radiation periods. This drawback can be rectified by coupling conventional thermal power plant with hybrid solar chimney-desalination system. The schematic of hybrid solar chimney-basin desalination system utilizing wind supercharger and flue gas from thermal power plant is shown in Fig. 11.9. The spiral heat exchanger installed inside the basin of desalination system to transfer heat energy of flue gas to the basin water is shown in Fig. 11.10. The hot flue gas after passing through the heat exchanger is allowed to flow through the chimney along with the ambient air stream heated in the solar collector. The mixed gas velocity (velocity of air stream and flue gas mixture entering chimney) and the average temperature rise of air inside the hybrid solar chimney system increase with an increase in incident solar radiation intensity thereby power and distillate output increase. The utilization of hot flue gas from the thermal power plant enhances distillate yield and turbine power output significantly even during low solar radiation periods.

The increased flue gas mass flow rate increases power output of the system due to the increased pressure difference between the chimney inlet and outlet. However, it has a negative impact on the yield of the system. This observation is attributed to the reduced temperature difference between the water and glass cover caused due to increased heat transfer between the flue gas and basin water which causes rapid steam generation and improper vapor condensation inside the basin. The depth of sea water in desalination system's basin also affects the performance of hybrid solar chimney system. The yield and power output drop with increase in seawater depth. This observation is attributed to the increased heat capacity of basin water at increased depth thereby incident solar energy and flue gas heat energy gets stored in water rather than initiating immediate evaporation. This phenomenon results in reduced gas mixture mass flow rate due to reduced glass cover temperature of basin desalination system resulting in reduced power output. The performance of the system remains nearly constant beyond sea water depth of 4.0 cm. The power generation and distillate yield of this system with wind supercharger is about 21.6% and 14.6% higher than the system without wind supercharger, respectively.

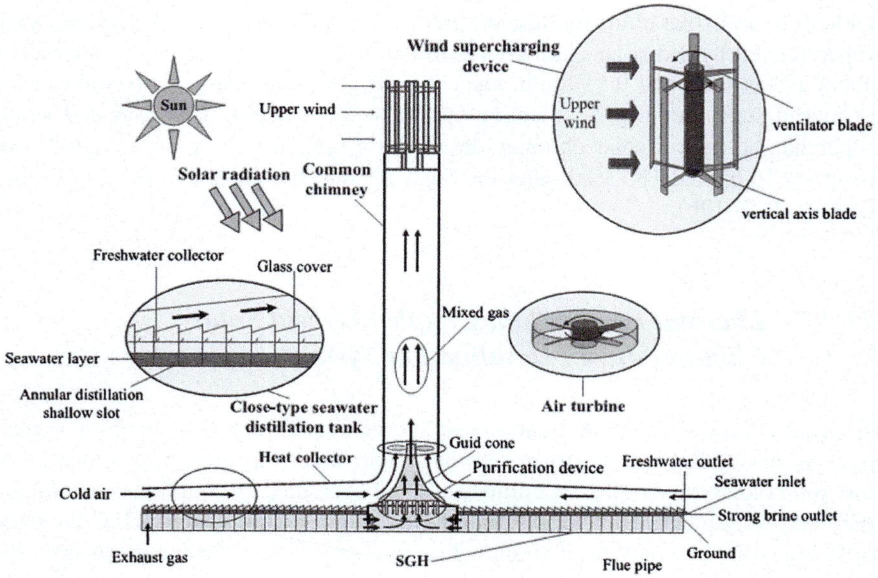


**Fig. 11.9** Schematic of hybrid solar chimney-desalination system (dimension similar to the Spanish prototype) with wind pressure ventilator at the top and spiral exhaust gas heat exchanger at the bottom (Zuo et al., 2020)

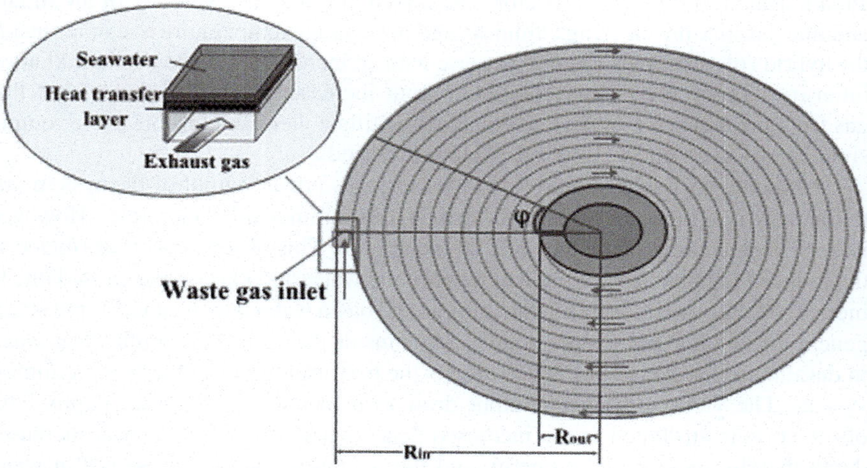


**Fig. 11.10** Schematic of spiral exhaust gas heat exchanger installed at bottom of solar chimney collector (Zuo et al., 2020)

### 11.2.8 Atmospheric Water Extraction Using Solar Chimney

Hot humid air inducing through a chimney where warm seawater is showered is shown in Fig. 11.11. The rise of hot humid air by convection happens due to the temperature difference inside and outside the chimney. The temperature of the rising humid air decreases as the height increases because its enthalpy gets converted into gravitational potential energy. In addition, the relative humidity of the moist air also increases. At some particular height, the temperature of moist air drops to dew point temperature thereby the vapors in the humid air start to condense over any solid surface it comes across. This process is similar to cloud and rain formation occurring in nature which is depicted in Fig. 11.12. This concept has been implemented to extract fresh water from atmospheric air using solar chimney by Ming et al. (2016).

The schematic of solar chimney system utilized for atmospheric water extraction is shown in Fig. 11.13. The solar glass/transparent collector in conventional solar chimney power plant are replaced by black plastic tube coils to absorb solar radiation. The tubes are filled with seawater that is pumped from the sea and it also serves as heat energy storage medium to facilitate operation even during night. The system is recommended to be installed especially in coastal regions due to favorable humidity, seawater availability, and high solar radiation insolation. The moist sea breeze flowing over the black tubes gets heated up by convection and flows towards the chimney inlet. The hot moist air continues to rise upwards inside the chimney by buoyancy

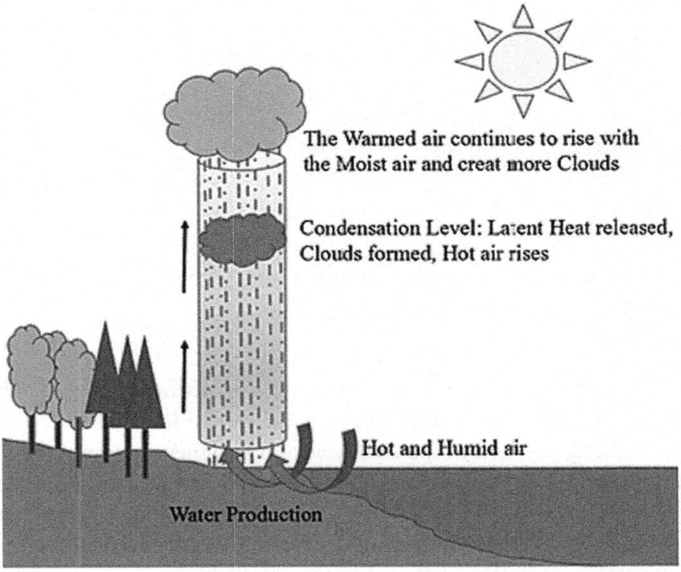


**Fig. 11.11** Concept of utilizing solar chimney for extracting freshwater from atmospheric air (Ming et al., 2016)

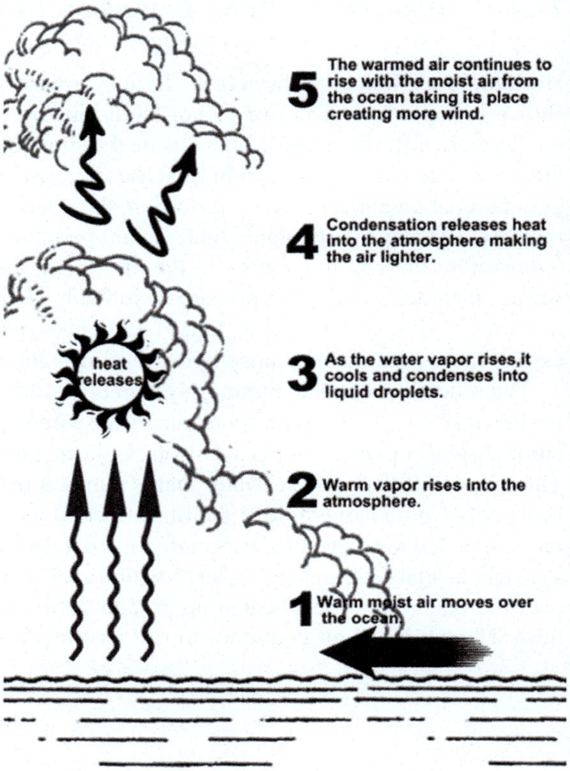


**Fig. 11.12** Natural water vapor condensation process occurring in nature (Wu et al., 2020)

and its temperature drops nearly to dew point temperature at some location inside the chimney resulting in condensation of vapors thereby freshwater is extracted.

### *11.2.9 Combined Power Generation and Atmospheric Water Extraction Solar Chimney System*

Solar chimney used for sole atmospheric water extraction has a low solar energy utilization factor. However, the utilization factor can be improved by incorporating wind turbine and hydraulic turbine in the system to generate power. The schematic of hybrid solar chimney with a special porous condenser, hydraulic turbine, and wind turbine is shown in Fig. 11.14. The solid porous condenser is located at the optimum condensation height to facilitate condensation of water vapor in the humid air. Wind turbine is installed at the chimney inlet to generate power from the hot moist air flowing over it. Hydraulic turbine is installed at the bottom to utilize the potential energy of condensate from the condenser to generate additional power. Optimum condensation height is inversely proportional to relative humidity but it

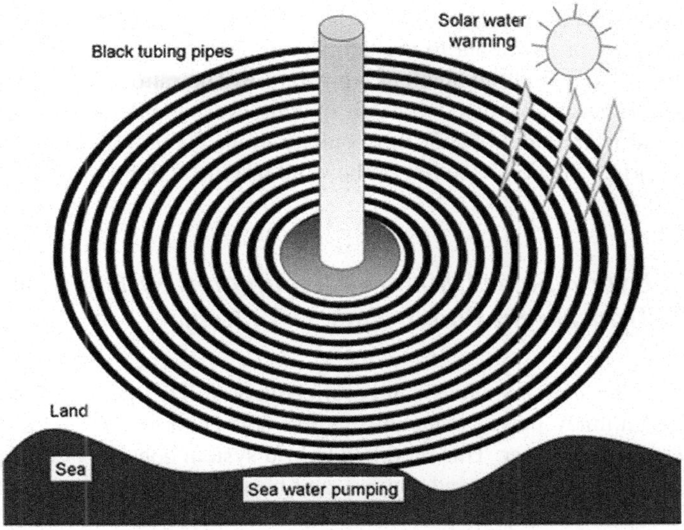


**Fig. 11.13** Schematic of solar chimney system used for atmospheric water extraction (Wu et al., 2020)

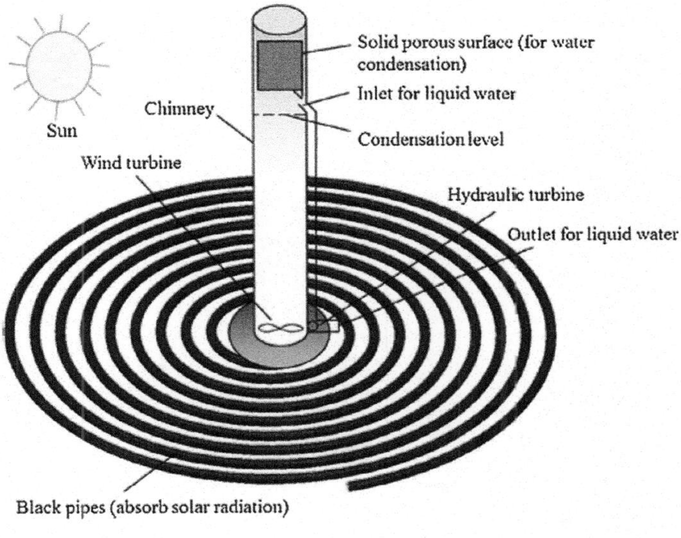


**Fig. 11.14** Schematic of hybrid solar chimney used for power generation and atmospheric water extraction (Ming et al., 2017a)

is nearly independent of the chimney height. However, the optimum condensation height is directly proportional to the temperature difference between the chimney inlet air temperature and the ambient temperature. The ratio of total power output to the thermal energy absorbed by the air flowing over the black tubes is the system efficiency and is around 7.0% for this configuration. The total power output of the system comprises the power output of the wind turbine and the hydraulic turbine.

### 11.2.10 Hybrid Solar Chimney-Desalination System with Transparent Photovoltaic (PV) Roof

Solar energy and land resources can be effectively utilized by modifying conventional solar chimney power plant into hybrid solar chimney-desalination system with transparent PV cover. This modified hybrid system consists of transparent PV roof, central chimney with wind trubine, and basin desalination system below the roof. The schematic of hybrid solar chimney-desalination system with transparent PV roof is shown in Fig. 11.15. Part of the incident solar radiation which is within

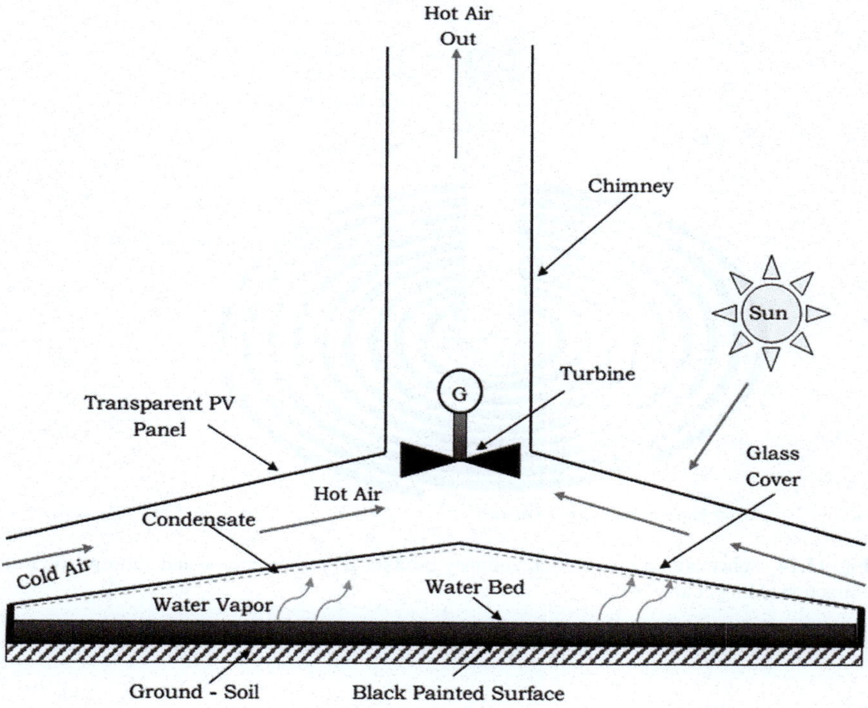


**Fig. 11.15** Schematic of hybrid solar chimney distillation system with transparent PV roof (Rahbar & Riasi, 2019)

the spectral response of PV cells in transparent PV roof is absorbed by the PV cells and the remaining radiation gets absorbed by the transparent cover, saline water, and basin liner of the desalination system. The water vapors generated inside the basin condense over the inner surface of transparent glass cover by releasing its latent heat. The released latent heat energy, convection, and radiation heat transfer from the water to glass cover along with the absorbed solar radiation energy by the glass cover, heats the cold air trapped between the glass cover of basin desalination system and transparent PV roof. The air thus heated rises up due to density difference and flows through the chimney with the aid of natural draft provided by the chimney. The flowing hot air makes the wind turbine on its way to rotate thereby power is generated. In this system, potable water is generated by basin desalination system and power is generated by both the transparent PV roof and generator coupled with wind turbine. Thus, both solar energy and land resource is effectively utilized.

Kiwan et al. (2020) studied the impact of incorporating bare and submerged PV panels as absorber section on the performance of hybrid solar chimney-desalination system. The schematic and layout of hybrid solar chimney-desalination system with submerged and bare PV panels is shown in Figs. 11.16, and 11.17, respectively. Both submerged PV and PV panels occupy only a fraction of the total absorber section of the system. The water bed above the PV panel cools the PV panel and

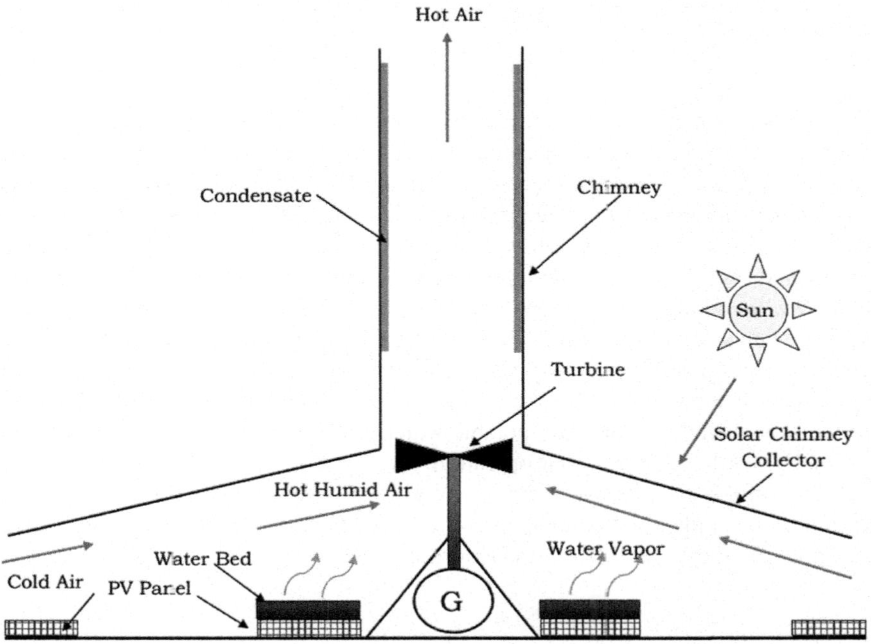


**Fig. 11.16** Schematic of hybrid solar chimney-desalination system with submerged and bare PV panels (Kiwan et al., 2020)

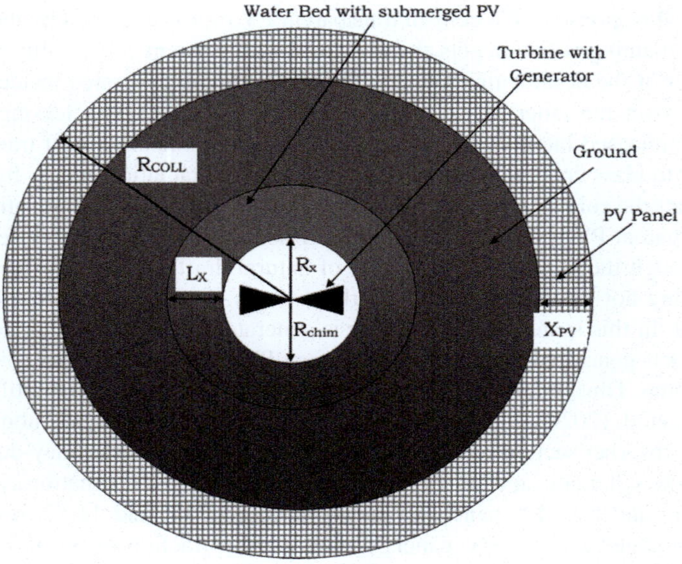


**Fig. 11.17** Schematic indicating the location of PV panels below the hybrid solar chimney-desalination system's collector roof (Chimney height 200m; chimney diameter 10 m; collector diameter 250 m; depth of water 0.03 m; PV efficiency 17.0% (STC); Solar radiation 800 W/m$^2$) (Kiwan et al., 2020)

also undergoes evaporation to generate potable water. The optimum length of bare PV panel measured as radial distance from collector entrance ($X_{PV}$) is about 2.0 m. The submerged PV panels are recommended to be installed closer to the outer radius of chimney and the optimum length of the submerged PV panel ($L_X$) is noticed to be about 15.0 m. The solar energy utilization factor of conventional solar chimney power plant is 0.51% while incorporating PV panels increased the utilization factor to 1.59%. In addition, the addition of water pool increases the utilization factor to 4.37% which is nearly 756.0% higher than the solar energy utilization factor of conventional solar chimney power plant. The annual electric power generated by the system is about 45.35% higher than the power generated by the standalone PV panel in the same location. The daily potable water production capacity of the system is about 160.0 kilotons for 6-hour operation.

The performance comparison of conventional solar chimney power plant, solar chimney power plant with tranparent PV roof and hybrid solar chimney-desalination system with transparent PV roof is shown in Table 11.2.

**Table 11.2** Performance comparison of different solar chimney power plant configurations (Rahbar & Riasi, 2019)

| Optimum input parameters | Conventional solar chimney power plant | Solar chimney power plant with tranparent PV roof | Hybrid solar chimney desalination system with transparent PV roof |
|---|---|---|---|
| Chimney radius (m) | 8.00 | 8.00 | 8.00 |
| Chimney height (m) | 400.00 | 400.00 | 399.79 |
| Roof radius (m) | 100.00 | 100.00 | 100.20 |
| Roof height (m) | 1.88 | 2.25 | 2.99 |
| Air mass flow rate (kg/s) | 799.90 | 800.00 | 799.58 |
| Output | | | |
| Collector efficiency (%) | 55.77 | 56.69 | 88.55 |
| Chimney efficiency (%) | 1.34 | 1.34 | 1.34 |
| Plant efficiency (%) | 0.63 | 0.65 | 1.01 |
| PV power output (kW) | – | 418.2 | 419.80 |
| Turbine power output (kW) | 97.93 | 122.06 | 120.20 |
| Distillate production (kg/h) | – | – | 11856.00 |

## 11.3 Features of Hybrid Solar Chimney Thermal Desalination Systems

(a) Electric power and potable water can be generated under one roof in a sustainable way.
(b) Enhanced utilization factor compared to sole solar chimney power plants.
(c) Chimneys of discarded coal power plants can also be utilized.
(d) Effective option to utilize waste heat rejected from conventional fossil fuel-based power plants for power production and water treatment.
(e) Improves microclimate of the locations if used for atmospheric water harvest.
(f) Large ground area occupancy.
(g) Requirement of high initial cost due to need of heavy and high structures.
(h) Transmission of electric power and potable water generated in the system to residential areas requires additional transmission and distribution systems due to its remoteness.
(i) No real-time studies have been conducted hence reliability is unknown.

## References

Asayesh, M., Kasaeian, A., & Ataei, A. (2017). Optimization of a combined solar chimney for desalination and power generation. *Energy Conversion and Management, 150*, 72–80.

Kiwan, S., Al-Nimr, M., & Salam, A. (2018). Solar chimney power-water distillation plant (SCP-WDP). *Desalination, 445*, 105–114.

Kiwan, S., Al-Nimr, M., & Salim, I. (2020). A hybrid solar chimney/photovoltaic thermal system for direct electric power production and water distillation. *Sustainable Energy Technologies and Assessments, 38*, 100680.

Ming, T., Gong, T., Richter, R. K., Liu, W., & Koonsrisuk, A. (2016). Freshwater generation from a solar chimney power plant. *Energy Conversion and Management, 113*, 189–200.

Ming, T., Gong, T., Richter, R. K., Wu, Y., & Liu, W. (2017a). A moist air condensing device for sustainable energy production and water generation. *Energy Conversion and Management, 138*, 638–650.

Ming, T., Gong, T., Richter, R. K., Cai, C., & Sherif, S. A. (2017b). Numerical analysis of seawater desalination based on a solar chimney power plant. *Applied Energy, 208*, 1258–1273.

Maia, C. B., Silva, F. V. M., Oliveira, V. I. C., & Kazmerski, L. L. (2019). An overview of the use of solar chimneys for desalination. *Solar Energy, 183*, 83–95.

Rahbar, K., & Riasi, A. (2019). Performance enhancement and optimization of solar chimney power plant integrated with transparent photovoltaic cells and desalination method. *Sustainable Cities and Society, 46*, 101441.

Wu, Y., Ming, T., Richter, R., & Hoffer, R. (2020). Large scale freshwater generation from the humid air using the modified solar chimney. *Renewable Energy, 146*, 1325–1336.

Zhou, X., Xiao, B., Liu, W., Guo, X., Yang, J., & Fan, J. (2010). Comparison of classical solar chimney power system and combined solar chimney system for power generation and seawater desalination. *Desalination, 250*, 249–256.

Zuo, L., Zheng, Y., Li, Z., & Sha, Y. (2011). Solar chimneys integrated with sea water desalination. *Desalination, 276*, 207–213.

Zuo, L., Yuan, Y., Li, Z., & Zheng, Y. (2012). Experimental research on solar chimneys integrated with seawater desalination under practical weather condition. *Desalination, 298*, 22–33.

Zuo, L., Ding, L., Chen, J., Zhou, X., Xu, B., & Liu, Z. (2018a). Comprehensive study of wind supercharged solar chimney power plant combined with seawater desalination. *Solar Energy, 166*, 59–70.

Zuo, L., Ding, L., Chen, J., Liu, Z., Qu, N., Zhou, X., & Yuan, Y. (2018b). The effect of different structural parameters on wind supercharged solar chimney power plant combined with seawater desalination. *Energy Conversion and Management, 176*, 372–383.

Zuo, L., Liu, Z., Zhou, X., Ding, L., Chen, J., Qu, N., & Yuan, Y. (2019a). Preliminary study of wind supercharging solar chimney power plant combined with seawater desalination by indirect condensation freshwater production. *Desalination, 455*, 79–89.

Zuo, L., Ding, L., Yuan, Y., Liu, Z., Qu, N., & Dai, P. (2019b). Research progress on integrated solar chimney system for freshwater production. *Global Energy Interconnection, 2*(3), 214–223.

Zuo, L., Liu, Z., Ding, L., Qu, N., Dai, P., Xu, B., & Yuan, Y. (2020). Performance analysis of a wind supercharging solar chimney power plant combined with thermal plant for power and freshwater generation. *Energy Conversion and Management, 204*, 112282.

# Chapter 12
# Solar Thermal Wastewater Treatment Systems

## 12.1 Wastewater Generation and Treatment Status

Rise in population, continuous growth in industries, and urbanization have resulted in large wastewater generation. This generated wastewater is generally dumped into surface water reserves even without any pre-treatment in most of the nations which results in increased pollution of those water bodies (Rajasulochana & Preethy, 2016). Apart from this, utilization of chemical-based pesticides and fertilizers in agriculture sector has also degraded groundwater quality in most of the countries (Rijsberman, 2006). For example in India, groundwater is highly polluted nearly in all the parts of the country (Yale Environment, 2017). According to the World Bank, around 21.0% of the spreadable diseases in highly populated developing nations are majorly due to unclean water (Rajasulochana & Preethy, 2016).

Wastewaters that are generally dumped into water bodies without any pre-treatment are mainly RO (reverse osmosis) reject and domestic sewage water. RO reject is a mixture of chemicals (used for feed water pre-treatment) and concentrated feed water from RO desalination systems. Domestic sewage water represents used water from households. Techniques like deep well injection and discharge into surface waters are widely followed for disposing RO reject but these methods have posed severe threats to the environment (Ahmed et al., 2000). Wastewater treatment and reuse is the only available option that can close water cycle, reduce water stress, and mitigate rising negative impacts on our environment (Vergine et al., 2017). Biological stabilization ponds (Rusan et al., 2007) and activated sludge-extended aeration plants (Al-Lahham et al., 2007) are widely used for treating domestic and industrial wastewaters. However, these techniques cannot remove toxic heavy metals, nitrogen, phosphorous, and organic and inorganic substances from the wastewater in a single step (Rajasulochana & Preethy, 2016).

Three steps in the wastewater treatment process are primary treatment, secondary treatment and tertiary treatment. Coagulation-flocculation, chemical precipitation, and microfiltration process fall under primary treatment. Aerobic and anaerobic biological methods fall under secondary treatment. Oxidation, distillation, electrochem-



H. Sharon and K. Srinivas Reddy, *Solar Thermal Desalination Technologies for Potable Water*, https://doi.org/10.1007/978-3-031-49155-9_12

ical precipitation, photocatalysis, ion exchange, adsorption, and membrane process fall under tertiary treatment. Tertiary treatment methods are capable of removing around 95% of the pollutants in wastewater (Arunkumar et al., 2022). Except distillation, other methods in general cannot tolerate high salinity and heavy metal concentration of RO reject. Hence, evaporation technique is widely recommended for treating RO reject (Giwa et al., 2017).

Open defecation is also a major problem in developing and under developing countries and it accounts for 10.0% in urban and 61.0% in rural areas of the developing nations like India. Moreover, open defecation also causes contamination of ground and surface water bodies by fecal bacteria (Tortajada, 2016). Around 500–1050 deaths/million per year are due to unclean water, improper sanitation, and poor hygiene (World Health Organization, 2017). Human urine is the largest contributor of nutrients in domestic wastewater. The quantity of urine excreted depends on how much a person drinks, sweats, and other factors such as diet, physical activity, and climate (Etter et al., 2011). The average excretion of urine varies between 1.5 and 2.0 L/person/day. Direct application of urine to the land is a very common practice in rural areas but it contaminates surface and groundwater sources significantly (Rose et al., 2015).

Current technologies available for urine treatment include physical treatments such as storage, distillation, filtration (Etter et al., 2011), chemical treatments such as precipitation, electrochemical oxidation, etc. (Zhao et al., 2013; Xie et al., 2016; Colon et al., 2015) biological treatments such as MABR (membrane aerated bio-film reactor) (Udert & Wĺachter, 2012) and advanced processes such as advanced oxidation, membrane filtration, reverse osmosis, electro-dialysis, and vacuum membrane distillation (VMD) (Zollig et al., 2015; Forbis-Stokes et al., 2016; Mehl et al., 2011). Ammonia volatilization from the stored urine, high nitrogen losses, and potential health risks to end-users due to the presence of pathogens are few issues associated with human urine treatment (Zollig et al., 2015). In addition, most of the urine treatment technologies are costly; require high energy, chemicals and are mostly inefficient in removing the salts present in the urine (Tilley et al., 2014; Asadi et al., 2013).

Average wastewater generated per person in urban region of developing nations varies between 98.0 and 220.0 L/d. Nearly, 90.0% of the water used for domestic purposes is returned back as wastewater which can be reused after proper treatment. On the contrary, water used for carrying out agricultural activities cannot be reused as it is consumed as food (Rijsberman, 2006). Moreover, per capita water required for food production is about 70 times higher than the per capita water required for domestic needs (Rijsberman, 2006). Hence, due to increased freshwater demand for irrigation, wastewaters generated from households and industries are directly sold to farmers for carrying out agricultural activities (Sato et al., 2013). Wastewater contains organic and inorganic substances, heavy metals, and toxic chemicals. Apart from this, pathogens like viruses, bacteria, and helminths are also present in it. Potential health risks are always possible with the continuous usage of wastewater for agricultural activities. Untreated wastewater is saline in nature and it reduces the quality of agricultural land and makes it unsuitable for further cultivation leading to chemo-desertification (Kaur et al., 2012; Rebhun, 2004).

Risks associated with utilization of raw wastewater for agriculture activities are listed below (Pereira et al., 2002)

(a) Health-related problems for field workers.
(b) Cumulative poisoning of agricultural land by heavy metals.
(c) Carcinogens and organic chemicals can enter the food chain of mankind during the consumption of crops cultivated using untreated wastewater.
(d) Chemo-desertification of agricultural lands due to salt accumulation.

The ratio of wastewater treated to wastewater generated is around 70.0% in highly developed countries. However, only 32.0% of the wastewater generated is treated in the Asian continent due to a lack of proper finance, infrastructures, and skilled workforce in wastewater treatment sectors (Sato et al., 2013). Energy consumption and $CO_2$ emission per $m^3$ of wastewater treated in biological sewage treatment plants are in the range of 0.40–4.87 kWh and 0.78–3.04 $kgCO_2eq$, respectively (Singh et al., 2016). Similarly, 27.4 kg of oil is required to produce 1.0 $m^3$ of distillate (treated water) by evaporation process (Kalogirou, 2005). Pollution and relatively high treatment cost of existing biological treatment plants and fossil fuel-based distillation systems can be reduced by utilizing renewable wind and solar energy for their operation (Haralambopoulos et al., 1997; Han et al., 2013; Halaby et al., 2017). Furthermore, it becomes necessary to develop renewable energy-powered wastewater treatment system that can treat wastewater in a single step. Moreover, it is also necessary to create awareness among the public about such systems thereby, the public may get attracted to implement such systems in their households, which in turn helps to achieve the sustainable development goal (SDG-6), which stresses availability and sustainable management of water and sanitation for all.

## 12.2 Solar Thermal Systems for Wastewater Treatment

Water-starved regions of the globe are blessed with abundant solar radiation potential and it can be tapped for wastewater treatment (Shatat et al., 2013) and water pasteurization (Duff & Hodgson, 2005). Basin solar distillation systems are capable of producing parasite and arsenic-free drinking water from polluted water (Onyegegbu, 1984; Jasrotia et al., 2013). Dewatering of wastewater sludge (Haralambopoulos et al., 2002) and recovery of antioxidants from oil mill wastewater (Sklavos et al., 2015) have been successfully carried out using basin solar distillation systems. The distillate obtained from basin solar distillation system has 80.0% lower COD (Chemical Oxygen Demand) and 90.0% lower TKN (Total Kjeldhal Nitrogen) compared to raw oil mill wastewater (Potoglou et al., 2003). Velmurugan et al. (2008, 2009) and Farahbod et al. (2013) have treated industrial effluent using basin solar distillation system. Asadi et al. (2013) used stepped solar distillation system for distilling kitchen and palm oil wastewaters. Sometimes, contamination of distillate (treated water) with polluted water or wastewater is highly possible in basin solar distillation systems during feed water addition to the basin (Hanson et al., 2004). However, this limitation can be completely avoided in tilted wick solar distillation systems because in wick-based systems the feed water flow is initiated only by the capillary effect of the wick (Reddy et al., 2018).

## 12.3 Quality Analyses of Wastewater and Distillate

Quality analyses of wastewater and distillate are essential to identify the suitability of the developed treatment system for further scaling up and commercialization. The various water quality parameters along with their measuring methods are tabulated in Table 12.1. The nutrients such as total nitrogen (TN), nitrate ($NO_{3^-}$-N), and total phosphate ($PO_{4^-}$) are analyzed after filtration of samples through the Whatman glass fiber filter paper of 0.45 μm pore diameter as described in standard method (APHA 4500). Heavy metals are measured using atomic absorption spectrometry (AAS) (Perkin Elmer Analyst 700, USA) by direct aspiration into an air-acetylene flame after filtration of wastewater sample. The elements such as percentage of carbon, nitrogen, hydrogen, and sulfur are estimated using the CHNS analyzer (Perkin Elmer, USA).

## 12.4 Reverse Osmosis (RO) Reject and Sewage Water Collection for Experimentation

RO reject is collected from Reverse Osmosis (RO) plant located in Brahmaputra Hostel Zone of Indian Institute of Technology Madras (IIT Madras), Chennai, India. The photograph of RO reject collection point is shown in Fig. 12.1. The capacity of RO plant is around 3000.0 kg/h and it purifies well water and metro water at a recovery ratio of 50.0%. Sewage water used for experimentation is collected from the mixing chamber (Fig. 12.2) of the biological sewage treatment plant of 5.0 ML/d capacity located inside IIT Madras. In the mixing chamber, kitchen and toilet wastewaters from the hostels, labs, and staff apartments are mixed together and finally supplied to the biological sewage treatment system for treatment. Sewage water needs to be filtered to remove floating particles before subjecting it to solar distillation. On contrary to RO reject, sewage water contains suspended particles which can be seen clearly in Fig. 12.2. Nearly, 60.0 L of RO reject and 60.0 L of sewage water are collected separately in high-density polyethylene cans a day before the start of the respective experiments (Reddy et al., 2018).

**Table 12.1** List of water quality parameters, measuring methods and instruments (Delhiraja et al., 2021)

| S.No | Quality parameter | Standards | Method/instrument |
|---|---|---|---|
| 1. | pH | – | Eutech cyberscan PCD 650 multi parameter kit (Thermo scientific, Singapore) |
| 2. | Electrical conductivity | – | Eutech cyberscan PCD 650 multi parameter kit (Thermo scientific, Singapore) |
| 3. | Salinity | – | Eutech cyberscan PCD 650 multi parameter kit (Thermo scientific, Singapore) |
| 4. | Turbidity | APHA-2130 B | Nephlometeric turbidity meter (Eutech cyber scan WL TB1000) |
| 5. | Total solids (TS) | APHA 2540 B-D | Gravimetric method |
| 6. | Total suspended solids (TSS) | APHA 2540 B-D | Gravimetric method |
| 7. | Total dissolved solids (TDS) | APHA 2540 B-D | Gravimetric method |
| 8. | Alkalinity | APHA 2320 B | 0.02N $H_2SO_4$ titrimetric method |
| 9. | Total hardness | APHA 2340 C | EDTA titrimetric method |
| 10. | Sulphate | APHA 4500-E | Nephelometric method |
| 11. | Chloride | APHA 4500 $Cl^-$ B | Argentometeric method |
| 12. | Biochemical oxygen demand (BOD) | APHA 5210 B | Winkler's method |
| 13. | Chemical Oxygen Demand (COD) | APHA 5220 B | closed reflux digestion method/HACH COD digester (Model No 45600, USA) |
| 14. | Total Organic Carbon (TOC) | APHA 5310 B | Total organic carbon analyzer V600 series (Shimadzu, Japan) |
| 15. | Total nitrogen (TN) | APHA 4500 N | TN analyzer (Shimadzu, Japan) |
| 16. | Nitrate ($NO_{3}$– N) | APHA 4500 $NO_3$ | UV-VIS spectrometer 800 (Shimandzu, Japan) by IN HCL method |
| 17. | Total phosphate ($PO_{4}$–) | APHA 4500 P | UV-VIS spectrometer 800 (Shimandzu, Japan) by stannous chloride method |
| 18. | Ammonia ($NH_3$-N) | HACH 8038 | |
| 19. | Coli forms | APHA 9221 | Multiple tube fermentation technique and most probable number (MPN) |
| 20. | Heavy metals | – | Perkin Elmer Analyst 700, Atomic absorption spectrometry |

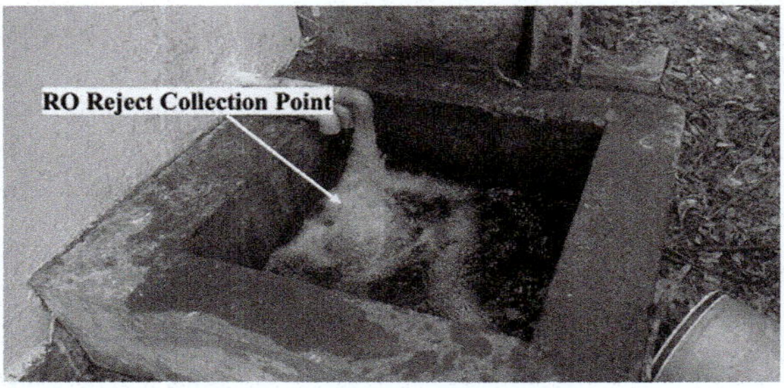


**Fig. 12.1** RO reject discharge point of well water desalting RO system at IIT Madras, Tamil Nadu, India

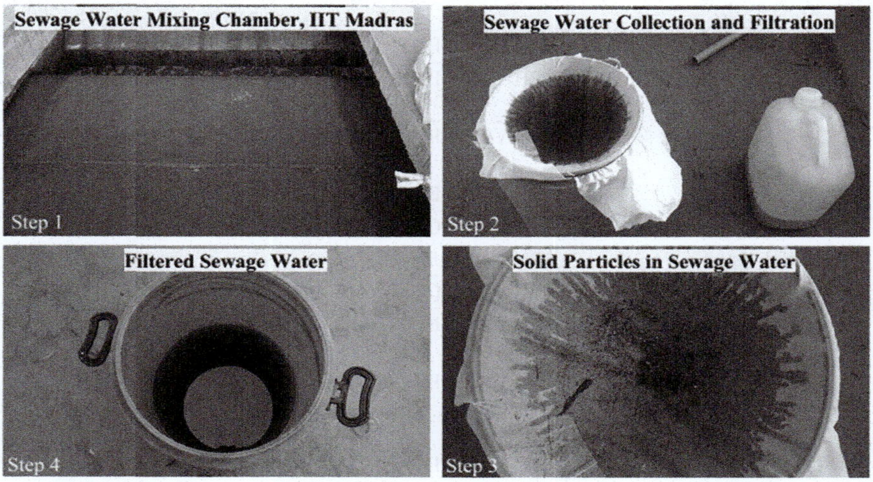


**Fig. 12.2** Sewage water collection and filtration

## 12.5 Performance of Tilted Wick Solar Distillation System Treating Reverse Osmosis (RO) Reject and Sewage Water

The overall summary of experimental results and climatic conditions during solar thermal treatment of RO reject and sewage water using tilted wick solar still is tabulated in Table 12.2. Daily average air-vapor mixture temperature is lower than the daily average wetted wick temperature for all the days irrespective of solar radiation intensity. The highest daily average hourly treated water production rate during RO reject and sewage water solar distillation is around 0.42 L/h and 0.39 L/h, respectively. For a cumulative solar radiation intensity of about 19.90 MJ/m$^2$d, higher wetted wick temperature and lower daily average treated water production rate are noticed during

**Table 12.2** Overall summary of experimental results of tilted wick solar still treating RO reject and sewage water (Reddy et al., 2018)

| Type of wastewater | RO reject | | | | Sewage water | | | |
|---|---|---|---|---|---|---|---|---|
| Experiment day no. | 1 | 2 | 3 | 4 | 1 | 2 | 3 | 4 |
| Daily average solar radiation intensity ($W/m^2$) | 471.0 | 474.0 | 511.0 | 490.0 | 401.0 | 491.0 | 526.0 | 421.0 |
| Cumulative solar radiation ($MJ/m^2$-d) | 19.08 | 19.20 | 20.70 | 19.86 | 16.29 | 19.90 | 21.33 | 17.09 |
| Daily average ambient temperature (°C) | 29.7 | 29.6 | 32.1 | 31.8 | 30.9 | 32.2 | 33.4 | 32.4 |
| Daily average wind velocity (m/s) | 2.6 | 3.0 | 2.3 | 2.4 | 2.7 | 2.6 | 2.9 | 3.6 |
| Daily average wetted wick temperature (°C) | 57.2 ± 3.5 | 57.7 ± 3.6 | 58.8 ± 3.7 | 57.6 ± 3.7 | 53.8 ± 3.7 | 61.0 ± 2.6 | 66.8 ± 4.4 | 60.2 ± 1.7 |
| Daily average Air-vapor mixture temperature (°C) | 55.8 ± 0.2 | 55.8 ± 0.2 | 55.9 ± 0.4 | 53.9 ± 0.5 | 50.3 ± 0.5 | 56.4 ± 0.8 | 61.9 ± 0.8 | 59.2 ± 1.2 |
| Daily average glass cover temperature (°C) | 46.5 ± 2.4 | 46.4 ± 2.3 | 46.0 ± 2.3 | 44.1 ± 2.0 | 40.8 ± 1.9 | 45.2 ± 2.1 | 47.3 ± 2.5 | 46.9 ± 1.8 |
| Daily average daily hourly treated water productivity (L/h) | 0.37 | 0.38 | 0.41 | 0.42 | 0.33 | 0.39 | 0.38 | 0.32 |

sewage water treatment (Day 2) in comparison to RO reject treatment (Day 4). This observation confirms poor wetting of wick due to clogging caused by the suspended particles in sewage water which leads to enhanced wetted wick temperature and reduced overall evaporation rate.

Treated water production rate and thermal efficiency of the tilted wick solar distillation system treating RO reject are in the range of 4.12–4.79 L/d and 43.4–48.5%, respectively. Nocturnal treated water production is only around 4.0% of the total treated water production because most of the heated feed water leaves the distillation system throughout the day leading to reduced availability of heated water mass inside the distillation system for further evaporation after sunset hours. Un-evaporated RO reject is re-circulated continuously during experimentation but this has not caused a significant drop in the performance of the distillation system and this may be due to the low salinity and low turbidity of RO reject (4452.43 mg/L and 1.34 NTU) used in this experimentation.

Highest and lowest diurnal treated water production during sewage water treatment are around 4.48 and 3.57 L/d on the second and the third day, respectively. During high solar intensity, enhanced evaporation of sewage water takes place, as a result, tiny suspended particles of re-circulated sewage water gets dried up and settles over the wick which lowers the capillary action and further wetting of the wick. Hence, further evaporation process slows down and leads to reduced overall treated water production and performance of the system. Higher thermal efficiency of 46.8% is noticed on Day 1 which has the lowest cumulative solar intensity of 16.29 MJ/m$^2$-d, while the lowest thermal efficiency of 40.5% is noticed on Day 3 which has the highest cumulative solar intensity of 21.33 MJ/m$^2$-d. This observation justifies the claim of wick clogging effect during sewage water distillation.

The wetted wick temperature of the system during sewage water treatment is higher than the water sterilization temperature (65.0 °C) for about 4–5 h in a day which indicates an efficient killing of bacteria and microbes even in un-evaporated sewage water (Hameed & Ahmad, 1997; Saitoh & El Ghetany, 1999). In order to prevent formation of dry patches due to the deposition of suspended particles over the evaporating surface, feed water flow rate must be at least 1.3 times higher than the evaporation rate (Chong et al., 2014). If wetting of the wick is achieved only by capillary effect, then feed water flow rate required to prevent the formation of dry patches will be difficult to achieve. In addition, liberation of trapped non-condensable gases like $CO_2$, $N_2$, $CH_4$, $NH_3$, and $H_2S$ (Sullivan & Krieger, 2001) also occurs during sewage water evaporation which in turn reduces distillation system's performance by increasing the air resistance within the distillation chamber. Hence, operation at lower temperature and evacuation at regular intervals (Al Hussaini & Smith, 1995) are necessary for enhanced performance of solar thermal distillation systems treating sewage water.

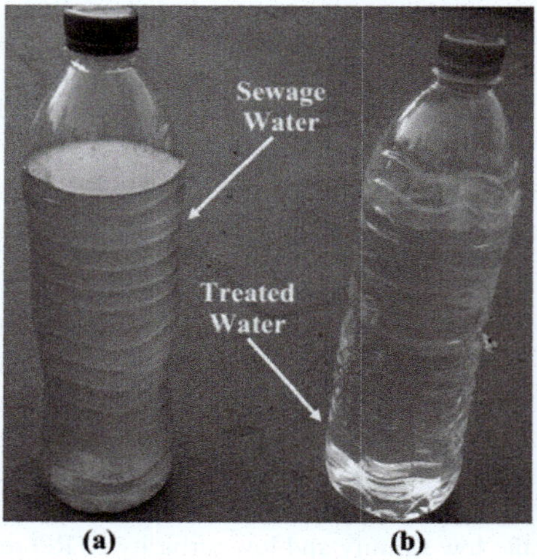


**Fig. 12.3** Sample of **a** raw sewage water and **b** distillate obtained through solar thermal distillation process

## 12.6 Quality Results of Treated RO Reject

The quality results of distillate obtained by solar thermal treatment of RO reject along with the necessary standards and removal efficiency is tabulated in Table 12.3. The pH value of distillate obtained by solar distillation of RO reject is within the drinking water standards. Salinity, chloride content, and electrical conductivity of the obtained distillate are very much lower than wastewaters which confirms the effective removal of ions by solar distillation process (Flendrig et al., 2009). According to WHO (World Health Organization) guidelines, BOD and COD values for safe drinking water are 2.0 and 10.0 mg/L (World Health Organization, 2004). BOD and COD contents of distillate obtained by solar distillation of RO reject are little bit higher than the drinking water standards, which stresses the need for some post-treatment processes to make the distillate more suitable for human consumption. Except BOD and COD, other qualities of distillate obtained by solar distillation of RO reject are well within the drinking water standards.

## 12.7 Quality Results of Treated Sewage Water

The quality results of distillate obtained by solar thermal treatment of sewage water along with the necessary standards and removal efficiency are tabulated in Table 12.4. Water quality of distillate obtained by solar distillation of sewage water is very much better than the standards proposed for safe disposal of effluent into inland surface waters (CPCB, 1986). BOD, COD, turbidity, and bacteria removal efficiency are around 97.3, 95.8, 98.3, and 100.0%, respectively, for tilted wick solar distillation system treating sewage water. COD removal efficiency of tilted wick solar distillation system is higher than the removal efficiency of basin solar distillation system studied by Potoglou et al. (2003). Distillate is very much clear which could be seen from Fig. 12.3, and can be justified by its turbidity value reported in Table 12.4. Both distillate and preheated un-evaporated sewage water are odor-free throughout the experiments. A similar observation has been recorded by Qasim (1978) while distilling sewage water using basin solar distillation system. Wang et al. (2007) noticed odor in biologically treated and disinfected wastewater which justifies the necessity and superiority of solar thermal distillation process for treating sewage water.

Except BOD and COD, all other quality values of distillate obtained by solar distillation of sewage water are within the drinking water standards hence the distillate can be used for domestic purposes like bathing, washing, and flushing (Hingorani, 2011). No health related issues have been noticed for persons handling biologically treated wastewater during a 14-month study conducted by Wang et al. (2007). Hence, superior distillate obtained by solar thermal distillation of sewage water is very safe and can be used for irrigation without any health risks. Heavy metal compositions of RO reject used for experimentations are within the standards for safe disposal. Lead concentration of sewage water is higher than the safe disposal standard. Heavy

**Table 12.3** Quality results of RO reject and distillate obtained from tilted wick solar distillation system (Reddy et al., 2018)

| S. No | Parameter | RO reject | Distillate | Indian standards (BIS, 2013) | Removal efficiency (%) |
|---|---|---|---|---|---|
| 1 | pH | 6.91 ± 0.05 | 7.21 ± 0.21 | 6.50 – 8.50 | – |
| 2 | EC (μS/cm) | 1127.00 ± 2.00 | 53.04 ± 0.19 | NA | 95.3 |
| 3 | Salinity (mg/L) | 4452.43 ± 5.86 | 19.29 ± 0.04 | NA | 99.6 |
| 4 | Turbidity (NTU) | 1.34 ± 0.06 | 0.92 ± 0.02 | 5.0 | 31.3 |
| 5 | Total hardness (as $CaCO_3$) (mg/L) | 806.70 ± 11.5 | 13.40 ± 5.77 | 300.00 | 98.3 |
| 6 | Chlorides (as Cl) (mg/L) | 2466.70 ± 5.86 | 10.67 ± 0.04 | 250.00 | 99.6 |
| 7 | Total solids (mg/L) | 1106.70 ± 20.2 | 52.50 ± 5.46 | NA | 95.3 |
| 8 | Total suspended solids (mg/L) | 104.40 ± 10.5 | 6.65 ± 5.34 | NA | 93.6 |
| 9 | Total dissolved solids (mg/L) | 1002.30 ± 0.58 | 46.88 ± 0.11 | 500.00 | 95.3 |
| 10 | Calcium (as $Ca^{+}$) (mg/L) | 23.54 ± 0.78 | 1.66 ± 0.82 | 75.00 | 92.9 |
| 11 | Sulphate (as $SO_4^{2-}$) (mg/L) | 823.40 ± 25.2 | 14.80 ± 1.28 | 200.00 | 98.2 |
| 12 | Nitrate (as $NO_{3-}$) (mg/L) | 56.60 ± 4.14 | 3.38 ± 0.53 | 45.00 | 94.0 |
| 13 | Alkalinity (as $CaCO_3$) (mg/L) | 420.00 ± 52.9 | 43.40 ± 5.77 | 200.00 | 89.7 |
| 14 | Magnesium (as Mg) | 26.48 ± 0.25 | 1.25 ± 0.88 | 30.00 | 95.3 |
| 15 | Phosphate (as $PO_4^{3-}$) (mg/L) | 5.14 ± 0.42 | 0.14 ± 0.06 | NA | 97.3 |
| 16 | Total nitrogen (mg/L) | 69.83 ± 0.81 | 10.22 ± 0.28 | NA | 85.4 |
| 17 | BOD (mg/L) | 95.60 ± 4.87 | 15.30 ± 3.05 | NA | 83.9 |
| 18 | COD (mg/L) | 353.40 ± 30.5 | 38.90 ± 1.10 | NA | 88.9 |
| 19 | Bacterial quality | | | | |
| | Total coliform (MPN/100mL) | 12.00 | <1.8 | Nil | 100.0 |
| | Fecal coliform (MPN/100mL) | <1.8 | <1.8 | Nil | 100.0 |
| Heavy metals | | | | | |
| 1 | Lead (Pb) (mg/L) | 0.069 ± 0.002 | 0.023 ± 0.005 | 0.05* | 66.7 |
| 2 | Zinc (Zn) (mg/L) | 0.140 ± 0.056 | 0.098 ± 0.023 | 5.00* | 32.9 |
| 3 | Manganese (Mn) (mg/L) | 0.078 ± 0.007 | 0.014 ± 0.007 | 0.05* | 82.1 |
| 4 | Copper (Cu) (mg/L) | 0.023 ± 0.005 | 0.009 ± 0.002 | 0.05* | 60.9 |

<1.8 less than detectable limit, *NA* not available
* Drinking Water Standards (Purdom, 1980)

**Table 12.4** Quality results of sewage water and distillate obtained from tilted wick solar distillation system (Reddy et al., 2018)

| S.No | Parameter | Sewage water | Treated water | Safe disposal standards (CPCB, 1986) | Removal efficiency (%) |
|---|---|---|---|---|---|
| 1 | pH | 7.23 ± 0.02 | 7.86 ± 0.04 | 5.50–9.00 | – |
| 2 | EC (μS/cm) | 1492.30 ± 2.08 | 173.50 ± 0.25 | NA | 88.4 |
| 3 | Salinity (mg/L) | 1371.30 ± 23.11 | 150.68 ± 0.45 | NA | 89.0 |
| 4 | Turbidity (NTU) | 55.10 ± 2.89 | 0.91 ± 0.09 | 5.0** | 98.3 |
| 5 | Chlorides (as Cl) (mg/L) | 759.71 ± 23.11 | 83.46 ± 0.45 | 250.00** | 89.0 |
| 6 | Total solids (mg/L) | 1916.70 ± 25.2 | 183.20 ± 4.97 | NA | 90.4 |
| 7 | Total suspended solids (mg/L) | 588.40 ± 21.9 | 28.80 ± 4.48 | 100.00 | 95.1 |
| 8 | Total dissolved solids (mg/L) | 1328.40 ± 3.24 | 154.40 ± 0.49 | 500.00** | 88.4 |
| 9 | Sulphate (as $SO_4^{2-}$) (mg/L) | 142.50 ± 5.37 | 19.26 ± 1.26 | 200.00** | 86.5 |
| 10 | Nitrate (as $NO_{3-}$) (mg/L) | 18.56 ± 2.66 | 2.11 ± 0.03 | 10.00 | 88.6 |
| 11 | Potassium (as K+) (mg/L) | 27.49 ± 0.45 | 2.80 ± 0.61 | NA | 89.8 |
| 12 | Phosphate (as $PO_4^{2-}$) (mg/L) | 28.23 ± 2.31 | 0.06 ± 0.02 | 5.00 | 99.8 |
| 13 | Total nitrogen (mg/L) | 60.80 ± 0.29 | 22.50 ± 1.12 | 50.00 | 62.9 |
| 14 | BOD (mg/L) | 506.70 ± 23.1 | 12.50 ± 0.42 | 30.00 | 97.3 |
| 15 | COD (mg/L) | 913.40 ± 30.6 | 38.70 ± 1.12 | 250.00 | 95.8 |
| 16 | Bacterial quality | | | | |
| | Total coliform (MPN/100Ml) | 40,000,000.00 | 2.00 | ≤100.00 | 99.9 |
| | Fecal coliform (MPN/100Ml) | 22,000.00 | Nil | ≤100.00 | 100.0 |
| Heavy metals | | | | | |
| 1 | Lead (Pb) (mg/L) | 0.102 ± 0.023 | 0.030 ± 0.002 | 0.10 | 70.6 |
| 2 | Zinc (Zn) (mg/L) | 0.092 ± 0.054 | 0.045 ± 0.010 | 5.00 | 51.1 |
| 3 | Manganese (Mn) (mg/L) | 0.029 ± 0.008 | 0.009 ± 0.007 | 2.00 | 68.9 |
| 4 | Copper (Cu) (mg/L) | 0.070 ± 0.047 | 0.030 ± 0.002 | 3.00 | 57.1 |

*NA* not available
** Drinking water standards as per (BIS, 2013)

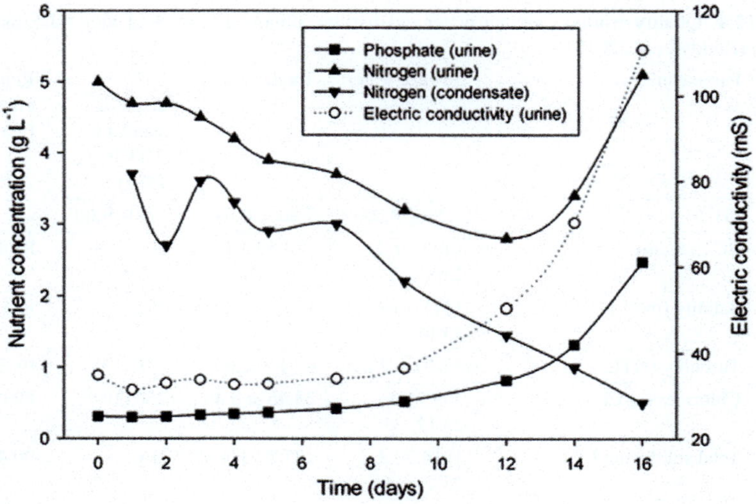


**Fig. 12.4** Quality results of undiluted human urine with days during evaporation in basin solar still (Antonini et al., 2012)

metals concentration of distillate obtained in a single step by solar thermal distillation of RO reject and sewage water are far better than the proposed safe drinking water and safe disposal standards.

## 12.8 Performance of Solar Thermal Distillation Systems Treating Human Urine

Human liquid waste (urine) is a complex solution consisting of 95% water, 3.5% organics and 1.5% inorganics (Rose et al., 2015). For experiments, simulated urine prepared as per the procedure given by Putnam (1971) is utilized. Tilted solar still with partitioned basin can hold the urine within the distillation chamber throughout the experiments thereby direct exposure of the researcher to the pungent smell of evaporating urine is avoided. In spite of this arrangement, a pungent smell was felt during experimentation. The distillation system is capable of holding around 10 L of urine (2.0 L of urine/d-person by considering five members in a family). The urine is allowed to evaporate continuously and the quality of distillate produced is measured daily while the quality of reject is measured at the end of the final experimental run such that the reject will be much concentrated. This approach is more practical because the liquid waste has to be treated/evaporated to the maximum level without the need for other disposal techniques. Maximum thermal efficiency of the system treating tap water and urine is around 52.02% and 48.9%, respectively. The specific energy consumption is inversely proportional to the thermal efficiency. The specific energy consumption of the tilted solar thermal distillation system with partition basin treating urine is about 1.35–2.77 $kWh_t/kg$. High specific energy consumption for treating urine is due to the presence of high dissolved salts within it.

Antonini et al. (2012) utilized concrete basin solar still (2 $m^2$ surface area) with black tiles as absorber and inner side walls covered with mirrors to recover nutrients from actual human urine collected from a dormitory in Vietnam. Fifty liters of undiluted human urine fed to the distillation system were allowed to evaporate continuously with the aid of solar radiation. Everyday urine sample from the basin and produced condensate was collected and analyzed for certain properties. Nearly, all the urine got evaporated by the end of the 26th day leaving behind 360 g of solid material inside the distillation chamber. Around 38.64 L of distillate was collected; 2.80 L of urine was taken as sample and 8.56 L of distillate was found to escape from the distillation system as vapor due to leakage during the 26-day study. Similarly, studies were also carried out by evaporating urine mixed with sulfuric acid (27.76 mL of $H_2SO_4$ per liter of urine) and urine mixed with phosphoric acid (31.46 mL of $H_3PO_4$ per liter of urine) to assess the impact of urine pH on nitrogen losses and nutrient content of the final solid product.

## 12.9 Quality Results of Treated Synthetic Human Urine

The quality results of raw urine, distillate, and reject are shown in Table 12.5. The slight pH increase in the reject is mainly due to the presence of ammonium salts (Antonini et al., 2012; Gulyas et al., 2014). Electrical conductivity of distillate ($569 \pm 20.6$ μS/cm) is lower than the raw urine ($21143 \pm 18.4$ μS/cm) and reject ($23502.5 \pm 60.1$ μS/cm). Salinity decreased from $10370 \pm 0.01$ to $498 \pm 130$ mg/L. TDS concentration has reduced from $14799.5 \pm 140$ to $466 \pm 18.7$ mg/L. The slight TDS concentration in distillate may be due to the condensation of ammonia generated along with vapors at high temperature during high solar radiation intensity periods. The presence of trace amount of salts such as sodium, potassium, nitrate etc., in the distillate might have also contributed to TDS.

Turbidity of the distillate is about $0.52 \pm 0.05$ NTU. Organics (measured in terms of COD and TOC) offer energy recovery potential whereas nutrients such as nitrogen, phosphorus, etc., plays a major role in fertilizer value of the material. Since the concentrations of heavy metals are generally low in human urine (Jönsson et al., 2004), they were not monitored. The COD and TOC concentration of distillate are $188.6 \pm 7.9$ mg/L (96.6% removal efficiency) and $53.7 \pm 2.8$ mg/L (94.7% removal efficiency), respectively. The organics removal efficiency is higher in tilted solar thermal distillation system in comparison to the stepped solar thermal distillation system treating domestic wastewater (Asadi et al., 2013) and industrial effluent (Velmurugan et al., 2009). The removal efficiencies of nutrients such as TN, $NO_3$-N, $PO_4$, $NH_4$-N and K, which are the major ingredients of urine, are 95.7%, 98.6%, 98.1%, 94.8%, and 98.5%, respectively. The quality of obtained distillate meets the USEPA reuse standard for hand washing, toilet flushing, and gardening. The overall recovery ratio

**Table 12.5** Quality analyses results of synthetic human urine, distillate and reject (Delhiraja et al., 2021)

| Parameters | Raw | Day 1 | Day 2 | Day 3 | Day 4 | Reject |
|---|---|---|---|---|---|---|
| pH | 7.83 ± 0.057 | 7.28 ± 0 | 7.1 ± 0.176 | 6.9 ± 0.13 | 6.89 ± 0.02 | 8.9 |
| EC (μS/cm) | 21143 ± 18.4 | 505.4 ± 22.2 | 539 ± 22.6 | 662.5 ± 16.3 | 571.5 ± 21.9 | 49300 |
| TDS (mg/L) | 14799.5 ± 140 | 378.3 ± 20.7 | 361 ± 49.2 | 474.8 ± 5.52 | 391 ± 1.41 | 34502 |
| Turbidity (NTU) | 8.37 ± 0.86 | 0.43 ± 0.09 | 0.87 ± 0.11 | 0.365 ± 0.01 | 0.4 ± 0.14 | 19 |
| NaCl (mg/L) | 10370 ± 0.01 | 481.6 ± 22.8 | 414.7 ± 19.7 | 689 ± 1.42 | 410 ± 14.1 | 23886 |
| COD (mg/L) | 5631 ± 46.7 | 148.5 ± 11.7 | 116.6 ± 6.51 | 200 ± 0 | 290.5 ± 13.4 | 13121 |
| TOC (mg/L) | 1466.5 ± 27.5 | 27.8 ± 1.52 | 92.87 ± 8.09 | 97.3 ± 0.92 | 88.2 ± 0.98 | 3379 |
| TN (mg/L) | 4761.5 ± 37.5 | 45.7 ± 1.08 | 47.77 ± 1.59 | 26.47 ± 0.60 | 20.19 ± 1.12 | 11213 |
| Nitrate (mg/L) | 611.5 ± 70 | 8.32 ± 0.91 | 10.6 ± 0.91 | 8.6 ± 1.94 | 6.46 ± 0.23 | 1435 |
| Phosphate (mg/L) | 245.4 ± 13.8 | 8.41 ± 0.09 | 2.96 ± 0.23 | 5.703 ± 0.39 | 1.729 ± 0.08 | 573.3 |
| Potassium (mg/L) | 910 ± 14 | 5.37 ± 0.11 | 4.23 ± 0.47 | 2.67 ± 0.32 | 2.82 ± 0.24 | 2148 |
| Ammonia (mg/L) | 185 ± 21.2 | 6.89 ± 0.01 | 10.3 ± 0.62 | 10.5 ± 0.63 | 10.5 ± 0.45 | – |

of the tilted solar still with partitioned basin treating human urine ranges between 50 and 60%. The percentage C, N, P, and K of the reject after drying is 10.9%, 12.8%, 3.4%, and 4.72%, respectively.

## 12.10 Quality Results of Treated Actual Human Urine

The variation of phosphate, nitrogen, and electrical conductivity of undiluted urine subjected to continuous evaporation in basin solar still is shown in Fig. 12.4. Electrical conductivity and phosphate content of urine inside the basin increased day by day due to accumulation of salts. However, its nitrogen content dropped from 5.0 g/L (day 1) to 2.8 g/L (day 12) and then rises to 5.1 g/L on day 16. The nitrogen content of distillate obtained at the beginning was high which then dropped from 3.8 g/L (day 1) to 0.5 g/L (day 16). The pH values of urine samples are in the range of 9.3–9.7, while the pH values of the distillate is slightly high and is in the range of 9.6–10.0. From X-ray diffraction spectroscopy, it was found that the sodium chloride, Aphthitalite, Niter, and Biphosammite content of the solid material obtained from the complete evaporation of undiluted actual human urine is about 90%, 6.0%, 4.0%, and 0%, respectively. For urine diluted with phosphoric acid, the sodium chloride, Aphthitalite, Niter, and Biphosammite content of the solid material obtained after evaporation is about 13.0%, 0%, 0%, and 87%, respectively.

**Table 12.6** Comparison of phosphorous, nitrogen and sulfur content of solid material obtained from various sources (Antonini et al., 2012)

| S. No. | Source from which solid material was recovered | Phosphate (%) | Nitrogen (%) | Sulfur (%) |
|---|---|---|---|---|
| 1. | Undiluted urine | 1.87 | 1.84 | 3.24 |
| 2. | Urine + Phosphoric acid | 17.55 | 8.33 | 1.04 |
| 3. | Urine + Sulfuric acid | 5.25 | 5.46 | 4.67 |
| 4. | Commercial fertilizer | 8.22 | 0.16 | 12.51 |

The comparison of nutrient content of the solid material derived from the human urine with conventional chemical fertilizer available in market is tabulated in Table 12.6. It could be seen that the addition of acid to urine has enhanced phosphorous and nitrogen content of the solid material obtained after evaporation. Acidification of urine with acids has prevented nitrogen loss but this step is not recommended for developing nations due to additional expenses and risks associated with handling acids (Antonini et al., 2012).

## 12.11 Wastewater Feed and Reject Management Strategies in Solar Thermal Distillation Systems (Arunkumar et al., 2022)

(a) Feed wastewater must be filtered to remove suspended particles to prevent clogging of wick surfaces.
(b) Microbe growth in wastewater is very intense at a temperature range of 30–40 °C which is a threat to the persons handling wastewater.
(c) Xylem filters derived from natural wood can be used to eliminate microbes from the wastewater (99% removal efficiency). These filters have also been effective in removing heavy metals
(d) High saline nature of wastewater can encourage corrosion of basin liners of the solar thermal treatment distillation systems and it can be prevented by corrosion inhibitor coatings over the basin components.
(e) Treatment of basin liner with acetic acid and sodium acetate is necessary to prevent fouling thereby life of the system can be enhanced.
(f) Salt fouling over basin liner can be delayed by coating micro/nanomaterial to reduce the chemical potential.
(g) The dried reject sediments of wastewater left after solar evaporation can be used as fertilizers.

## 12.12 Features of Solar Thermal Wastewater Treatment Systems

(a) Treatment of wastewater in a single step which is not possible in biological treatment systems.
(b) High pollutant removal efficiency (greater than 95%) compared to other available treatment systems.
(c) Effective removal of heavy metals in a single step.
(d) Suitable and sustainable option to treat and reuse human urine, especially in rural and remote locations to improve sanitation.
(e) Clogging of wick due to suspended particles present in sewage water is a limiting factor. However, this limitation can be rectified by using tilted partitioned basins instead of wicks.
(f) Selective removal of volatile components from the wastewater stream is not possible without the aid of vacuum pumps and sophisticated controls.
(g) Long-term studies on reliability of the system are needed to induce public acceptability.

## References


Ahmed, M., Shayya, W. H., Hoey, D., Mahendran, A., Morris, R., & Al-Handaly, J. (2000). Use of evaporation ponds for brine disposal in desalination plants. *Desalination, 130*, 155–168.

Al-Hussaini, H., & Smith, I. K. (1995). Enhancing of solar still productivity using vacuum technology. *Energy Conversion and Management, 36*(11), 1047–1051.

Al-Lahham, O., El-Assi, N. M., & Fayyad, M. (2007). Translocation of heavy metals to tomato (Solanum lycopersicom L.) fruit irrigated with treated water. *Scientia Horiculture, 113*, 250–254.

Antonini, S., Nguyen, P. T., Arnold, U., Eichert, T., & Clemens, J. (2012). Solar thermal evaporation of human urine for nitrogen and phosphorus recovery in Vietnam. *Science of the Total Environment, 414*, 592–599.

Arunkumar, T., Sathyamurthy, R., Denkenberger, D., & Lee, S. J. (2022). Solar distillation meets the real world: A review of solar stills purifying real wastewater and seawater. *Environmental Science and Pollution Research*. https://doi.org/10.1007/s11356-022-18720-2

Asadi, R. Z., Suja, F., & Ruslan, M. H. (2013). The application of a solar still in domestic and industrial wastewater treatment. *Solar Energy, 93*, 63–71. https://doi.org/10.1016/j.solener.2013.03.024

BIS (2012). http://cgwb.gov.in/Documents/WQ-standards.pdf.

Chong, Tze-Ling., Huang, Bin-Juine., Wu, Po-Hsien., & Kao, Yeong-Chuan. (2014). Multiple-effect diffusion solar still coupled with a vacuum-tube collector and heat pipe. *Desalination, 347*, 66–76.

Colon, t' J., Forbis-Stokes, A. A., & Deshusses, M. A. (2015). Anaerobic digestion of undiluted simulant human excreta for sanitation and energy recovery in less-developed countries. *Energy for Sustainable Development, 29*, 57–64. https://doi.org/10.1016/j.esd.2015.09.005

CPCB (1986). Government of India. http://cpcb.nic.in/GeneralStandards.pdf. Accessed 16 May 2017

Delhiraja, K., Sharon, H., Reddy, K. S., & Philip, L. (2021). Performance evaluation of solar thermal systems as an alternative for human waste treatment. *Sustainable Energy Technologies and Assessments, 47*, 101393.

Duff, W. S., & Hodgson, D. A. (2005). A simple high efficiency solar water purification system. *Solar Energy, 79*, 25–32.

Etter, B., Tilley, E., Khadka, R., & Udert, K. M. (2011). Low-cost struvite production using source-separated urine in Nepal. *Water, 45*(2), 852–62. https://doi.org/10.1016/j.watres.2010.10.007

Farahbod, F., Mowla, D., Jafari Nasr, M. R., & Soltanieh, M. (2013). Experimental study of a solar desalination pond as second stage in proposed zero discharge desalination process. *Solar Energy, 97*, 138–146.

Flendrig, L. M., Shah, B., Subrahmaniam, N., & Ramakrishnan, V. (2009). Low cost thermoformed solar still water purifier for D&E countries. *Physics and Chemistry of the Earth, 34*, 50–54.

Forbis-Stokes, A. A., O'Meara, P. F., Mugo, W., Simiyu, G. M., & Deshusses, M. A. (2016). On-site fecal sludge treatment with the anaerobic digestion pasteurization latrine. *Environmental Engineering Science, 33*(11), 898–906. https://doi.org/10.1089/ees.2016.0148

Giwa, A., Dufour, V., Al-Marzooqi, F., Al-Kaabi, M., & Hasan, S. W. (2017). Brine management methods: Recent innovations and current status. *Desalination, 407*, 1–23.

Gulyas, H., Zhang, S., & Otterpohl, R. (2014). Pre-treating stored human urine for solar evaporation by low-technology ammonia stripping. *Journal of Environmental Protection, 5*(11), 962–969.

Halaby, A., Ghoneim, W., & Helal, A. (2017). Sensitivity analysis and comparative studies for energy sustainability in sewage treatment. *Sustainable Energy Technologies and Assessments, 19*, 42–50.

Hameed, S. M. K., & Ahmad, I. (1997). Solar sterilization of water. *Renewable Energy, 12*(3), 321–324.

Hanson, A., Zachritz, W., Stevens, K., Mimbela, L., Polka, R., & Cisneros, L. (2004). Distillate water quality of a single-basin solar still: Laboratory and field studies. *Solar Energy, 76*, 635–645.

Han, C., Liu, J., Liang, H., Guo, X., & Li, L. (2013). An innovative integrated system utilizing solar energy as power for the treatment of decentralized wastewater. *Journal of Environmental Sciences, 25*(2), 274–279.

Haralambopoulos, D., Pantelakis, I., Paraskevas, P., & Lekkas, Th. (1997). Waste-water treatment and renewable energy potential in the Aegean islands. *Energy, 22*(7), 725–733.

Haralambopoulos, D. A., Biskos, G., Halvadakis, C., & Lekkas, T. D. (2002). Dewatering of wastewater sludge through a solar still. *Renewable Energy, 26*, 247–256.

Hingorani, P. (2011). The economics of municipal sewage water recycling and reuse in India. *India Infrastructure Report, 2011*, 312–322.

Jasrotia, S., Kansal, A., & Kishore, V. V. N. (2013). Application of solar energy for water supply and sanitation in Arsenic affected rural areas: A study for Kaudikasa village, India. *Journal of Cleaner Production, 60*, 102–106.

Jönsson, H., Stintzing, A. R., Vinnerås, B., & Salomon, E. (2004). Guidelines on the use of urine and faeces in crop production. *Eco San Res Programme*.

Kalogirou, S. (2005). Sea water desalination using renewable energy sources. *Progress in Energy and Combustion Science, 31*, 242–281.

Kaur, R., Wani, S. P., Singh, A. K., & Lal, K. (2012). Wastewater production, treatment and use in India. Presented at the Second Regional Workshop of the Project 'Safe Use of Wastewater in Agriculture' 16–18 May 2012, New Delhi.

Mehl, J., Kaiser, J., Hurtado, D., Gibson, D. A., Izurieta, R., & Mihelcic, J. R. (2011). Pathogen destruction and solids decomposition in composting latrines: Study of fundamental mechanisms and user operation in rural Panama. *Journal of Water and Health, 9*(1), 187–99. https://doi.org/10.2166/wh.2010.138

Onyegegbu, S. O. (1984). Solar distillation of pond (muddy) water. *Energy Conversion and Management, 24*(1), 1–4.

Pereira, L. S., Oweis, T., & Zairi, A. (2002). Irrigation management under water scarcity. *Agricultural Water Management, 57*, 175–206.

Potoglou, D., Kouzeli-Katsiri, A., & Haralambopoulos, D. (2003). Solar distillation of olive mill wastewater. *Renewable Energy, 29*, 569–579.

Purdom, P.W. (1980). Environmental health (2nd ed.). London: Academic Press Inc.

Putnam, D. F. (1971). Composition and concentrative properties of human urine.

Qasim, S. R. (1978). Treatment of domestic sewage by using solar distillation and plant culture. *Journal of Environmental Science and Health. Part A: Environmental Science and Engineering: Toxic/Hazardous Substances and Environmental Engineering, 13*(8), 615–627.

Rajasulochana, P., & Preethy, V. (2016). Comparison on efficiency of various techniques in treatment of waste and sewage water - A comprehensive review. *Resource-Efficient Technologies, 2*, 175–184.

Rebhun, M. (2004). Desalination of reclaimed wastewater to prevent salinization of soils and groundwater. *Desalination, 160*, 143–149.

Reddy, K. S., Sharon, H., Krithika, D., & Philip, L. (2018). Performance, water quality and enviro-economic investigations on solar distillation treatment of reverse osmosis reject and sewage water. *Solar Energy, 173*, 160–172.

Rijsberman, F. R. (2006). Water scarcity: Fact or fiction? *Agricultural Water Management, 80*, 5–22.

Rose, C., Parker, A., Jefferson, B., & Cartmell, E. (2015). The characterization of faeces and urine: A review of the literature to inform advanced treatment technology. *Critical Reviews in Environmental Science and Technology, 45*(17), 1827–79. https://doi.org/10.1080/10643389.2014.1000761

Rusan, M. J. M., Hinnawi, S., & Rousan, L. (2007). Long term effect of wastewater irrigation of forage crops on soil and plant quality parameters. *Desalination, 215*, 143–152.

Saitoh, T. S., & El-Ghetany, H. H. (1999). Solar water-sterilization system with thermally-controlled flow. *Applied Energy, 64*, 387–399.

Sato, T., Qadir, M., Yamamoto, S., Endo, T., & Zahoor, A. (2013). Global, regional and country level need for data on wastewater generation, treatment and use. *Agricultural Water Management, 130*, 1–13.

Shatat, M., Worall, M., & Riffat, S. (2013). Ecocnomic study for an affordable small scale solar water desalination system in remote and semi-arid region. *Renewable and Sustainable Energy Reviews, 25*, 543–551.

Singh, P., Kansal, A., & Carliell-Marquet, C. (2016). Energy and carbon footprints of sewage treatment methods. *Journal of Environmental Management, 165*, 22–30.

Sklavos, S., Gatidou, G., Stasinakis, A. S., & Haralambopoulos, D. (2015). Use of solar distillation for olive mill wastewater drying and recovery of polyphenolic compounds. *Journal of Environmental Management, 162*, 46–52.

Sullivan, J. B., & Krieger, G. R. (2001). Clinical environmental health and toxic exposures, 2nd edn, LWW 2001. (accessed through Google books)

Tilley, E., Strande, L., Luthi, C., Mosler, H. J., Udert, K. M., Gebauer, H., & Hering, J. G. (2014). Looking beyond technology: An integrated approach to water, sanitation and hygiene in low income countries. https://doi.org/10.1021/es501645d

Tortajada, C. (2016). Policy dimensions of development and financing of water infrastructure: The cases of China and India. *Environmental Science & Policy, 64*, 177–187.

Udert, K. M., & Wĺachter, M. (2012). Complete nutrient recovery from source-separated urine by nitrifiation and distillation. *Water Research, 46* (2), 453–64. https://doi.org/10.1016/j.watres.2011.11.020

Velmurugan, V., Deenadayalan, C. K., Vinod, H., & Srithar, K. (2008). Desalionation of effluent using fin type solar still. *Energy, 33*, 1719–1727.

Velmurugan, V., Kumar, K. N., Haq, T. N., & Srithar, K. (2009). Performance analysis in stepped solar still for effluent desalination. *Energy Journal, 34*(9), 1179–1186. https://doi.org/10.1016/j.energy.2009.04.029

Vergine, P., Salerno, C., Libutti, A., Beneduce, L., Gatta, G., Berardi, G., & Pollice, A. (2017). Closing the water cycle in the agro-inductrial sector by reusing treated wastewater for irrigation. *Journal of Cleaner Production, 164*, 587–596.

Wang, Jun-feng, Wang, Gen-xu, & Wanyan, Hua. (2007). Treated wastewater irrigation effect on soil, crop and environment: Wastewater recycling in the loess area of China. *Journal of Environmental Sciences, 19*, 1093–1099.

World Health Organization (WHO) (2004). http://14.139.186.108/jspui/bitstream/123456789/16170/7/6.1. Annexures.pdf. Accessed 2 July 2017.

World Health Organization (WHO) (2017). http://www.who.int/heli/risks/water/water/en/. Accessed 19 June 2017.

Xie, M., Shon, H. K., & Gray, S. R. (2016). Elimelech M. Membrane-based processes for wastewater nutrient recovery: Technology, challenges, and future direction. *Water Research, 89*, 210–21. https://doi.org/10.1016/j.watres.2015.11.045

Yale Environment 360 (2017). http://e360.yale.edu/digest/global_pesticide_map_shows_large_areas_of_high_water_pollution_risk. Accessed 19 June 2017

Zhao, Z. P., Xu, L., Shang, X., & Chen, K. (2013). Water regeneration from human urine by vacuum membrane distillation and analysis of membrane fouling characteristics. *Separation and Purification Technology, 118*, 369–76. https://doi.org/10.1016/j.seppur.2013.07.021

Zollig, H., Fritzsche, C., Morgenroth, E., & Udert, K. M. (2015). Direct electrochemical oxidation of ammonia on graphite as a treatment option for stored source-separated urine. *Water Research, 69*, 284–94. https://doi.org/10.1016/j.watres.2014.11.031

# Chapter 13
# Environmental Footprints, Economics, and Sustainability of Solar Thermal Desalination Systems

## 13.1 Environmental Footprints

Environmental footprints refer to the impacts caused by an activity on the environment. During development, deployment, and operation of solar thermal desalination/wastewater treatment systems, there may be certain positive impacts and adverse impacts on the environment which need to be evaluated to assess the true sustainability of the developed systems. Stainless steel, mild steel, copper, glass, aluminum, rubber, glass wool, rock wool, paints, plastic, etc., are used in the development of solar thermal desalination/wastewater treatment systems. These materials are developed from minerals extracted from the ores by utilizing various energy-consuming chemical/mechanical/thermal processes. Starting from ore extraction to fabrication of the system a lot of energy is consumed and this consumed energy is called the **energy density or embodied energy (EE)** (Kumar & Kurmaji, 2013).

In the current scenario, the energy spent for these processes is mainly from conventional coal or fossil fuel-based power generation units which contribute to significant greenhouse gas emissions. The embodied energy of various components generally used in solar thermal desalination/wastewater treatment systems is listed in Table 13.1. The summation of the energy density of each component of the desalination/wastewater treatment system gives the embodied energy (EE) of the entire system. The embodied energy of certain solar thermal desalination/wastewater treatment systems along with their lifetime and $CO_2$ emission mitigation potential is tabulated in Table 13.2. The embodied energy estimation procedure for a solar thermal desalination system can be easily understood from an example depicted in Table 13.3.



H. Sharon and K. Srinivas Reddy, *Solar Thermal Desalination Technologies for Potable Water*, https://doi.org/10.1007/978-3-031-49155-9_13

**Table 13.1** Embodied energy of common materials used in solar thermal desalination/wastewater treatment systems (Sharon & Reddy, 2015; Reddy & Sharon, 2016; Sharon et al., 2017; Reddy & Sharon, 2017; Sharon et al., 2021; Hilarydoss, 2022).

| S.no | Material | Embodied energy (MJ/kg) |
|---|---|---|
| 1. | Rubber gasket | 11.83 |
| 2. | Glass wool insulation[a] | 139.00 |
| 3. | Black paint | 90.40 |
| 4. | Stainless steel components | 56.70 |
| 5. | Aluminum components | 170.00 |
| 6. | Tempered glass[a] | 66020.00 |
| 7. | Brass components | 62.00 |
| 8. | Mild steel component | 34.20 |
| 9. | Plastic tank—HDPE | 103.00 |
| 10. | Felt sheet (wick) | 18.00 |
| 11. | Fiber-reinforced plastic | 92.20 |
| 12. | Polystyrene insulation | 117.00 |
| **S.no** | **Component** | **Cumulative embodied energy (MJ)** |
| 1. | DC pump (50 W) | 45.23 |
| 2. | Vacuum pump (180 W) | 165.09 |

[a]Embodied energy is given in $MJ/m^3$

**Table 13.2** Embodied energy of common materials used in solar thermal desalination/wastewater treatment systems (Sharon & Reddy, 2015; Reddy & Sharon, 2016, 2017; Sharon et al., 2017, 2021; Hilarydoss, 2022)

| S.no | Solar Thermal Desalination/ Wastewater Treatment System | Annual Capacity[a] ($M_Y$) ($L/m^2Yr$) | Embodied Energy (EE) (kWh) | Life time (LT) (Year) | $CO_2$ emission mitigated during life time[b] (tons) |
|---|---|---|---|---|---|
| 1. | Commercial fiber-reinforced plastic still (Sharon et al., 2021) | 590 | 599.00 | 10 | 1.55 |
| 2. | Insulated commercial fiber-reinforced plastic still (Sharon et al., 2021) | 1030 | 655.90 | 10 | 7.93 |
| 3. | Commercial fiber-reinforced still with black dye (Sharon et al., 2021) | 893 | 586.47 | 10 | 7.95 |
| 4. | Insulated commercial fiber-reinforced plastic still with black dye (Sharon et al., 2021) | 1288 | 643.35 | 10 | 11.90 |
| 5. | Conventional basin still (Shoeibi et al., 2021) | 579.00 | 319.00 | 30 | 20.70 |

(continued)

**Table 13.2** (continued)

| S.no | Solar Thermal Desalination/ Wastewater Treatment System | Annual Capacity[a] ($M_Y$) (L/m$^2$Yr) | Embodied Energy (EE) (kWh) | Life time (LT) (Year) | $CO_2$ emission mitigated during life time[b] (tons) |
|---|---|---|---|---|---|
| 6. | Basin still with pin fin absorber (Shoeibi et al., 2021) | 1394.00 | 318.00 | 10 | 14.40 |
| 7. | Basin still with circular fin absorber (Shoeibi et al., 2021) | 1097.00 | 390.00 | 20 | 20.50 |
| 8. | Basin still with square fin absorber (Shoeibi et al., 2021) | 1168.00 | 379.00 | 20 | 22.10 |
| 9. | Basin still with aluminum heat sink with pin fins (Shoeibi et al., 2021) | 1556.00 | 553.00 | 10 | 17.15 |
| 10. | Double slope still with PV/T collector (Shoeibi et al., 2021) | 4757.00 | 8957.00 | 50 | 368.08 |
| 11. | Basin still with flat plate solar collector (Shoeibi et al., 2021) | 1724.00 | 1583.00 | 20 | 20.10 |
| 12. | Basin still with evacuated tube collector (Shoeibi et al., 2021) | 1045.00 | 1138.00 | 30 | 77.20 |
| 13. | Tilted wick still (Sharon et al., 2017) | 800 | 1397.15 | 20 | 14.75 |
| 14. | Tilted still with partitioned basin (Sharon et al., 2017) | 965 | 1820.81 | 20 | 17.65 |
| 15. | Double sided vertical convection still (Sharon, 2020) | 785 | 1590.00 | 20 | 69.36 |
| 16. | Hybrid basin-vertical diffusion still (Sharon, 2021) | 1150 | 1915.67 | 10–25 | 23.40–63.10 |
| 17. | Evaporative cooled hybrid inclined multi-effect vertical diffusion still with reflector (Hilarydoss, 2022) | 1780 | 2577.91 | 10 | 37.02–65.05 |
| 18. | Active double sided vertical still (Sharon & Reddy, 2015) | 1505 | 4389.73 | 20 | 69.85 |
| 19. | Evacuated active multi-effect vertical convection still (Reddy & Sharon, 2016) | 5323 | 3800.80 | 20 | 81.81 |
| 20. | Wetted wick-cooled evacuated active multi-stage series flow system (Reddy & Sharon, 2017) | 5300 | 6824.94 | 20 | 221.80 |
| 21. | Passive multi-chamber vertical diffusion still (Sharon et al., 2020a) | 2783 | 1035.00 | 20 | 67.93 |
| 22. | Solar chimney–basin desalination system (Zuo et al., 2011) | 1536 | NA | 25 | 6725–8950 |

[a] Annual capacity is based on number of clear days of the site and per m$^2$ area of the system
[b] $CO_2$ emission mitigated during lifetime is based on cumulative capacity of the system and $CO_2$ emission factor of the type of fossil fuel-based power generation system

**Table 13.3** Estimation of embodied energy of active multiple stage series flow solar thermal desalination system (Reddy & Sharon, 2017)

| S.no | List of components | Dimension/capacity | Quantity | Mass (kg) | Embodied energy (MJ(kWh)) |
|---|---|---|---|---|---|
| 1 | SS sheet for distillation chamber | 1.6 × 1.0 × 0.002 m | 4 | 101.12 | 5733.50(1592.64) |
| | | 1.0 × 1.0 × 0.002 m | 1 | 15.80 | 895.86(248.85) |
| | | 1.0 × 0.5 × 0.002 m | 1 | 7.90 | 447.93(124.43) |
| 2 | SS evaporator-condenser tray | 1.1 × 1.0 × 0.001 m | 5 | 43.45 | 2463.62(684.34) |
| 3 | SS distillation collection tray (L Shaped) | 1.0 × 0.03 × 0.03 m (0.001 m thickness) | 5 | 2.37 | 134.38(37.33) |
| 4 | SS hollow square pipe-support structure | 2.6 m (0.04 × 0.04 × 0.0026 m) | 4 | 30.37 | 1721.98(478.33) |
| | | 1.0 m (0.04 × 0.04 × 0.0026 m) | 16 | 46.72 | 2649.02(735.84) |
| 5 | Plastic tank | (960 × 720 × 350 mm)/200 L | 1 | 3.25 | 334.75(92.99) |
| 6 | Al cladding sheet | 1.6 × 1.1 × 0.0005 m | 4 | 13.09 | 2225.30(618.14) |
| | | 1.0 × 1.1 × 0.0005 m | 2 | 4.09 | 695.30(193.14) |
| 7 | SS vacuum pipe line | 2.1 m (12.7 mm dia) | 1 | 1.03 | 58.40(16.22) |
| 8 | SS pipe from distillation unit to collector | 6.2 m (19.05 mm dia) | 1 | 7.87 | 446.23(123.95) |
| 9 | SS pipe for distillate transport | 4.1 m (12.7 mm dia) | 1 | 2.00 | 113.40(31.50) |
| 10 | SS pipe for reject transport | 1.0 m (12.7 mm dia) | 1 | 0.49 | 27.78(7.71) |
| 11 | Brass valve | 19.05 mm dia | 5 | 1.05 | 65.10(18.08) |
| | | 12.7 mm dia | 3 | 0.32 | 19.84(5.51) |
| 12 | SS elbow | 19.05 mm dia | 8 | 0.84 | 47.63(13.23) |
| | | 12.7 mm dia | 5 | 0.30 | 17.01(4.73) |

(continued)

**Table 13.3** (continued)

| S.no | List of components | Dimension/capacity | Quantity | Mass (kg) | Embodied energy (MJ(kWh)) |
|---|---|---|---|---|---|
| 13 | SS Tee joint | 12.7 mm dia | 8 | 0.72 | 40.82(11.34) |
| | | 19.05 mm dia | 1 | 0.21 | 11.91(3.31) |
| 14 | SS bolts and nuts | 6.0 mm dia (20.0 mm length) | 35 | 0.88 | 49.90(13.86) |
| 15 | Glass wool insulation[a] | 0.25 $m^3$ | 1 | – | 34.75(9.65) |
| 16 | Gasket | 5.2 m | 1 | 1.22 | 14.43(4.00) |
| 17 | Solar flat plate collector | 2.0 $m^2$ aperture area | 2 | 88.00 | 5892.52(1636.81) |
| 18 | Vacuum pump | 0.25 HP | 1 | 6.30 | 165.09(45.86) |
| 19 | Hot water circulation pump | 0.10 HP | 1 | 2.40 | 120.00(33.33) |
| Total Embodied energy of air cooled evacuated solar thermal desalination system (kWh) | | | | | 6785.12 |
| Total Embodied energy air cooled non-evacuated solar thermal desalination system (kWh) | | | | | 6698.02 |
| Total Embodied energy wetted wick-cooled evacuated solar thermal desalination system (kWh) | | | | | 6824.94 |
| Total embodied energy wetted wick-cooled non-evacuated solar thermal desalination system (kWh) | | | | | 6737.84 |

[a]Embodied energy in $MJ/m^3$

The carbon dioxide ($CO_2$), sulfur dioxide ($SO_2$), and nitrogen oxide (NO) emission associated with the embodied energy of the system is estimated by (Kumar & Kurmaji 2013)

$$\begin{pmatrix}\text{CO}_2 \text{ emission during} \\ \text{life time of the system}\end{pmatrix} = \begin{pmatrix}\text{Embodied energy } \times \text{ CO}_2 \text{ emission factor} \\ \text{of the conventional power generation unit}\end{pmatrix} \tag{13.1}$$

$$\begin{pmatrix}\text{SO}_2 \text{ emission during} \\ \text{life time of the system}\end{pmatrix} = \begin{pmatrix}\text{Embodied energy } \times \text{ SO}_2 \text{ emission factor} \\ \text{of the conventional power generation unit}\end{pmatrix} \tag{13.2}$$

$$\begin{pmatrix}\text{NO emission during} \\ \text{life time of the system}\end{pmatrix} = \begin{pmatrix}\text{Embodied energy } \times \text{ NO emission factor} \\ \text{of the conventional power generation unit}\end{pmatrix} \tag{13.3}$$

In the above formulations, it has been assumed that the developed system utilizes only solar energy for its operation during its entire life term. If fossil fuel is utilized for system operation, it must be included in the above relations to properly estimate the emissions associated with the system. The $CO_2$, $SO_2$ and NO emission factor is about 1.58, 0.012, and 0.0046 kg/kWh of electricity generated from Indian coal power plants. These factors are country-specific, plant-specific, and fuel-specific (Mittal et al., 2014).

## 13.2 Environmental Benefits Estimation

The solar thermal desalination/wastewater treatment system uses only solar energy for it operation during its life span. Hence, it can mitigate a significant amount of greenhouse gas emissions into the environment during its lifetime (LT). The mitigated quantity can be estimated by subtracting the energy spent for evaporating the water using solar energy and overall system embodied energy and multiplying this result with the corresponding greenhouse gas emission factor of the conventional power source that is generally used for treating/desalting water. Overall system embodied energy ($EE_{overall}$) refers to the system's embodied energy estimated by taking into account the embodied energy of the replaced items in the system during its life span (Hilarydoss, 2022).

The net $CO_2$ emission mitigation (NCEM) potential of the system in kg is estimated by

$$NCEM = \left(\left[\left(\frac{M_Y \times h_{fg}}{3600}\right) \times LT\right] - (EE)_{overall}\right) \times \text{CO}_2 \text{ emission factor} \tag{13.4}$$

The net $SO_2$ emission mitigation (NSEM) potential of the system in kg is estimated by

$$NSEM = \left(\left[\left(\frac{M_Y \times h_{fg}}{3600}\right) \times LT\right] - (EE)_{overall}\right) \times \text{SO}_2 \text{ emission factor} \quad (13.5)$$

The net NO emission mitigation (NNEM) potential of the system in kg is estimated by

$$NNEM = \left(\left[\left(\frac{M_Y \times h_{fg}}{3600}\right) \times LT\right] - (EE)_{overall}\right) \times \text{NO emission factor} \quad (13.6)$$

### 13.2.1 Energy Payback Time

Energy payback time (EPBT in year) is also an important indicator usually used in renewable energy-based systems to assess the time taken to regain the energy spent on the development of the system. EPBT of the solar thermal desalination/wastewater treatment system is estimated by

$$EPBT = \frac{(EE)_{overall}}{\left(\frac{M_Y \times h_{fg}}{3600}\right)} \quad (13.7)$$

### 13.2.2 Greenhouse Gas Payback Time

Greenhouse gas payback time (in year) is defined as the ratio of greenhouse emission associated with the embodied energy of the system to its annual greenhouse gas emission mitigation potential (Tirmikci & Yavuz, 2020).

$$\text{Greenhouse gas payback time} = \frac{\text{Embodied greenhouse gas emissions}}{\text{Annual greenhouse gas emission savings}} \quad (13.8)$$

### 13.2.3 Life Cycle Conversion Efficiency

Life cycle conversion efficiency (LCCE in %) is defined as the ratio of net desired energy output of the system over its life span to the cumulative solar energy incidence over its aperture/solar energy collection device over its life span. It will be lower than the thermal or energy efficiency of the system due to the reduction of system's embodied energy from its desired energy output. Desired energy output of the system refers to the product of distillate production rate by evaporation and latent heat of evaporation. Life cycle conversion efficiency of the system is estimated by the following relation (Sharon et al., 2020b):

$$LCCE = \frac{\left(\frac{\sum_{i=1}^{i=LT} M_{Y(i)} \times h_{fg}}{3600}\right) - (EE)_{overall}}{\left(\frac{\sum_{i=1}^{i=LT} (I_s \times \Delta t)_{(i)}}{3600}\right) \times A_s} \tag{13.9}$$

Where,

$I_s$ - Incident solar radiation intensity in $W/m^2$

$\Delta t$ - Time interval between each intensity measurement in Seconds

## 13.3 Adverse Environmental Impacts Estimation

NO emitted into the atmosphere gets converted in $NO_X$ (Oxides of Nitrogen). The global warming potential of $NO_X$ is about 30–33 times higher than $CO_2$. $SO_2$ is responsible for acid rains which affects agricultural activities, metallic structures, and causes death of infants. Financial damage caused by $SO_2$ emission is about 350 times higher than the damage caused by $CO_2$ emission. Acidification potential, global warming potential, photochemical oxidants creation potential, abiotic resource depletion potential, eutrophication potential, freshwater aquatic eco-toxicity potential, human toxicity potential, ozone depletion potential, and marine aquatic eco-toxicity potential have been used to indicate the environmental impacts of any system (Greening & Azapagic, 2014; Stamford & Azapagic, 2018). Solar thermal desalination/wastewater treatment system uses solar thermal energy to produce desalinated/treated water thereby burning of fossil fuel/biomass is eliminated.

### *13.3.1 Global Warming Potential (GWP)*

Global warming potential of solar thermal desalination/wastewater system per liter of desalinated/treated water is defined as the ratio of total global warming potential associated with the embodied energy of the system throughout its lifetime to the total desalinated/treated water produced during its lifetime (Ott et al., 2020) and is expressed mathematically as

$$\text{GWP of the system} = \frac{\begin{pmatrix}(\text{CO}_2 \text{ emission factor} \times \text{GWP of CO}_2)+ \\ (\text{NO emission factor} \times \text{GWP of NO}_x)\end{pmatrix} \sum_{i=1}^{i=LT} EE}{M_Y \times LT} \tag{13.10}$$

## 13.3.2 *Acidification Potential (AP)*

Acidification potential of solar thermal desalination/wastewater system per liter of desalinated/treated water is defined as the ratio of total AP of $SO_2$ emission & $NO_X$ emission associated with the embodied energy of the system throughout its life time to the total desalinated/treated water produced during its lifetime (Ott et al., 2020) and is expressed mathematically as

$$\text{AP of the system} = \frac{\begin{pmatrix}(\text{SO}_2 \text{ emission factor} \times \text{AP of SO}_2)+ \\ (\text{NO emission factor} \times \text{AP of NO}_x)\end{pmatrix} \sum_{i=1}^{i=LT} EE}{M_Y \times LT} \tag{13.11}$$

## 13.3.3 *Photochemical Oxidants Creation Potential (POCP)*

Photochemical oxidants creation potential of solar thermal desalination/wastewater system per liter of desalinated/treated water is defined as the ratio of total POCP of $NO_X$ emission associated with the embodied energy of the system throughout its lifetime to the total desalinated/treated water produced during its lifetime and is expressed mathematically as

$$\text{POCP of the system} = \frac{(\text{NO emission factor} \times \text{POCP of NO}_x) \sum_{i=1}^{i=LT} EE}{M_Y \times LT} \tag{13.12}$$

## 13.3.4 *Eutrophication Potential (EP)*

Eutrophication potential of solar thermal desalination/wastewater system per liter of desalinated/treated water is defined as the ratio of total EP of $NO_X$ emission associated with the embodied energy of the system throughout its lifetime to the total desalinated/treated water produced during its lifetime and is expressed mathematically as

**Table 13.4** Characterization factors of emissions considered for environmental impact assessment (Heijungs et al., 1992; Environdec, 2021)

| Emissions | GWP | AP | POCP | EP | HTP |
|---|---|---|---|---|---|
| | (kg $CO_2$ eq./kg) | (kg $SO_2$ eq./kg) | (kg $C_2H_4$ eq./kg) | (kg $PO_4$ eq./kg) | (kg DCB eq./kg) |
| $CO_2$ | 1.0 | – | – | – | – |
| $NO_x$ | 33.0 | 0.70 | 1.0 | 0.13 | 0.78 |
| $SO_2$ | – | 1.0 | – | – | 1.20 |

GWP—Global Warming Potential; AP—Acidification Potential; POCP—Photochemical Oxidants Creation Potential; EP—Eutrophication Potential; HTP—Human Toxicity Potential

$$\text{EP of the system} = \frac{(\text{NO emission factor} \times \text{EP of NO}_x) \sum_{i=1}^{i=LT} EE}{M_Y \times LT} \quad (13.13)$$

## 13.3.5 *Human Toxicity Potential (HTP)*

Human toxicity potential of solar thermal desalination/wastewater system per liter of desalinated/treated water is defined as the ratio of total HTP of $SO_2$ emission & $NO_X$ emission associated with the embodied energy of the system throughout its lifetime to the total desalinated/treated water produced during its lifetime and is expressed mathematically as

$$\text{HTP of the system} = \frac{\begin{pmatrix} (\text{SO}_2 \text{ emission factor} \times \text{HTP of SO}_2)+ \\ (\text{NO emission factor} \times \text{HTP of NO}_x) \end{pmatrix} \sum_{i=1}^{i=LT} EE}{M_Y \times LT} \quad (13.14)$$

Copper, aluminum, and stainless steel used for system construction results in heavy metal, halon 1211, and hydrogen fluoride emissions (Greening & Azapagic, 2014; Whiting & Azapagic, 2014). Hence, abiotic resource depletion potential, freshwater aquatic eco-toxicity potential, ozone depletion potential, and marine aquatic eco-toxicity potential need to be considered for assessing the adverse environmental impacts of the system in a complete form. The characterization factor of the associated emissions considered for assessing the adverse environmental impacts of the system is tabulated in Table 13.4.

## 13.4 Economics of Solar Thermal Desalination Systems

Solar energy is freely available in nature. However, in order to utilize it for desalination or wastewater treatment proper systems need to be designed, fabricated, and installed which requires significant cash investment. Even though these developed systems are eco-friendly they can be accepted by the general public only if they are economical in comparison to the currently available commercial conventional fossil fuel-based systems. In general, solar thermal desalination/wastewater treatment systems have high initial cost and low maintenance & operation cost. The following components are essential to estimate the economics of solar thermal desalination/wastewater treatment systems (Sharon & Reddy, 2015).

### 13.4.1 Capital Cost (CC)

It includes design cost, fabrication cost, and installation cost of the system. It also includes land cost for large capacity systems. For rooftop and household-based systems land cost is generally assumed to be nil.

### 13.4.2 Life Span (LT)

Life span of the system also affects the economics. Use of low-quality materials for fabrication may reduce the system's initial cost but, at the same time, it reduces the system's life span which may lead to frequent replacement of system components. Life span of commercial thermal desalination systems is about 50 years however for small-scale solar desalination systems it can range between 5 and 20 years.

### 13.4.3 Salvage Value (SV)

Salvage value is also called the residual value or scrap value. It is defined as the cost of the solar thermal desalination/wastewater treatment system at the end of its life span or at the time of disposal. There is also some cost associated with disposing of the system in some countries. Salvage value is about 15–20% of the useful material cost (UMC) of the system in general.

### 13.4.4 Operation and Maintenance Cost (OMC)

It indicates the cost associated with the pretreatment of feed water, pumping of feed water, operation of electrical accessories like vacuum pumps, circulation pumps, etc., cleaning of glass covers of desalination/wastewater treatment systems, solar collectors, salt removal from the basin, periodic black coating, replacement of wicks, etc.

### 13.4.5 Interest Rate (IR)

If the capital amount is taken as a loan from bank or any other financial body, it has to be paid back with interest. The interest amount that has to be paid every year along with principal amount depends on the interest rate fixed by the lending body. Interest rate can vary between 5 and 15%.

### 13.4.6 Fixed Annual Cost (FAC)

It is the product of capital cost and capital recovery factor. Capital recovery factor is the factor using which annual payment has to be made to cover the capital cost for the given time by considering the interest rate. It is expressed mathematically as

$$FAC = CC \times \left[ \frac{IR(1 + IR)^{LT}}{(1 + IR)^{LT} - 1} \right] \tag{13.15}$$

### 13.4.7 Annual Salvage Value (ASV)

It is the product of salvage value and sinking fund factor. Sinking fund factor refers to the factor that is used to calculate the future value of a series of equal annual cash flow. It is expressed mathematically as

$$ASV = SV \times \left[ \frac{IR}{(1 + IR)^{LT} - 1} \right] \tag{13.16}$$

$$ASV = 0.20 \times CC \times \left[ \frac{IR}{(1 + IR)^{LT} - 1} \right] \tag{13.17}$$

### 13.4.8 Annual Operation and Maintenance Cost (AOMC)

It is assumed to be about 10–15% of the fixed annual cost in general. It is mathematically expressed as

$$AOMC = 0.10 \text{ to } 0.15 \times FAC \tag{13.18}$$

### 13.4.9 Total Annualized Cost (TAC)

It is the annualized value of the total net present cost. It is expressed mathematically as

$$TAC = FAC + AOMC - ASV \tag{13.19}$$

The economics of the desalination/wastewater treatment system is determined based on the cost per liter of desalinated/treated water produced by the system and system's finance payback period.

### 13.4.10 Cost Per Liter (CPL)

Cost per liter is determined by dividing the total annualized cost by annual desalinated/treated water production ($M_Y$) capacity of the system. It is mathematically expressed as

$$CPL = \frac{TAC}{M_Y} \tag{13.20}$$

Another simplest formula used for estimation of cost per liter in a rough way is (Omara et al., 2014)

$$CPL = \frac{\text{Fixed cost } + \text{ (yearly variable cost } \times \text{LT)}}{M_Y \times LT} \tag{13.21}$$

Yearly variable cost is taken about 30% of the fixed cost.

The amount of desalinated water/treated water produced by the system per unit USD or Euro or Indian Rupee invested on the system is estimated by taking reciprocal of cost per liter estimated using Eq. (13.20).

### 13.4.11 Net Carbon Credit Earning Potential

The net carbon credit earning potential of the solar thermal desalination/wastewater treatment system can be estimated by multiplying the net $CO_2$ emission mitigated by the system during its lifetime with the international carbon price. The international carbon price is about 14.5 USD/ton of $CO_2$ as of the year 2015.

### 13.4.12 Finance Payback Period (FPP)

It represents the time taken to gain back the financial investment made on the solar thermal desalination/wastewater treatment system. In other words, it can be defined as the time duration between the initial investments on the system to the breakeven point. At breakeven point, the cumulative returns made from the system equals the investment made on the system.

The finance payback period (in year) can be estimated in many ways. In simple form, it can be estimated by

$$FPP = \frac{\text{Captial cost of the desalination system}}{\text{Yearly drinking water cost offset by the system}} \tag{13.22}$$

$$FPP = \frac{\text{Captial cost of the wastewater treatment system}}{\text{Yearly conventional wastewater treatment cost offset by the system}} \tag{13.23}$$

If the system is used for commercial packaged water production or for production of double distilled water for application in battery, industrial, and chemical processes then finance payback period can be estimated by

$$FPP = \frac{\text{Captial cost of the system}}{\text{Yearly desalinated/treated water production} \times \text{Selling price per liter}} \tag{13.24}$$

If interest rate is taken into account, then the finance payback period can be estimated by

$$FPP = \frac{\ln\left(\frac{M_Y \times S_p}{(M_Y \times S_p) - (CC \times IR)}\right)}{\ln(1 + IR)} \tag{13.25}$$

## 13.4.13 Equated Monthly Installment (EMI)

Subsidy for solar systems is given either by state and federal governments in most of the countries thereby, it covers some portion of the capital cost. The remaining amount "$P$" can be taken as loan from financial institutions which must be paid back with some interest rate for a loan period of some years. EMI refers to the amount payable to the financial institution every month until all the loan amount is paid off. The EMI that has to be paid for "$N$" monthly installments can be estimated by (Njgroup, 2021)

$$EMI = \frac{P \times MIR \times (1 + MIR)^N}{(1 + MIR)^N - 1} \tag{13.26}$$

Loan amount is given as

$$P = \text{Capital Cost} - \text{Subsidy} \tag{13.27}$$

### 13.4.14 Monthly Interest Rate (MIR)

Monthly interest rate can be found by dividing the yearly interest rate ($IR$) by 12 (i.e.) no. of months in a year. It is mathematically as

$$\text{MIR} = \frac{\text{Year Interest Rate}}{\text{Number of months in a year}} \tag{13.28}$$

The economics of some of the solar thermal desalination systems and commercial desalination systems is tabulated in Table 13.5.

## 13.5 Sustainability Index (SI) Estimation

Sustainable development of a society can be achieved only if sustainable/clean energy sources are utilized effectively. Effective utilization of resources for solar thermal desalination/wastewater treatment leads to no/minimal harmful emissions into the environment. Therefore, effective utilization of resources leads to minimal adverse environmental impact and higher sustainability. Exergy efficiency ($\eta_{ex}$) is an indicator that tells whether the clean energy resource is utilized effectively or not (Kanoglu et al., 2009). Higher exergy efficiency indicates lower exergy destruction and enhanced sustainability. Exergy destruction ($Ex_{des}$) indicates irreversibility associated with the process. The sustainability of the system can be estimated by the sustainability index (SI) which is the inverse of the depletion number. The depletion number is given as the ratio of exergy destruction in the system to the exergy input ($Ex_{in}$) to the system (Rosen et al., 2008). The relation between sustainability index, exergy input and exergy destruction is given as (Rosen et al., 2008)

$$SI = \frac{1}{\left(\frac{Ex_{des}}{Ex_{in}}\right)} \tag{13.29}$$

The relation between exergy efficiency and depletion number is given as (Rosen et al., 2008)

$$\eta_{ex} = 1 - \left(\frac{Ex_{des}}{Ex_{in}}\right) \tag{13.30}$$

Hence, the sustainability index (SI) of solar thermal desalination/wastewater treatment system can be estimated by (Caliskan, 2017; Fudholi et al., 2019)

$$SI = \frac{1}{1 - \eta_{ex}} \tag{13.31}$$

**Table 13.5** Economics of certain solar thermal and commercial desalination systems

| S.no | Desalination/Wastewater Type | Capital cost (USD) | Life time (year) | Annual capacity (L) | Interest rate (%) | Cost per litre (USD/L) |
|---|---|---|---|---|---|---|
| 1. | Basin still (Kabeel et al., 2010) | 275 | 10 | 1511 | 12 | 0.035 |
| 2. | Basin still with solar collector (Kabeel et al., 2010) | 1144 | 10 | 1203 | 12 | 0.180 |
| 3. | Basin still with solar concentrator (Kabeel et al., 2010) | 300 | 10 | 990 | 12 | 0.058 |
| 4. | Basin still with sun tracking system (Kabeel et al., 2010) | 300 | 10 | 250 | 12 | 0.230 |
| 5. | Pyramid still (Kabeel et al., 2010) | 250 | 10 | 1533 | 12 | 0.031 |
| 6. | Pyramid still with solar collector (Kabeel et al., 2010) | 450 | 10 | 844 | 12 | 0.103 |
| 7. | Basin still with fins (Kabeel et al., 2010) | 200 | 10 | 720 | 12 | 0.054 |
| 8. | Basin still with wick and fins (Kabeel et al., 2010) | 250 | 10 | 731 | 12 | 0.065 |
| 9. | Hemispherical still (Kabeel et al., 2010) | 958 | 10 | 1026 | 12 | 0.180 |
| 10. | Stepped still with fins and sponges (Kabeel et al., 2010) | 180 | 10 | 546 | 12 | 0.064 |
| 11. | Stepped still with sun tracking (Kabeel et al., 2010) | 350 | 10 | 958 | 12 | 0.071 |
| 12. | Weir type still (Kabeel et al., 2010) | 280 | 10 | 1001 | 12 | 0.054 |
| 13. | Basin still with sponge and solar pond (Kabeel et al., 2010) | 350 | 10 | 837 | 12 | 0.080 |
| 14. | Basin still with shallow solar pond (Kabeel et al., 2010) | 320 | 10 | 1183 | 12 | 0.052 |
| 15. | Basin still with separate condenser (Kabeel et al., 2010) | 350 | 10 | 1116 | 12 | 0.060 |
| 16. | Commercial fiber-reinforced plastic still (Kabeel et al., 2010) | 170 | 10 | 690 | 15 | 0.075 |
| 17. | Insulated commercial fiber-reinforced plastic still (Kabeel et al., 2010) | 177 | 10 | 1205 | 15 | 0.039 |
| 18. | Commercial fiber-reinforced still with black dye (Kabeel et al., 2010) | 170 | 10 | 1045 | 15 | 0.047 |
| 19. | Insulated commercial fiber-reinforced plastic still with black dye (Kabeel et al., 2010) | 177 | 10 | 1507 | 15 | 0.027 |
| 20. | Tilted wick still (Sharon et al., 2017) | 277 | 10 | 1080 | 12 | 0.065 |
| 21. | Tilted still with partitioned basin (Sharon et al., 2017) | 299 | 10 | 1303 | 12 | 0.059 |

(continued)

**Table 13.5** (continued)

| S.no | Desalination/Wastewater Type | Capital cost (USD) | Life time (year) | Annual capacity (L) | Interest rate (%) | Cost per litre (USD/L) |
|---|---|---|---|---|---|---|
| 22. | Double sided vertical convection still (Sharon, 2020) | 469 | 10 | 3140 | 12 | 0.027 |
| 23. | Hybrid basin-vertical diffusion still (Sharon, 2021) | 537 | 10 | 3450 | 15 | 0.042 |
| 24. | Evaporative cooled hybrid inclined multi-effect vertical diffusion system with reflector (Hilarydoss, 2022) | 638 | 10 | 5340 | 10 | 0.017 |
| 25. | Active double sided vertical still (Sharon & Reddy, 2015) | 825 | 10 | 6020 | 12 | 0.038 |
| 26. | Evacuated active multi-effect vertical system (Reddy & Sharon, 2016) | 870 | 10 | 5323 | 12 | 0.040 |
| 27. | Wetted wick-cooled evacuated active multi-stage series flow system (Reddy & Sharon, 2017) | 1828 | 20 | 21200 | 12 | 0.026 |
| 28. | Passive multi-chamber vertical diffusion still (Sharon et al., 2020a) | 249 | 10 | 2783 | 12 | 0.015 |
| 29. | CPC-Tubular desalination still (Arunkumar et al., 2013) | 206 | 15 | 894 | 6 | 0.018 |
| 30. | CPC-Tubular desalination still with water cooling (Arunkumar et al., 2013) | 279 | 15 | 1490 | 6 | 0.015 |
| 31. | PTC-Tubular desalination still (Elashmawy, 2020) | 172 | 15 | 1153 | 6 | 0.0189 |
| 32. | PTC-Tubular desalination still with gravel (Elashmawy, 2020) | 174 | 15 | 1317 | 6 | 0.0167 |
| 33. | PTC-Tubular desalination still with PCM tubes (Elashmawy et al., 2021) | 190 | 15 | 1621 | 6 | 0.01485 |
| 34. | Tubular desalination still with wick covered hollow rotating drum (Essa et al., 2021) | 272 | 10 | 1927 | 15 | 0.024 |
| 35. | Modular-Arrayed tubular desalination system (Xie et al., 2016) | 2555 | 20 | 23625 | 5 | 0.0061 |
| 36. | Two effect tubular desalination system with Fresnel reflector (Gang et al., 2011) | 1602 | 20 | 19104 | 5 | 0.0062 |

(continued)

**Table 13.5** (continued)

| S.no | Desalination/Wastewater Type | Capital cost (USD) | Life time (year) | Annual capacity (L) | Interest rate (%) | Cost per litre (USD/L) |
|---|---|---|---|---|---|---|
| 37. | Three effect tubular desalination system under low pressure operation (40–60 kPa) (Xie et al., 2020) | 345 | 20 | 1350 | 5 | 0.0082 |
| 38. | Solar chimney-basin desalination system (Zuo et al., 2011) | 3567655 | 25 | 69462000 | 6.2 | 0.0027 |
| 39. | Stepped desalination still with internal and external reflectors (Omara et al., 2014) | 160 | 10 | 2040 | NA | 0.031 |
| 40. | Conventional RO system | – | – | Few $m^3$/d | – | 0.011 |
| 41. | Conventional MED system | – | – | <100 $m^3$/d | – | 0.007 |
| 42. | Solar PV-RO system | – | – | <100 $m^3$/d | – | 0.014 |
| 43. | Solar PV-ED system | – | – | – | – | 0.010 |
| 44. | Solar Membrane Distillation system (MD) | – | – | 10 $m^3$/d | – | 0.017 |

The thermodynamic performance and sustainability index of various solar thermal desalination systems is tabulated in Table 13.6.

## References

Arunkumar, T., Jayaprakash, R., Ahsan, A., Denkenberger, D., & Okundamiya, M. S. (2013). Effect of water and air flow on concentric tubular solar water desalting system. *Applied Energy, 103*, 109–115.

Bait, O. (2019). Exergy, environeeconomic and economic analyses of a tubular solar water heater assisted solar still. *Journal of Cleaner Production, 212*, 630–646.

Balachandran, G. B., David, P. W., Chellam, P. V., Ali, M. N. A., Radhakrishnan, V., Balamurugan, R., & Manokar, A. M. (2020). Rehash of cooked oil for the palatable water production using single slope solar still. *Fuel, 271*, 117613.

Caliskan, H. (2017). Energy, exergy, environmental, enviroeconomic, exergoenvironmental (EXEN) and exergoenviroeconomic (EXENEC) analyses of solar collectors. *Renewable and Sustainable Energy Reviews, 69*, 488–492.

Environdec. (2021). https://www.environdec.com/resources/impact-categories

Elashmawy, M. (2020). Improving the performance of a parabolic concentrator solar tracking tubular solar still (PCST-TSS) using gravel as a sensible heat storage material. *Desalination, 473*, 114182.

Elashmawy, M., Alhadri, M., & Ahmed, M. M. Z. (2021). Enhancing tubular solar still performance using novel PCM-tubes. *Desalination, 500*, 114880.

El-Said, E. M. S., Elshamy, S. M., & Kabeel, A. E. (2020). Performance enhancement of a tubular solar still by utilizing wire mesh packing under harmonic motion. *Desalination, 474*, 114165.

Essa, F. A., Abdullah, A. S., & Omara, Z. M. (2021). Improving the performance of tubular solar still using rotating drum-Experimental and theoretical investigation. *Process Safety and Environmental Protection, 148*, 579–589.

Fudholi, A., Zohri, M., Rukman, N. S. B., Nazri, N. S., Mustapha, M., Yen, C. H., Mohammad, M., & Sopian, K. (2019). Exergy and sustainability index of photovoltaic thermal (PVT) air collector: A theoretical and experimental study. *Renewable and Sustainable Energy Reviews, 100*, 44–51.

Greening, B., & Azapagic, A. (2014). Domestic solar thermal water heating: A sustainable option for the UK? *Renewable Energy, 63*, 23–36.

Hassan, H., Yousef, M. S., Fathy, M., & Ahmed, M. S. (2020). Assessment of parabolic trough solar collector assisted solar still at various saline water mediums via energy, exergy, exergoeconomic, and enviroeconomic approaches. *Renewable Energy, 155*, 604–616.

Gang, W., Qichang, Y., Hongfei, Z., Yi, Z., Hui, F., Rihui, J., (2019). Direct utilization of solar linear Fresnel reflector on multi-effect eccentric horizontal tubular still with falling film. *Energy 170*, 170–184.

Heijungs, R., Guinée, J. B., Huppes, G., Lankreijer, R. M., Udo de Haes, H. A., Wegener Sleeswijk, A., Ansems, A. M. M., Eggels, P. G., Duin, R., & Goede, van, de, H. P. (1992). *Environmental life cycle assessment of products: Background and guide*. Leiden: Multicopy.

**Table 13.6** Thermodynamic performance and sustainability index of certain solar thermal desalination systems

| S.no | Type of solar thermal desalination system | Energy or Thermal efficiency (%) | Exergy efficiency (%) | Sustainability index |
|---|---|---|---|---|
| 1. | Passive basin solar still (Ranjan et al., 2016) | 30.42 | 4.93 | 1.05 |
| 2. | Tubular water heater-assisted basin solar still (Bait, 2019) | 14.0–27.0 | 1.5–3.0 | 1.02 |
| 3. | Basin solar still PCM-loaded absorber plate (Yousef & Hassan, 2019) | 34.8 | 2.33 | 1.02 |
| 4. | Basin solar still with sand as energy storage material (Hassan et al., 2020) | 35.82–40.73 | 1.48–1.69 | 1.02 |
| 5. | Commercial FRP solar still (Sharon et al., 2021) | 22.71 | 2.11 | 1.02 |
| 6. | Un-insulated FRP solar still with black dye (Sharon et al., 2021) | 38.38 | 4.61 | 1.05 |
| 7. | Insulated FRP solar still with black dye (Sharon et al., 2021) | 61.73 | 9.31 | 1.10 |
| 8. | Evacuated tube collector-assisted basin solar still (Singh et al., 2013) | 33.0 | 2.50 | 1.03 |
| 9. | Parabolic trough collector assisted basin solar still (Hassan et al., 2020) | 20.03–23.04 | 0.95–1.21 | 1.01 |
| 10. | Parabolic trough collector-assisted basin solar still with sand as energy storage material (Hassan et al., 2020) | 21.96–26.31 | 1.08–1.58 | 1.01 |
| 11. | Basin solar still with graphite nanoparticles dispersed basin water (Sharshir et al., 2018) | 41.18 | 4.32 | 1.05 |
| 12. | Basin solar still with CuO nanoparticles dispersed basin water (Sharshir et al., 2018) | 38.61 | 3.78 | 1.04 |

(continued)

**Table 13.6** (continued)

| S.no | Type of solar thermal desalination system | Energy or Thermal efficiency (%) | Exergy efficiency (%) | Sustainability index |
|---|---|---|---|---|
| 13. | Basin solar still with cooked oil as heat storage medium (Balachandran et al., 2020) | 24.35 | 0.69 | 1.01 |
| 14. | Basin solar still with external air-cooled condenser (Rahmai et al., 2021) | 23.2 | 1.9 | 1.02 |
| 15. | Tilted wick solar still (Sharon et al., 2017) | 33.83 | 2.88 | 1.03 |
| 16. | Tilted partition basin solar still (Sharon et al., 2017) | 41.06 | 3.06 | 1.03 |
| 17. | Tubular solar still with vibrating wire mesh (El-Said et al., 2020) | 50.3 | 4.64 | 1.05 |
| 18. | Tubular solar still with carbon-coated mushrooms (Sharshir et al., 2021) | 68.0 | 5.25 | 1.06 |
| 19. | Tubular solar still with wick-lined open-ended rotating drum (Essa et al., 2021) | 56.4 | 3.45 | 1.04 |
| 20. | Tubular solar still with wick-lined closed rotating drum (Essa et al., 2021) | 61.0 | 3.60 | 1.04 |
| 21. | Double sided vertical convection solar still (Sharon, 2020) | 62.16 | 4.73 | 1.05 |
| 22. | Passive triple effect vertical diffusion solar still without tracking (Sharon et al., 2020a) | ~2.0 (performance ratio) | 9.14 | 1.01 |
| 23. | Passive triple effect vertical diffusion solar still with manual tracking (Sharon et al., 2020a) | 2 (performance ratio) | 13.85 | 1.16 |
| 24. | Hybrid basin-vertical diffusion solar still (Sharon, 2021) | 56.17 | 6.93 | 1.07 |
| 25. | Evaporative cooled hybrid inclined multi-effect diffusion solar still with bottom reflector (Hilarydoss, 2022) | 1.05 (performance ratio) | 10.44 | 1.12 |


Hilarydoss, S. (2022). Techno-enviro-economic assessment of novel hybrid inclined-multi-effect vertical difFusion solar still for sustainable water distillation. *Environmental Science and Pollution Research*. https://doi.org/10.1007/s11356-022-23286-0

Kabeel, A. E., Hamed, A. M., & El-Agouz, S. A. (2010). Cost analysis of different solar still configurations. *Energy, 35*, 2901–2908.

Kanchan, S., Kumarankandath, A. (2015). The Indian power sector: Need of sustainable energy access. Socio-Ecol Narratives 126–141.

Kanoglu, M., Dincer, I., & Cengel, Y. A. (2009). Exergy for better environment and sustainability. *Environment, Development and Sustainability, 11*, 971–988. https://doi.org/10.1007/s10668-008-9162-3.

Kumar, S., Kurmaji, K. T. (2013). Carbon credit earned by some designs of solar stills. *Desalination and Water Treatment*, 1–10. https://doi.org/10.1080/19443994.2013.770269.

Mittal, M. L., Sharma, C., & Singh, R. (2014). Decadal emission estimates of carbondioxide, sulphur dioxide and nirtic oxide emissions from coal burning in electric power generation plants in India. *Environmental Monitoring and Assessment, 186*, 6857–66.

Njgroup. (2021). Retrieved December 24, 2021, from https://www.njgroup.in/app/pdf/equated-monthly-installments-(emi).pdf.

Omara, Z. M., Kabeel, A. E., & Younes, M. M. (2014). Enhancing the stepped solar still performance using internal and external reflectors. *Energy Conversion and Management, 78*, 876–881.

Ott, E. M., Kabus, C. A., Baxter, B. D., Hannon, B., & Celik, I. (2020). *Environmental analysis of agrivoltaic systems. reference module in earth systems and environmental sciences*. Elsevier. ISBN 9780124095489. https://doi.org/10.1016/B978-0-12-819727-1.00012-1.

Ranjan, K. R., Kaushik, S. C., & Panwar, N. L. (2016). Energy and exergy analysis of passive solar distillation systems. *International Journal of Low-Carbon Technologies, 11*, 211–221.

Rahmani, A., Khemmar, F., & Saadi, Z. (2021). Experimental investigation on the negative effect of the external condenser on the conventional solar still performance. *Desalination, 50*, 114914.

Reddy, K. S., & Sharon, H. (2016). Active multi-effect vertical solar still: Mathematical modeling, performance investigation and enviro-economic analyses. *Desalination, 395*, 99–120.

Reddy, K. S., & Sharon, H. (2017). Energy-environment-economic investigations on evacuated active multiple stage series flow solar distillation unit for potable water production. *Energy Conversion and Management, 151*, 259–285.

Rosen, M. A., Dincer, I., & Kanoglu, M. (2008). Role of exergy in increasing efficiency and sustainability and reducing environmental impact. *Energy Policy, 36*, 128–137.

Sharon, H., & Reddy, K. S. (2015). Performance investigation and enviro-economic analysis of active vertical solar distillation units. *Energy, 84*, 794–807.

Sharon, H., Reddy, K. S., Krithika, D., & Philip, L. (2017). Experimental performance investigation of tilted solar still with basin and wick for distillate quality and enviro-economic aspects. *Desalination, 410*, 30–54.

Sharon, H., Reddy, K. S., & Gorjian, S. (2020). Parametric investigation and year round performance of a novel passive multi-chamber vertical solar diffusion still: Energy, exergy and enviro-economic aspects. *Sol Energy, 211*, 831–846.

Sharon, H., Reddy, K. S., Krithika, D., & Philip, L. (2020). Viability assessment of solar distillation for desalination in coastal locations of Indian sub- continent - Thermodynamic, condensate quality and enviroeconomic aspects. *Solar Energy, 197*, 84–98.

Sharon, H. (2020). Energy, exergy and enviro-economic assessment of productivity enhanced passive double sided vertical convection solar distiller for fresh water production. *Sustainable Energy Technologies and Assessments, 42*, 100846.

Sharon, H. (2021). Energy, exergy, environmental benefits and economic aspects of novel hybrid solar still for sustainable water distillation. *Process Safety and Environmental Protection, 150*, 1–21.

Sharon, H., Prabha, C., Vijay, R., Niyas, A. M., & Gorjian, S. (2021). Assessing suitability of commercial fire reinforced plastic solar still for sustainable potable water production in rural India through detailed energy-exergy-economic analyses and environmental impacts. *Journal of Environmental Management, 295*, 113034.

Sharshir, S. W., Pend, G., Elsheikh, A. H., Edreis, E. M. A., Eltawil, M. A., Abdelhamid, T., Kabeel, A. E., Zang, J., & Yang, N. (2018). Energy and exergy analysis of solar stills with micro/nano particles: A comparative study. *Energy Conversion and Management, 177*, 363–375.

Sharshir, S. W., Hamada, M. A., Kandeal, A. W., El-Said, E. M. S., Elsaid, A. M., Rashad, M., & Abdelaziz, G. B. (2021). Augmented performance of tubular solar still integrated with cost-effective nano-based mushrooms. *Solar Energy, 228*, 27–37.

Shoeibi, S., Rahbar, N., Esfahlani, A. A., & Kargarsharifabad, H. (2021). A comprehensive review of Enviro-Exergo-economic analysis of solar stills. *Renewable and Sustainable Energy Reviews, 149*, 111404.

Singh, R. V., Kumar, S., Hasan, M. M., Khan, M. E., & Tiwari, G. N. (2013). Performance of a solar still integrated with evacuated tube collector in natural mode. *Desalination, 318*, 25–33.

Stamford, L., & Azapagic, A. (2018). Environmental impacts of photovoltaics: The effects of technological improvements and transfer of manufacturing from Europe to China. *Energy Technology, 6*, 1148–1160.

Tirmikçi, C. A., & Yavuz, C. (2020). Environmental impact of a 290.4 kWp grid-connected photovoltaic system in Kocaeli, Turkey. *Clean Technologies and Environmental Policy, 22*, 1943–1951.

Whiting, A., & Azapagic, A. (2014). Life cycle environmental impacts of generating electricity and heat from biogas produced by anaerobic digestion. *Energy, 70*, 181–193.

Xie, G., Sun, L., Mo, Z., Liu, H., Du, M., 2016. Conceptual design and experimental investigation involving a modular desalination system composed of arrayed tubular solar stills. *Applied Energy 179*, 972–984.

Xie, G., Chen, W., Yan, T., Tang, J., Liu, H., & Cao, S. (2020). Three-effect tubular solar desalination system with vacuum operation under actual weather conditions. *Energy Conversion and Management, 205*, 112371.

Yousef, M. S., & Hassan, H. (2019). Assessment of different passive solar stills via exergoeconomic, exergoenvironmental, and exergoenviroeconomic approaches: A comparative study. *Solar Energy, 182*, 316–331.

Zuo, L., Zheng, Y., Li, Z., & Sha, Y. (2011). Solar chimneys integrated with sea water desalination. *Desalination, 276*, 207–213.

# Appendix A
# First Law Analysis of Solar Thermal Desalination Systems

The first law of thermodynamics deals with the principle of conservation of energy. The first law is quantitative. In simpler terms, the first law can be stated as "Energy can neither be created nor be destroyed but it can be transformed from one form to another form".

For an open system, which involves mass transfer in addition to heat transfer across the boundary, the energy balance equation based on first law under steady-state condition can be written as

$$\text{Energy In} = \text{Energy Out}$$

$$\sum E_i + \sum_{j=1}^{n} Q_j + W_j = \sum E_o + \sum_{j=1}^{n} Q_k + W_o \tag{A.1}$$

where

$E_i$ and $E_o$ are the energy associated with the mass entering and leaving the system, respectively.
$Q_j$ represents the heat energy transferred to the system from the source which is at a temperature of "$T_j$".
$Q_k$ represents the heat energy transferred from the system to the surrounding which is at a temperature of "$T_k$".
$W_i$ represents the work input to the system like energy for operating circulation pumps, vacuum pumps, etc.
$W_o$ represents the work output of the system.

Solar thermal desalination systems discussed in this book do not produce any form of work output hence its work output is nil. Similarly, for simplicity, the energy utilized for circulation pumps and vacuum pumps can be neglected in the analysis as their magnitude will be nearly negligible in comparison to the solar radiation intensity utilized by the desalination systems.



H. Sharon and K. Srinivas Reddy, *Solar Thermal Desalination Technologies for Potable Water*, https://doi.org/10.1007/978-3-031-49155-9

Equation (A.1) can be used for evaluating energy balance in each component of the solar still like basin liner, water mass, wetted wick, glass cover, etc. In this book, unsteady state has been considered in all of the solar still components. Hence, a time dependent first-order derivate will be included on the right-hand side of the Eq. (A.1).

First law efficiency is also called energy efficiency or thermal efficiency. It is the ratio of the desired output energy to the input energy. In solar thermal desalination systems, the desired output energy is the energy actually utilized to produce potable water from the feed water. This is nothing but the product of daily distillate produced and latent heat of evaporation. Input energy is the daily cumulative solar radiation intensity incident over the aperture of solar still or solar collector (in case of active mode operation). First law efficiency of solar thermal desalination systems is given as (Ranjan et al., 2016)

$$\eta_I(\%) = \frac{\sum_{6:00\ AM}^{6:00\ AM\ next\ day} \dot{m}_d \times h_{fg}}{\sum_{6:00\ AM}^{18:00\ PM} I_s \times A_a} \times 100 \tag{A.2}$$

In the case of multi-effect/stage desalination systems, the supplied heat energy is reused thereby; their first law efficiency will be higher than 100%. In general, performance of multi-stage/effect desalination systems is assessed by their performance ratio (PR) which has already been discussed in Chap. 7.

# Appendix B
# Second Law Analysis of Solar Thermal Desalination Systems

The second law of thermodynamics talks about the exergy destruction or entropy generation in an irreversible process. The second law also indicates the direction of energy transfer which is not possible with the first law. Exergy represents the maximum possible useful work that can be extracted from a system at a particular state undergoing a reversible process with the ambient environment as dead state. Exergy represents the quality of energy (Gakkhar et al., 2016). Hence, the second law of thermodynamics is qualitative. In reversible processes, exergy is conserved (i.e.) no entropy generation and no exergy destruction. However, all real processes are irreversible in nature thereby there is a need to analyze exergy destruction to identify scope for further process enhancements (Gude et al., 2012). From exergy analysis of solar thermal desalination systems, the magnitude of exergy destruction occurring in its individual components can be evaluated using which the particular component's improvement potential is calculated. Higher improvement potential of a component indicates more scope for performance enhancement of the component. Moreover, from exergy analysis, how effectively solar radiation has been utilized for desalination process can also be quantified (Sharon, 2021).

For an open system involving mass transfer in addition to heat transfer, the exergy balance under steady state is given as

$$\text{Exergy In} = \text{Exergy Out} + \text{Exergy Destructed}$$

$$\sum X_i + \sum_{j=1}^{j=n} X(Q_j) + W_j = \sum X_o + \sum_{k=1}^{k=n} X(Q_k) + W_o + X_{des} \tag{B.1}$$

where

$X_i$ indicates exergy of the inflowing mass.
$X_o$ indicates exergy of the outflowing mass.



H. Sharon and K. Srinivas Reddy, *Solar Thermal Desalination Technologies for Potable Water*, https://doi.org/10.1007/978-3-031-49155-9

$X(Q_j)$ indicates exergy associated with heat transfer to the system from the source which is at a temperature of "$T_j$".
$X(Q_k)$ indicates exergy associated with heat transfer from the system to the ambient which is at a temperature of "$T_k$".
$W_i$ and $W_o$ indicate work input to the system and work output of the system, respectively.

Different forms of energy have different capacities to do work. Potential, kinetic energy, and work energy are of high quality and they can be converted completely into useful work. "Work" is of high quality hence exergy magnitude of work is equivalent to the magnitude of the work itself (Gude et al., 2012). In solar thermal desalination systems discussed in this book, work input term is neglected due to its low magnitude and work output is nil.

In the case of heat energy, only a fraction of it can be converted into useful work, while the remainder gets rejected to the sink or surroundings. Hence, heat energy is considered to have low quality. The maximum possible useful work that can be extracted from the convection, conduction, and evaporation heat transfer is evaluated by multiplying the corresponding heat transfer quantity with the Carnot efficiency factor of the component ($\eta_{c,component}$).

$$\eta_{c,component} = \left(1 - \frac{T_a}{T_{component}}\right) \tag{B.2}$$

For example, exergy of convection heat transfer occurring between the basin water and glass cover ($Q_{c(bw-gc)}$) can be estimated by

$$X(Q_{c(bw-gc)}) = Q_{c(bw-gc)}\left(1 - \frac{T_a}{T_{basin\ water}}\right) \tag{B.3}$$

The exergy of solar radiation is calculated by multiplying the solar radiation intensity incident over the aperture with solar radiation exergy factor ($\psi$)

$$\psi = \left(1 + \frac{1}{3}\left(\frac{T_a}{6000}\right)^4 - \frac{4}{3}\left(\frac{T_a}{6000}\right)\right) \tag{B.4}$$

Exergy of solar radiation incident over the aperture area is estimated by,

$$X_{S,a} = I_s A_a \psi \tag{B.5}$$

where

$X_{S,a}$ indicates exergy of solar radiation incident over the aperture.
$I_s$ indicates solar radiation intensity over the aperture.
$A_a$ indicates aperture area.

The maximum possible useful work that can be extracted from the radiation heat transfer is evaluated by multiplying the corresponding radiation heat transfer quantity with the Carnot efficiency factor associated with the radiation heat transfer of the component ($\eta_{c,r,component}$)

$$\eta_{c,r,component} == \left(1 + \frac{1}{3}\left(\frac{T_a}{T_{component}}\right)^4 - \frac{4}{3}\left(\frac{T_a}{T_{component}}\right)\right) \quad \text{(B.6)}$$

The values of temperature must be in Kelvin.

Exergy associated with the inflowing feed water can be estimated by (Gude et al., 2012)

$$X_i = (h_i - h_a) - T_a(s_i - s_a) \quad \text{(B.7)}$$

$$(h_i - h_a) = m_{fwi}c_p(T_{fwi} - T_a) \quad \text{(B.8)}$$

$$(s_i - s_a) = m_{fwi}c_p \ln\left(\frac{T_{fwi}}{T_a}\right) \quad \text{(B.9)}$$

where

$h_i$ indicates specific enthalpy of inflowing feed water (kJ/kg).
$h_a$ indicates specific enthalpy of inflowing feed water at ambient temperature (kJ/kg).
$s_i$ indicates specific entropy of inflowing feed water (kJ/kgK).
$s_a$ indicates specific entropy of inflowing feed water at ambient temperature (kJ/kgK).
$c_p$ indicates specific heat capacity of inflowing feed water (kJ/kgK).
$m_{fwi}$ indicates mass flow rate of inflowing feed water (kg/s).

A similar kind of equation can be formulated for the outflowing reject stream. Substituting, all these calculated exergy values in Eq. (B.1) the exergy destruction of the component can be estimated. This process must be repeated for estimating the exergy destruction of other components.

Second law efficiency or exergy efficiency is defined as the ratio of desired exergy output to the exergy input. Desired exergy output is the exergy of evaporation heat transfer (Ranjan & Kaushik, 2013).

$$\eta_{II}(\%) = \frac{\text{Desired exergy output}}{\text{Exergy input}} \times 100 \quad \text{(B.10)}$$

$$\eta_{II}(\%) = \left(1 - \frac{\text{Exergy destruction}}{\text{Exergy input}}\right) \times 100 \quad \text{(B.11)}$$

Exergy efficiency will always be less than 100%.

Exergy of solar radiation is very high and is about 93% of its energy content. However, the exergy efficiency of solar energy-based systems is very much lower.

The main reasons for low exergy efficiency of solar thermal desalination systems are (Ranjan & Kaushik, 2013):

(a) Degradation of solar energy quality—High temperature solar radiation (6000 K) is degraded to the low temperature of the absorber surface (which is not much greater than the atmospheric conditions).
(b) Impossible full absorption of the incident solar insolation by the absorber surface.
(c) Large exergy loss takes place during absorption of the solar radiation by the absorber surface which is at a lower temperature than the sun. Solar radiation absorption must take place at high temperature of the absorber to have less exergy destruction.

The improvement potential of a component in solar thermal desalination system is estimated by Sharon (2021)

$$IP_{component} = \left(1 - \frac{\eta_{ex,component}}{100}\right)\left(\sum X_{des,component}\right) \tag{B.12}$$

Detailed equations for estimating exergy efficiency of solar thermal desalination system's component can be found in Sharon (2021).

# Appendix C
# Correlations for Estimating Nano-fluid Properties

**Density (kg/m$^3$)**—mass of the nano-fluid per unit volume (Said et al., 2021)

$$\rho_{nf} = \varphi\rho_{np} + (1 + \varphi)\rho_{bf} \tag{C.1}$$

The subscript "$nf$", "$np$", and "$bf$" indicates nanofluid, nanoparticle, and base fluid, respectively.

Volume concentration ($\varphi$) of nanoparticle in the base fluid is given by (Ali & Salam, 2020)

$$\varphi = \left[\frac{\frac{m_{np}}{\rho_{np}}}{\frac{m_{np}}{\rho_{np}} + \frac{m_{bf}}{\rho_{bf}}}\right] \tag{C.2}$$

$$\varphi(\%) = \left[\frac{\frac{m_{np}}{\rho_{np}}}{\frac{m_{np}}{\rho_{np}} + \frac{m_{bf}}{\rho_{bf}}}\right] \times 100 \tag{C.3}$$

**Specific heat capacity (kJ/kgK)**—Amount of heat energy required to raise the temperature of 1 kg of nanofluid by 1 °K.

By neglecting density effect (Said et al., 2021),

$$c_{nf} = \varphi c_{np} + (1 - \varphi)c_{bf} \tag{C.4}$$

By including density effect (Said et al., 2021),

$$c_{nf} = \frac{\varphi\rho_{np}c_{np} + (1 - \varphi)\rho_{bf}c_{bf}}{\rho_{nf}} \tag{C.5}$$

**Thermal Expansion Coefficient (1/K)** Change in nanofluid volume per 1 degree temperature rise of the nanofluid.


H. Sharon and K. Srinivas Reddy, *Solar Thermal Desalination Technologies for Potable Water*, https://doi.org/10.1007/978-3-031-49155-9

By neglecting density effect (Said et al., 2021),

$$\beta_{nf} = \varphi\beta_{np} + (1-\varphi)\beta_{bf} \tag{C.6}$$

By including density effect (Said et al., 2021),

$$\beta_{nf} = \frac{\varphi_{np}\beta_{np} + (1-\varphi)\rho_{bf}\beta_{bf}}{\rho_{nf}} \tag{C.7}$$

**Thermal Conductivity (W/mK)** It represents the ability of the nano-fluid to conduct heat (Said et al., 2021).

$$k_{nf} = k_{bf} + \frac{3\varphi(k_{np}-k_{bf})}{k_{np}+k_{bf}-\varphi(k_{np}-k_{bf})} \tag{C.8}$$

If ratio of thermal conductivity of nano-particle to base fluid is greater than 100 then (Said et al., 2021),

$$k_{nf} = k_{bf}\left[\frac{k_{np}+(n-1)k_{bf}+(n-1)\varphi(k_{np}-k_{bf})}{k_{np}+(n-1)k_{bf}-\varphi(k_{np}-k_{bf})}\right] \tag{C.9}$$

where

$$n = \frac{3}{\psi} \tag{C.10}$$

$$\psi = \frac{\text{surface area of sphere with a volume equal to the nano-particle}}{\text{surface area of the nano-particle}} \tag{C.11}$$

If the nanoparticle is spherical in shape, then its particle sphericity ($\psi$) is 1.

Correlation incorporating temperature and nanoparticle diameter is given as (Iqbal et al., 2021)

$$k_{nf} = k_{bf}\left(0.9808 + 0.0142\varphi - 0.2718\left(\frac{T_{bf}}{70}\right) + 0.1020\left(\frac{d_{np}}{150}\right)\right) \tag{C.12}$$

Valid for,

$$0\% \le \varphi \le 20\%$$
$$20^\circ C \le T_{bf} \le 70\,^\circ C$$
$$13\ nm \le d_{np} \le 150\ nm$$

**Dynamic Viscosity (Pa.s)** It refers to the resistance offered by the nanofluid to flow when an external force is applied (Khanafer & Vafai, 2011).

$$\mu_{nf} = -0.4491 + \frac{28.837}{T_{nf}} + 0.574\varphi - 0.1634(\varphi^2) + 23.053\left(\frac{\varphi}{T_{nf}}\right)^2 + 0.0132(\varphi^3) - 2354.735\left(\frac{\varphi}{T_{nf}^3}\right) + 23.498\left(\frac{\varphi}{d_{np}}\right)^2 - 3.0185\left(\frac{\varphi^3}{d_{np}^2}\right) \quad \text{(C.13)}$$

Valid for,

$$1\% \leq \varphi \leq 9\%$$
$$20\,^\circ C \leq T \leq 70\,^\circ C$$
$$13\ nm \leq d_{np} \leq 131\ nm$$

**Kinematic Viscosity ($m^2/s$)** It refers to the resistance offered by the nanofluid to flow when no external force other than gravity is applied.

$$v_{nf} = \frac{\mu_{nf}}{\rho_{nf}} \quad \text{(C.14)}$$

# Appendix D
# Correlations for Estimating Base Fluid (Water) Properties

**Density (kg/m³)**

$$\rho_w = 999.79684 + 0.068317355(T_w) - 0.010740248(T_w)^2 + 0.00082140905(T_w)^{2.5} - 2.3030988 \times 10^{-5}(T_w)^3 \quad \text{(D.1)}$$

**Specific heat capacity (kJ/kgK)**

$$c_{p,w} = 4.2174356 - 0.0056181625(T_w) + 0.0012992528(T_w)^{1.5} - 0.00011535353(T_w)^2 + 4.14964 \times 10^{-6}(T_w)^{2.5} \quad \text{(D.2)}$$

**Dynamic viscosity (Pa.s)**

$$\mu_w = \frac{1}{557.82468 + 19.408782(T_w) + 0.1360459(T_w)^2 - 3.1160832 \times 10^{-4}(T_w)^3} \quad \text{(D.3)}$$

**Thermal conductivity (W/mK)**

$$k_w = 0.5650258 + 0.0026363895(T_w) - 0.0012516934(T_w)^{1.5} - 1.5154918 \times 10^{-6}(T_w)^2 - 0.0009412945(T_w)^{0.5} \quad \text{(D.4)}$$

All the above correlations are valid for 0 to 150 °C (Popiel & Wojtkowiak, 1998).

**Latent heat of evaporation (kJ/kg)** (Tsilingiris, 2012)

$$h_{fg} = 2503.94 - (2.4515 \times T_w) \quad \text{(D.5)}$$

Valid for $10 < T_w < 100$ °C.



H. Sharon and K. Srinivas Reddy, *Solar Thermal Desalination Technologies for Potable Water*, https://doi.org/10.1007/978-3-031-49155-9

# Appendix E
# Correlations for Estimating Humid Air Properties

**Specific heat capacity (J/kgK)**

$$c_{pha} = 999.2 + 0.1343(T_{av}) + (1.01 \times 10^{-4})(T_{av})^2 - (6.758 \times 10^{-8})(T_{av})^3 \tag{E.1}$$

**Thermal conductivity (W/mK)**

$$k_{ha} = 0.0244(0.7673 \times 10^{-4})(T_{av}) \tag{E.2}$$

**Dynamic viscosity (Pa.s)**

$$\mu_{ha} = (1.718 \times 10^{-5}) + (4.62 \times 10^{-8})(T_{av}) \tag{E.3}$$

**Density (kg/m$^3$)**

$$\rho_{ha} = \frac{353.44}{(T_{av} + 273.15)} \tag{E.4}$$

**Thermal expansion coefficient (1/K)**

$$\beta = \frac{1}{(T_{av} + 273.15)} \tag{E.5}$$

where

$$T_{av} = \frac{T_{\text{evaporating surface}} + T_{\text{condensing surface}}}{2} \tag{E.6}$$

$T_{av}$ in °C in the above correlations. The above-listed correlations are from (Aggarwal & Tiwari, 1998)



H. Sharon and K. Srinivas Reddy, *Solar Thermal Desalination Technologies for Potable Water*, https://doi.org/10.1007/978-3-031-49155-9

**Saturation vapor pressure of water vapor at temperature "$T$" (N/m$^2$)** (Somwanshi and Tiwari, 2014)

$$P_{wv,T} = \exp\left(25.317 - \left[\frac{5144}{T + 273.15}\right]\right) \tag{E.7}$$

# Appendix F
# Water Quality Parameters

**pH** It is the negative logarithm of hydrogen ion concentration in water. pH of water can range between 0 and 14. pH value less than 7 indicates the acidic nature of water, while pH value greater than 7 indicates the alkaline nature. Acidic water and alkaline water contain extra $H^+$ and $OH^-$ ions, respectively. Pure water is neutral with a pH of 7 at 25°C. pH of rainwater is about 5.6. pH of potable water should be in the range of 6.5–8.5 (Omer, 2019; FAO, 2023).

**Turbidity** It indicates the cloudiness of water caused by suspended materials like clay, slit, organic material, planktons, and other particulate matters. Turbidity greater than 5 NTU is visible to human eye (Omer, 2019).

**Electrical Conductivity** Electrical conductivity indicates the ability of water to conduct electricity. Electrical conductivity increases with increase in ion concentration in water. It is represented in deciSimens/m (dS/m). The electrical conductivity of drinking water is about 0.005–0.05 S/m (Omer, 2019).

**Total Solids** Total solids represents total dissolved and total suspended solids in water (Omer, 2019). **Dissolved solids** consists of calcium, nitrate, phosphorous, iron, sulfur, and other ion particles that can pass through a filter with pores of around 2.0 microns in size. **Suspended solids** include slit, clay, plankton, algae, organic matter, and other particulates that cannot pass through a filter with pores of size around 2.0 microns (EPA, 2023a). Total dissolved solids can be estimated from electrical conductivity using the relation given below (Omer, 2019):

$$\text{Total Dissolved Solids} = \text{Electrical conductivity}(\text{ds/m}) \times (0.55 \text{ to } 0.70) \quad (\text{F.1})$$

**Alkalinity** It indicates the capacity of water to neutralize acids. Alkalinity is due to the presence of bicarbonates, carbonates, and hydroxides in water (EPA, 2023b). It is represented in terms of equivalents of $CaCO_3$.

**Salinity** It represents the dissolved salt content of water. Electrical conductivity is directly proportional to salinity. Salinity affects the thermal conductivity, heat capacity, density, and surface tension of water (EPA, 2023c). Sodium chloride gives



H. Sharon and K. Srinivas Reddy, *Solar Thermal Desalination Technologies for Potable Water*, https://doi.org/10.1007/978-3-031-49155-9

salty taste to water at a concentration of 250 mg/L. The salty taste of magnesium chloride is detectable only around a concentration of 1000 mg/L (Omer, 2019).

**Hardness** It indicates the ability of water to precipitate soap. For achieving same level of cleaning, high amount of soap is required if hardness of water is increased. Hardness is chemically defined as the sum of polyvalent cations like $Ca^{2+}$ $Mg^{2+}$, $Fe^{2+}$, $Sr^{2+}$ and $Mn^{2+}$. Contributions of $Fe^{2+}$, $Sr^{2+}$ and $Mn^{2+}$ ions in hardness is very less, hence they are generally neglected while estimating hardness. **Total hardness** is the sum of carbonate hardness and non-carbonate hardness. Carbonate hardness is due to $CaCO_3$ and $MgCO_3$. Non-carbonate hardness is due to $CaCl_2$, $MgCl_2$, and $MgSO_4$. Total hardness is usually represented in terms of equivalents of $CaCO_3$ (Edis, 2023).

**Nitrogen** Organic nitrogen, ammonia nitrogen, nitrite nitrogen, and nitrate nitrogen are the four forms of nitrogen that are usually present in water and wastewater. Summation of ammonia nitrogen and organic nitrogen is **Total Kjeldahl Nitrogen** (WQ, 2023a). Contamination of water with sewage water transforms organic and ammonia nitrogen into nitrites and nitrates due to the activities of microbes present in wastewater. High concentration of nitrate in water causes rapid growth of algae resulting in poor water quality. In addition, concentration of nitrate beyond 10 mg/L in drinking water can lead to severe health threats to infants (Omer, 2019).

**Sulfate** Sulfate is the highest oxidation state of sulfide. Metal sulfides are oxidized by dissolved oxygen in water to form sulfate during the natural weathering process. Sulfates are generally non-toxic; however, high concentrations in water may result in an objectionable taste. Sometimes, it can also lead to calcium deposition and indirect poisoning of crops with sodium (WQ, 2023b; Omer, 2019).

**Total Organic Carbon** It represents the milligrams of carbon in the organic pollutants per liter of water (WQ, 2023c). **Total carbon** is the summation of total organic and inorganic carbon in water. **Inorganic carbon** includes dissolved $CO_2$ gas and carbonic salts (MEL, 2023).

**Total Phosphate** It is composed of orthophosphate, polyphosphate, and organic phosphorous. Phosphorous in water exists in the form of phosphate and it is essential for plants and animals. Higher amount of phosphorous in water lead to rapid algae growth and death of other organisms. Death and decay of algae depletes the dissolved oxygen content of water and leads to eutrophication (WQ, 2023d).

**Heavy metals** It is a very broad term used to describe metallic elements having density and molecular weight higher than water. Heavy metals like iron, copper, zinc, and manganese at low concentrations are essential for human beings but the same becomes toxic at higher concentrations. Arsenic, mercury, cadmium, lead, thallium, and mercury are toxic even at lower concentrations (Hero, 2023). Heavy metals like cadmium, chromium, and lead get easily dissolved in acidic water and increase the toxicity of water (Omer, 2019).

**Biological Oxygen Demand (BOD)** It indicates the amount of oxygen required by the micro-organisms in water to remove waste organic matter from the water in the process of decomposition under aerobic conditions. Higher BOD indicates organic pollution and less availability of oxygen to support life in water (FAO, 2023; USGS, 2023).

**Chemical Oxygen Demand (COD)** It indicates the amount of oxygen required to chemically oxidize the organic material and inorganic nutrients (ammonia or nitrate) in water (Proteus-Instruments, 2023).

**Coliform bacteria** These are short rod-shaped bacteria that normally live in intestines of humans and other animals. Coliform group bacteria are also found in feces and are called **fecal coliforms**. If a large amount of coliform is detected, it indicates that the water is very recently polluted by feces. Coliforms have a lifetime larger than other pathogenic bacteria. Hence, if coliforms are not detected in water, it indicates low probability of the presence of other pathogenic bacteria in water (WQ, 2023e). It has been estimated that at least 3 million E.Coli bacteria will be present in 100 mL of untreated sewage water (Omer, 2019).

## References

Aggarwal, S., & Tiwari, G. N. (1998). Convective mass transfer in a double condensing chamber and a conventional solar still. *Desalination, 115*, 181–188.

Ali, A.R.I., & Salam, B. (2020). A review on nanofuid: Preparation, stability, thermophysical properties, heat transfer characteristics and application. *SN Applied Sciences 2*, 1636 (https://doi.org/10.1007/s42452-020-03427-1).

Edis (2023). https://edis.ifas.ufl.edu/publication/SS540.

EPA (2023a). https://archive.epa.gov/water/archive/web/html/vms58.html#:~:text=Total%20solids%20are%20dissolved%20solids,(0.002%20cm)%20in%20size.

EPA (2023b). https://archive.epa.gov/water/archive/web/html/vms510.html.

EPA (2023c). https://www.epa.gov/national-aquatic-resourcesurveys/indicators-salinity.

FAO (2023). https://www.fao.org/3/x5624e/x5624e05.htm.

Gakkhar, N., Soni, M. S., & Jakhar, S. (2016). Second law thermodynamic study of solar assisted distillation system: A review. *Renewable and Sustainable Energy Reviews, 56*, 519–535.

Gude, V. G., Nirmalakhandan, N., Deng, S., & Maganti, A. (2012). Desalination at low temperatures: An exergy analysis. *Desalination and Water Treatment, 40*(1–3), 272–281.

Hero (2023). https://hero.epa.gov/hero/index.cfm/reference/details/reference_id/7077818.

Iqbal, A., Mahmoud, M. S., Sayed, E. T., Elsaid, K., Abdelkareem, M. A., Alawadhi, A., & Olabi, A. G. (2021). Evaluation of the nanoflid-assisted desalination through solar stills in the last decade. *Journal of Environmental Management, 277*, 111415.

Khanafer, K., & Vafai, K. (2011). A critical synthesis of thermophysical characteristics of nanofluids. *International Journal of Heat and Mass Transfer, 54*, 4410–4428.

MEL (2023). https://maineenvironmentallaboratory.com/?p=1095.

Omer, N. H. (2019). *Water Quality Parameters*. Book: Water quality science, assessments and policy, IntechOpen.

Popiel, C. O., & Wojtkowiak, J. (1998). Simple formulas for thermophysical properties of liquid water for heat transfer calculations (from 0 °C to 150 °C). *Heat Transfer Engineering, 19*(3), 87–101.

Proteus-Instruments (2023). https://proteus-instruments.com/parameters/chemical-oxygen-demand-cod-sensors/.

Ranjan, K. R., & Kaushik, S. C. (2013). Energy, exergy and thermo-economic analysis of solar distillation systems: A review. *Renewable and Sustainable Energy Reviews, 27*, 709–723.

Ranjan, K. R., Kaushik, S. C., & Panwar, N. L. (2016). Energy and exergy analysis of passive solar distillation systems. *International Journal of Low-Carbon Technologies, 11*, 211–221.

Said, Z., Hachicha, A. A., Aberoumand, S., Yousef, B. A. A., Sayed, E. T., & Bellos, E. (2021). Recent advances on nanofluids for low to medium temperature solar collectors: Energy, exergy, economic analysis and environmental impact. *Progress in Energy and Combustion Science, 84*, 100898.

Sharon, H. (2021). Energy, exergy, environmental benefits and economic aspects of novel hybrid solar still for sustainable water distillation. *Process Safety and Environmental Protection, 150*, 1–21.

Somwanshi, A., Tiwari, A.K. (2014). Performance enhancement of a single basin solar still with flow of water from an air cooler on the cover. *Desalination 352*, 92–102.

Tsilingiris, P. T. (2012). Combined heat and mass transfer analyses in solar distillation systems—The restrictive conditions and a validity range investigation. *Solar Energy, 86*, 3288–3300.

USGS (2023). https://www.usgs.gov/special-topics/water-scienceschool/science/biological-oxygen-demand-bod-and-water.

WQ (2023a). https://wq.epa.gov.tw/EWQP/en/Encyclopedia/NounDefinition/Pedia_30.aspx.

WQ (2023b). https://wq.epa.gov.tw/EWQP/en/Encyclopedia/NounDefinition/Pedia_20.aspx.

WQ (2023c). https://wq.epa.gov.tw/EWQP/en/Encyclopedia/NounDefinition/Pedia_13.aspx.

WQ (2023d). https://wq.epa.gov.tw/EWQP/en/Encyclopedia/NounDefinition/Pedia_32.aspx.

WQ (2023e). https://wq.epa.gov.tw/EWQP/en/Encyclopedia/NounDefinition/Pedia_03.aspx.

The manufacturer's authorised representative in the EU is Springer Nature Customer Service Centre GmbH, Europaplatz 3, 69115 Heidelberg, Germany. If you have any concerns regarding our products, please contact ProductSafety@springernature.com

Printed and bound by CPI Group (UK) Ltd, Croydon, CR0 4YY

01/10/2026

02225853-0006